International Human Resources
Development Corporation
Boston

Enders A. Robinson
Distinguished Professor of Geophysics
University of Tulsa

Seismic velocity analysis and the convolutional model

Cover & interior design by Outside Designs

Line drawings by Jennifer Card

Library of Congress Cataloging in Publication Data

Robinson, Enders A.
Seismic velocity analysis and the convolutional model.

Bibliography: p.
Includes index.
1. Seismic prospecting. 2. Seismic waves.
3. Convolutions (Mathematics) I. Title.
TN269.R554 1983 622'.159 83-12631
ISBN 0-934634-63-7

Printed in the United States of America

To Norman Ricker (1897–1980)

Contents

Preface

This book is an introductory text for undergraduate majors in geology and geophysics. Although other books in exploration geophysics cover the entire area of exploration geophysics and give a balanced overall picture, this book concentrates on the fundamental aspects of the seismic method used in petroleum exploration. Velocity analysis and the convolutional model form the backbone of exploration seismology and knowledge of these lead to an understanding of the entire rationale of seismic data acquisition and processing. This book is for the student and includes exercises. Answers to selected exercises are given at the end of the book.

During the past two decades, exploration geophysics has moved into the forefront of scientific and technological thought. Oil companies and geophysical companies are great users of computers. In the digital signal processing area of geophysics there has been much activity, accomplishment, and promise of things to come. These technological developments have enabled geophysicists to make more precise measurements than before. Advanced computational analyses have been applied to high quality data and elaborate theoretical models have been devised to interpret them. Hopefully, after reading this book, the student will find access to this exciting scientific endeavor easy and enjoyable.

I owe much to my class in Geophysics 4053 at the University of Tulsa during the spring semester of 1983: Bruce M. Day, Allison A. Dupigny, Kris Anne Erickson, Karen B. Goodfellow, Eduardo P. Jaramillo, Marcelo A. Morales, Susan M. Riley, Brenda L. Roberts, Timothy J. Smith, Scott J. Spaunhorst, and Wadja S. Tampubolon. I want to thank Professor Eduard Douze, Chairman of the Department of Geosciences, for his encouragement and help. Special thanks are due to Wayne Igo for his advice and to Donald Straley for the preparation of the bibliography. I would also like to thank the many authors whose illustrations I have reprinted in the text. Of course the whole book was made possible by Terry Saunders; she did an excellent job in preparing a beautiful manuscript on the word processor. Carolyn Yoder and Mike Hays of IHRDC provided the professional know-how to produce this book. Their work is unequalled by any other publisher in my experience.

Enders A. Robinson
July 1983
Tulsa, Oklahoma

1. The reflection seismic method

Geophysical methods

The ultimate goal of exploration geophysics is to increase our geologic understanding—to allow us to understand more clearly the nature of the subsurface structure of the earth. Information acquired by exploration is used to locate deposits of oil and gas, minerals, geothermal reservoirs, or groundwater. In addition, geophysics can be used to determine safe sites for buildings, dams, nuclear power plants, and other construction. All geophysical exploration methods are involved with looking for anomalies—that is, locations where there are departures from regularity. An example of such an anomaly would be a geologic anticline that serves as a trap for hydrocarbons (see fig. 1.1). Geophysical information is conveyed by means of signals. These signals usually occur in the presence of noise that obscures the signal to varying degrees. It is more difficult to detect features deep within the earth than shallow ones. All geophysical methods have more difficulty defining small features than larger ones; some features are so small that they are beyond the limits of resolution. Because there are always several possible explanations for each set of observations, each geophysical method suffers from some type of ambiguity. Additional data are always helpful in reducing the ambiguity.

The three principal geophysical methods used in petroleum exploration are the seismic method, the magnetic method, and the gravity method. The magnetic method, which is the oldest geophysical method, is based on the measurement of variations in the magnetic field due to changes of structure or magnetic susceptibility of the rocks (see fig. 1.2). Sedimentary rocks generally have a smaller susceptibility than igneous or metamorphic rocks, so an interpretation of the recorded anomalies can yield the maximum depth values for a sedimentary basin. Today, magnetic surveys for hydrocarbon exploration are usually carried out from the air (aeromagnetics) or from a ship.

The gravity method is based on the measurements of the variations in the pull of gravity from rocks in the upper layers of the earth's surface (see fig. 1.3). Denser rocks have greater gravitational attraction than less dense rocks. A structural uplift of a denser rock, for example, will appear as an anomaly on the gravity map. Gravity surveys for hydrocarbons are carried out on land, in the air on helicopters, and at sea on ships.

The most widely used geophysical method is the seismic method. In a report to the Royal Irish Academy in 1846, Robert Mallet made the first proposal to obtain subsurface geologic information by recording the seismic waves from a man-made source. In the next few years, he developed a mercury-bowl seismograph and measured the velocity of seismic waves through the granites of

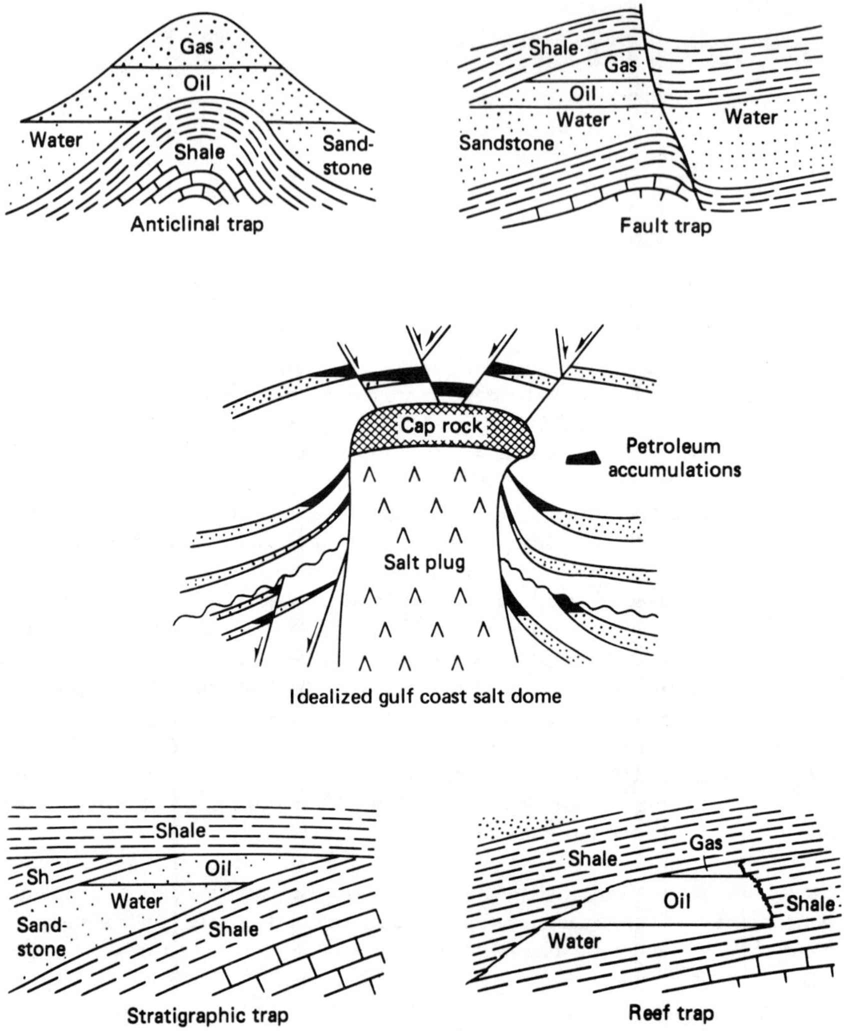

Figure 1.1. Typical structural configurations for trapping hydrocarbons. Reprinted by permission of W. H. Freeman and Company from Levorsen 1967 and IHRDC from Dix 1981.

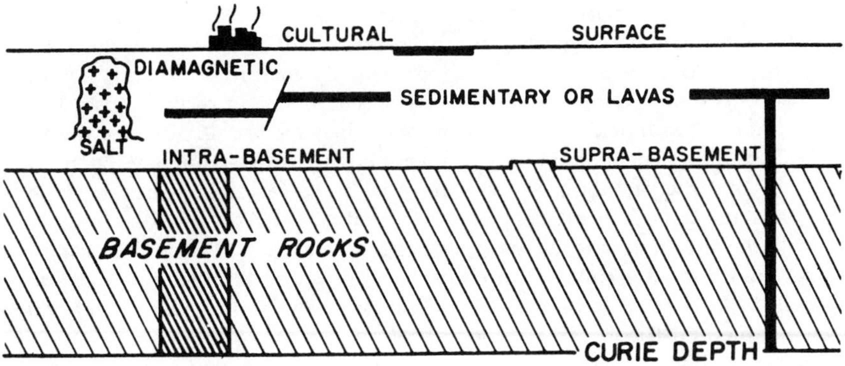

Figure 1.2. Types of magnetic sources. Reprinted by permission of IHRDC from Sheriff 1978.

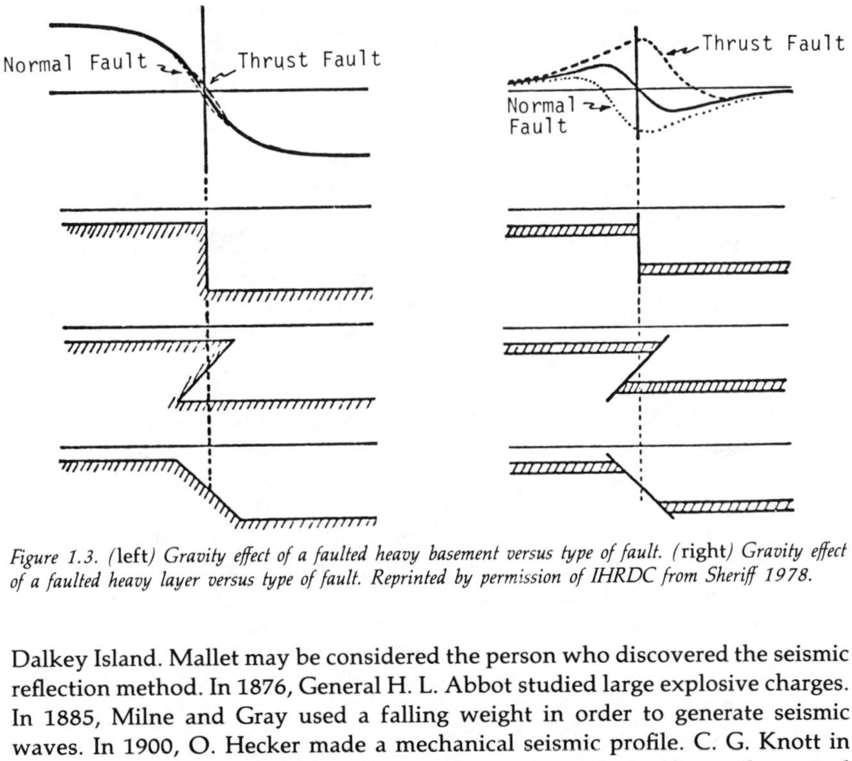

*Figure 1.3. (*left*) Gravity effect of a faulted heavy basement versus type of fault. (*right*) Gravity effect of a faulted heavy layer versus type of fault. Reprinted by permission of IHRDC from Sheriff 1978.*

Dalkey Island. Mallet may be considered the person who discovered the seismic reflection method. In 1876, General H. L. Abbot studied large explosive charges. In 1885, Milne and Gray used a falling weight in order to generate seismic waves. In 1900, O. Hecker made a mechanical seismic profile. C. G. Knott in 1899 and Wiechert and Zoeppritz in 1907 performed significant theoretical studies. All this work represented important steps in the early history of exploration seismic methods.

In 1914, Reginald Fessenden made the first use of reflected energy while using his sonic sounder to find ocean depths. In 1917, W. P. Haseman and J. P. Karcher of the National Bureau of Standards studied the feasibility of using reflected seismic waves to locate potential oil structures. The first successful reflection seismograms resulted from their experiments in a rock quarry in Washington,

D.C. In 1921, Haseman and Karcher, with D. W. Ohern, Irving Perrine, and W. C. Kite, gathered together the necessary equipment to test the seismic reflection method for petroleum exploration. Their initial attempt, however, turned out to be a failure on the structure that was later recognized to be the Oklahoma City oil field. Not giving up, they then successfully defined the flank of a known dome near Dougherty, Oklahoma, by the reflection method. Significant in the interpretation of these records is the fact that Karcher properly migrated the seismic record section they had obtained. As we will see, migration is still an important step in the processing of seismic reflection records.

At about the same time as this early reflection work in Oklahoma was taking place, L. Mintrop in Germany was applying for patent coverage for a refraction seismic method for locating the depth of subsurface formations. In 1924, he brought his instruments to Texas, and, working for the Gulf Oil Corporation, he located the Ochard Dome of Bend County, Texas, which was the first refraction discovery. Amerada Oil Company formed the Geophysical Research Corporation in Tulsa which, under DeGolyer, discovered the Maude Poole in Oklahoma in 1927; this was the first reflection discovery. These discoveries were the first of many discoveries that established seismic methods as the essential petroleum exploration tool. Over the succeeding years, the reflection method, because of its remarkable depth of penetration, has eclipsed the earlier refraction method in the search for oil and gas. Every geophysicist knows, however, the importance to the exploration picture of the complete integration of all geologic and geophysical techniques and information.

Exploration seismology is divided into two branches: reflection seismology and refraction seismology. Most of petroleum exploration is done by the reflection seismic method. Reflection seismology is a method of mapping the subsurface sedimentary rock layers from measurements of the arrival times of events reflected from the subsurface layers. The technology of collecting and processing reflection seismic data is based on a fundamental concept. By generating seismic energy that penetrates the earth's layered media, reflections of the seismic waves at the interfaces are measured and recorded by receiving devices at or near the earth's surface. Basically, the concept is the same for both land and marine surveys; only the mechanics vary. On land, seismic energy is generated at or near the earth's surface by arrays of small chemical explosions or vibrating machines or thumping devices. Preplaced detectors on the surface run in a line, as illustrated in figure 1.4. The seismic waves resulting from the downward propagation of this source energy are reflected from the various interfaces and received by the detectors (see fig. 1.5). A central recording unit then digitizes the analog signals and records them on magnetic tape for subsequent analysis (see fig. 1.6). At sea, a source, such as an array of air guns (see fig. 1.7), is actuated every few seconds as the ship moves over a predetermined course. The seismic waves are picked up by detectors embedded in a cable (called a streamer) trailing the ship (fig. 1.8). As in land surveys, the data are transmitted to a central recording unit and recorded in digital form on magnetic tape.

The essence of a seismic data gathering system as implemented in the field by a seismic crew may thus be summarized as follows. Elastic waves (called seismic waves) created by man-made (artificial) sources propagate through the earth. Interfaces between geologic rock strata reflect the spreading wavefronts.

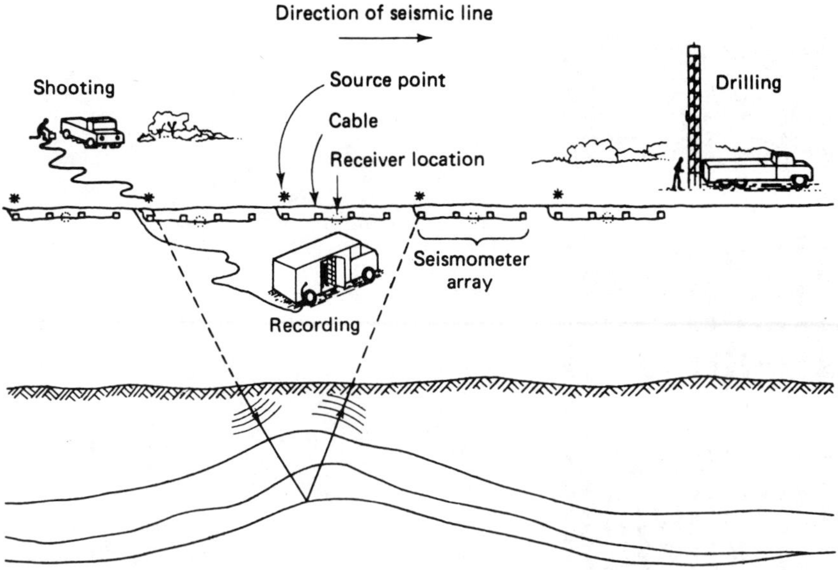

Figure 1.4. Configuration of source points and receiver points in seismic exploration. Reprinted by permission of Prentice-Hall, Inc. from Robinson and Treitel 1980.

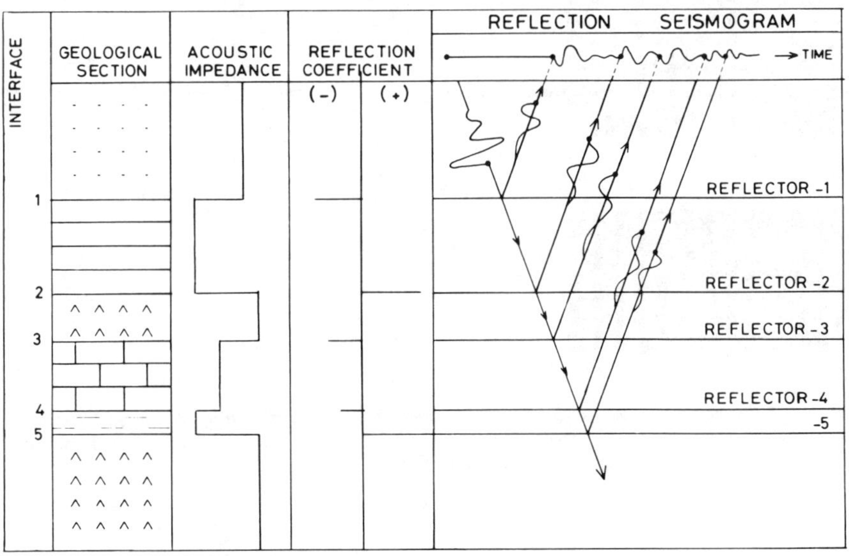

Figure 1.5. Reflection seismogram, showing primary reflections only. Reprinted by permission of Birkhauser-Verlag from Al-Sadi 1980.

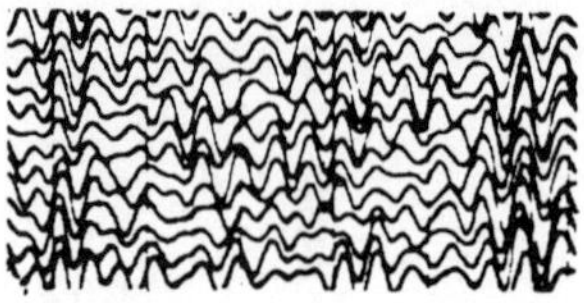

(a)

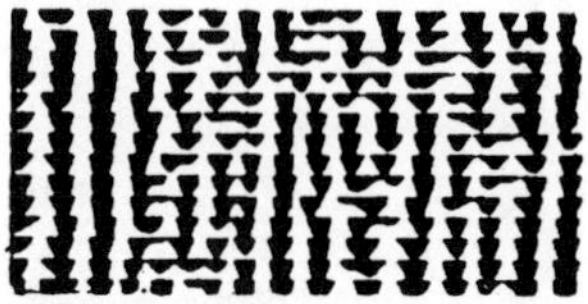

(b)

(c)

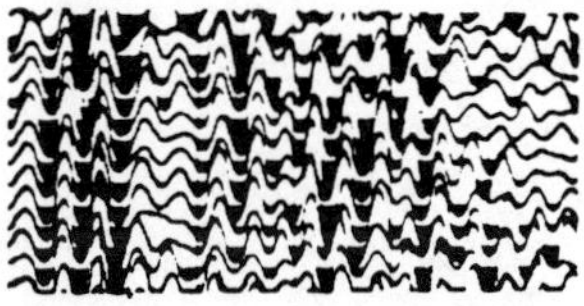

(d)

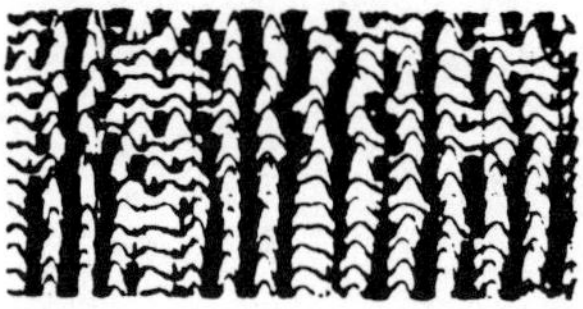

(e)

Figure 1.6. Display modes: (a) wiggle trace; (b) variable area; (c) variable density; (d) wiggle trace superimposed on variable area; (e) wiggle trace superimposed on variable density. Reprinted by permission of the Society of Exploration Geophysicists from Sheriff 1973.

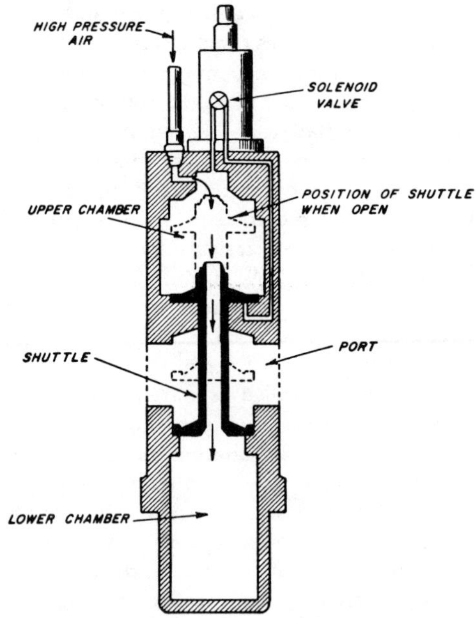

Figure 1.7. Air gun—a marine seismic source that injects a bubble of highly compressed air into the water. Oscillations of the bubble as it alternately expands and contracts generate a sonic wave whose frequency depends on the amount of air in the bubble, its pressure, and the water depth (or water pressure). Arrays of guns of different sizes are sometimes used so that a broader frequency spectrum will be generated. High-pressure air flows continuously into the upper chamber and through the shuttle into the lower chamber. Opening the solenoid valve puts high-pressure air under the upper shuttle seat, causing the shuttle to move upward, opening the lower chamber and allowing its air to flow out through ports to form a bubble of high-pressure air in the water. The size of a gun is the size of its lower chamber. Courtesy Bolt Associates. Reprinted by permission of the Society of Exploration Geophysicists from Sheriff 1973.

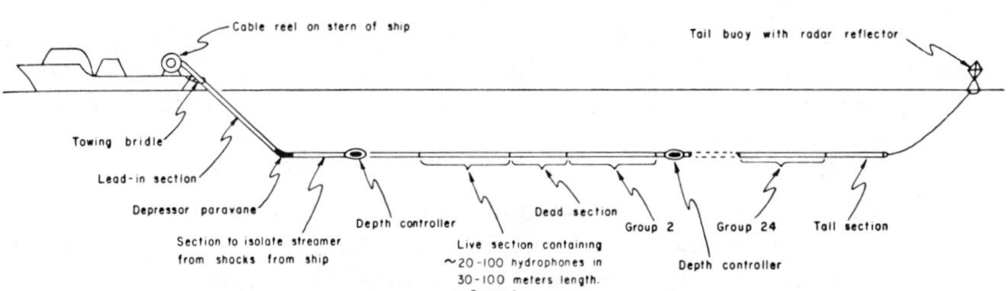

Figure 1.8. Marine seismic streamer—a marine cable incorporating pressure hydrophones internally, designed for continuous towing through the water. A marine streamer is typically made up of active, or live, sections that contain hydrophone arrays separated by spacer, or dead, sections. Usually the streamer is nearly neutrally buoyant, and depressors, or depth controllers, are attached to depress the streamer to the proper towing depth. The entire streamer may be 5000 to 9000 ft long. Reprinted by permission of the Society of Exploration Geophysicists from Sheriff 1973.

Receivers laid out on the surface of the ground, or in a cable slightly under the surface of the sea, record the returning seismic waves. Each receiver actually consists of a group (i.e., array) of transducers, which are geophones on land (fig. 1.9) and hydrophones at sea. Each group may consist of up to 100 individual transducers, arranged in various linear or spatial patterns. The distance between receiver positions ranges from 50 to 900 ft. It is common practice to use either 24, 48, or 96 receivers, and the seismic master cable joining the groups ranges from 1 to 3 mi long. The returned signal that is recorded by each receiver is called a seismic trace (fig. 1.10). Each time a source is activated, either 24, 48, or 96 traces are recorded on a digital tape (see fig. 1.11). A seismic trace is a graph (seismogram) of the amplitude of the received signal versus time. A strong reflecting horizon within the earth causes a strong echo (reflection), which exhibits itself as a high-amplitude event on the seismic trace at the arrival time of the echo. Arrival times of single-bounce echoes (called primary reflections or simple reflections) permit the determination of depths and inclination angles of the reflecting horizons (reflectors), as we shall see in later discussions. After each time the source is activated and the resulting seismic traces (called the common source point, or CSP, profile or gather) are recorded (fig. 1.12), the source point is moved to the next location. Seismic surveys are conducted along straight or nearly straight lines, and individual straight lines may extend for distances up to 100 miles. To cover an area of interest, parallel survey lines are used, so that the source points fall on a two-dimensional (2-D) grid on the surface of the earth.

By itself, a seismic source may be too weak to penetrate deep within the earth. Weak sources, therefore, are laid out in arrays similar to the arrays used for the receivers (see fig. 1.13). Such arrays used for both source and receiver have many good properties. Arrays can cancel unwanted ambient noise, attenuate surface waves (called ground roll), smooth time variations caused by surface irregularities, and discriminate against harmful scattered energy. The optimum design of an array is an important problem that must be solved for each new prospect in order to obtain satisfactory reflection quality.

The data acquisition system used by a geophysical crew records an accurate facsimile of the ground's motion to incoming seismic energy. In land surveys, the geophone converts the particle velocity of the ground to electrical voltages, which are then recorded. In marine work, the hydrophones convert pressure variations to voltage. Because each receiver point may have up to 100 transducers, the outputs of the transducers are added together and recorded as a single seismic trace (one channel). If there are 48 receiver positions, then the resulting 48 seismic traces (48 channels) are multiplexed and recorded digitally on magnetic tape in an instrumentation truck or ship's cabin. These 48 traces make up a single seismic record, called a common source point (CSP) profile. Each seismic trace is about 6 sec long. The reason seismic traces seldom exceed 6 sec is that hydrocarbon reservoirs (oil and gas traps) rarely occur below 30,000 ft in geologic basins; thus, at an average velocity of 15,000 ft/sec, the two-way traveltime (down and up) would be only 4 sec. Admittedly, there is much deep information (below 30,000 ft) on the seismograms from 4 to 6 sec that is never utilized by oil companies but would be of great interest to geologists interested in the growth of continents. One exception to the 6 sec record length is important. When a low-amplitude vibration is used as a source, as in

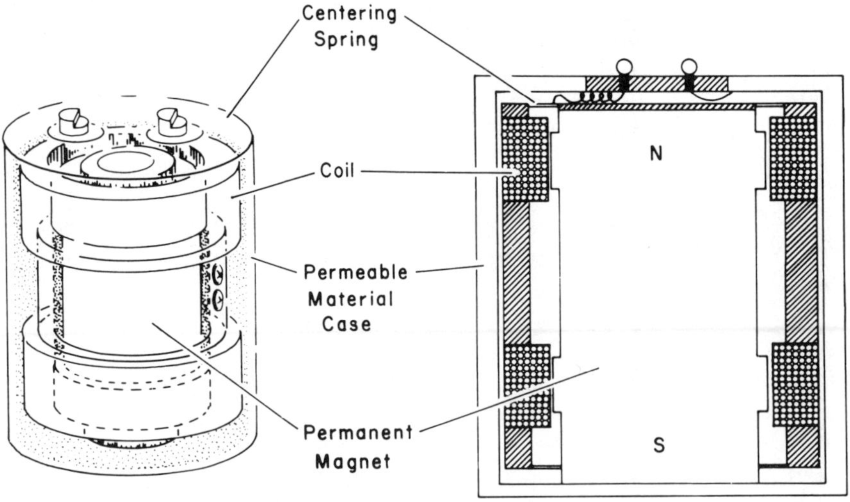

Figure 1.9. Schematic of a moving-coil geophone element. (Electrical connections are diagrammatic; not drawn to scale.) Reprinted by permission of John Wiley & Sons, Inc. from Reflection Seismology by K.H. Waters 1978. Copyright © 1978 John Wiley & Sons, Inc.

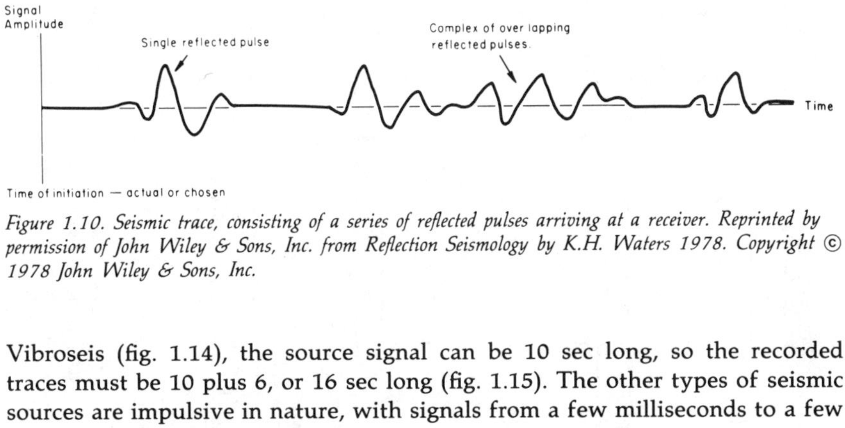

Figure 1.10. Seismic trace, consisting of a series of reflected pulses arriving at a receiver. Reprinted by permission of John Wiley & Sons, Inc. from Reflection Seismology by K.H. Waters 1978. Copyright © 1978 John Wiley & Sons, Inc.

Vibroseis (fig. 1.14), the source signal can be 10 sec long, so the recorded traces must be 10 plus 6, or 16 sec long (fig. 1.15). The other types of seismic sources are impulsive in nature, with signals from a few milliseconds to a few hundred milliseconds long.

The reflected signals received at the surface of the earth contain all frequencies, but what are considered the useful frequencies range from a few hertz to a few hundred hertz. In digitizing the seismic traces, samples are taken at 1, 2, or 4 msec rates, depending on the survey, that have alias frequencies (equal to half the sample frequencies) of 500, 250, or 125 Hz, respectively. The digital seismic recording systems have dynamic range of about 80 dB. In exploration geophysics, a decibel is defined as 20 log (A/A_0) where A/A_0 is the amplitude ratio. Some very strong signals may rise 100 dB over the ambient noise levels, however, so such signals will not be entirely recorded. The amplitude of the reflected waves decay about 100 dB in the first 4 sec of recording, because of the spherical spreading of the wavefront and attenuation losses to heat and to multiple reflections. As a result, the amplifier gain levels must be changed many

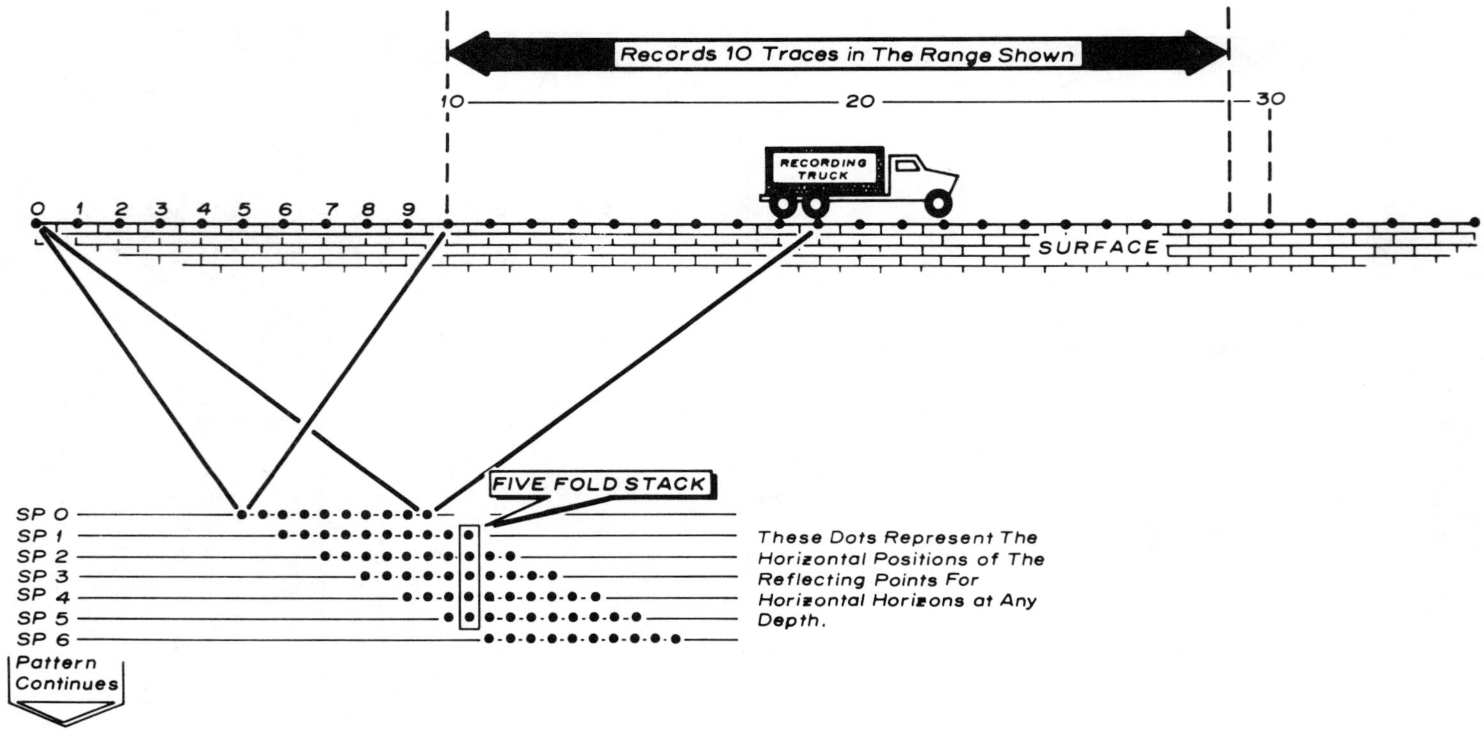

Figure 1.11. The basic arrangement of a typical (fivefold) reflection spread. The recording truck contains a switch that allows selection of the correct sequence of geophones as the shot point is moved up. As convenient, portions of cable and geophones are moved from the left end of the cable and connected to the right end. Reprinted by permission of John Wiley & Sons, Inc. from Reflection Seismology by K.H. Waters 1978. Copyright © 1978 John Wiley & Sons, Inc.

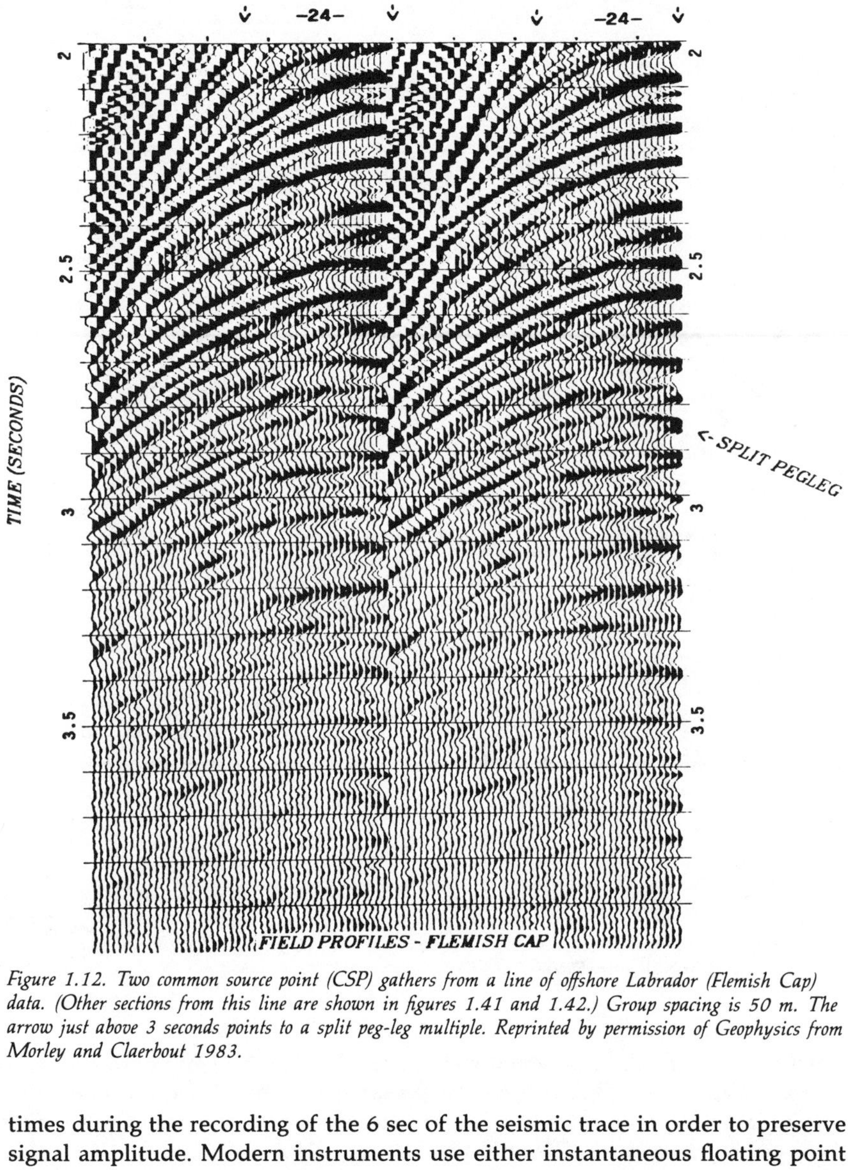

Figure 1.12. Two common source point (CSP) gathers from a line of offshore Labrador (Flemish Cap) data. (Other sections from this line are shown in figures 1.41 and 1.42.) Group spacing is 50 m. The arrow just above 3 seconds points to a split peg-leg multiple. Reprinted by permission of Geophysics from Morley and Claerbout 1983.

times during the recording of the 6 sec of the seismic trace in order to preserve signal amplitude. Modern instruments use either instantaneous floating point or binary gain control. The binary gain amplifiers record the times at which the gain is changed, so that in subsequent processing, the true signal can be reconstructed. The field recording system records 16 bits for each amplitude value on the digitized seismic trace. In subsequent computer processing, however, usually only 12 bits are used, and when the final results are displayed the plotter uses only the most significant 8 bits.

Let us now consider a typical ship that would be used in a marine seismic survey. The ship would be a steel-hulled vessel, 55 m long, with a 12 m beam and a draft of 4 m. There would be twin-screw propulsion and a stabilization

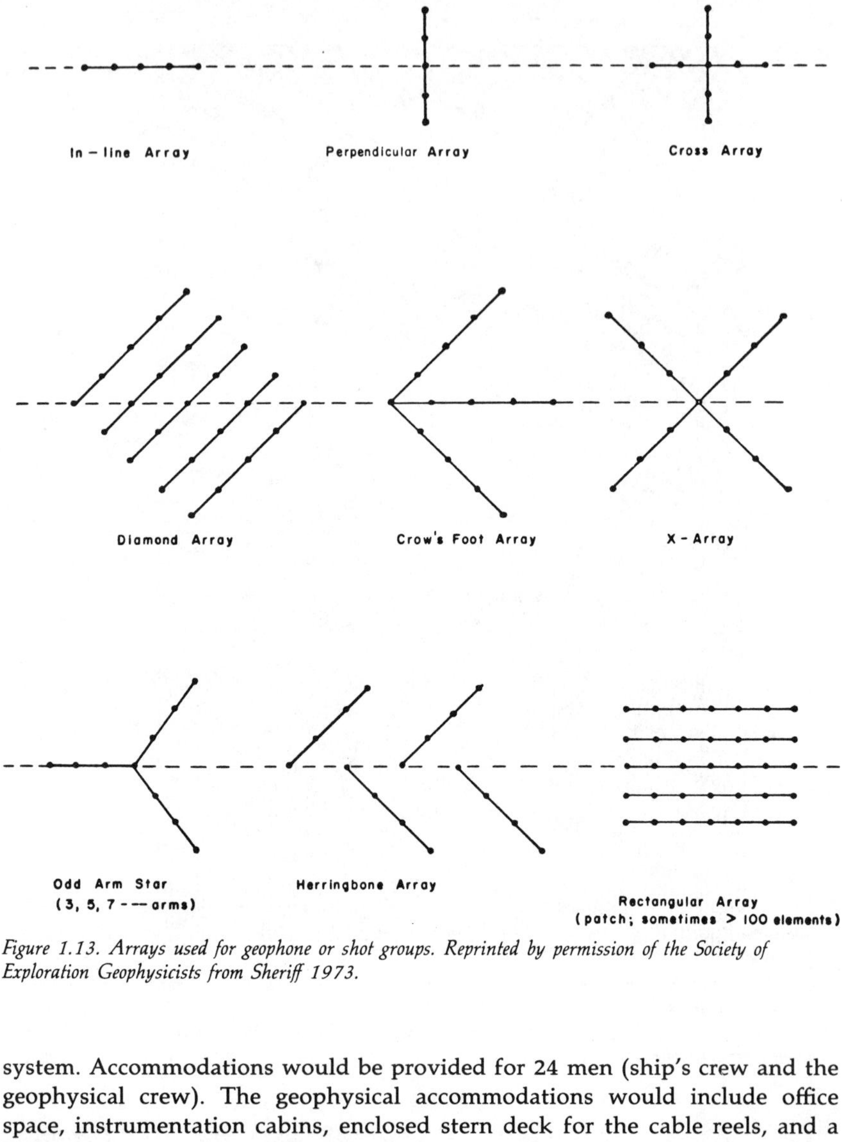

Figure 1.13. Arrays used for geophone or shot groups. Reprinted by permission of the Society of Exploration Geophysicists from Sheriff 1973.

system. Accommodations would be provided for 24 men (ship's crew and the geophysical crew). The geophysical accommodations would include office space, instrumentation cabins, enclosed stern deck for the cable reels, and a heliport superstructure. The ship would have a cruising speed of 10 nautical miles (nmi) per hour and a towing speed of 4 to 6 nmi per hour. Marine navigational facilities would include dual radars with a 24 mi range and a fully integrated satellite navigation system, with auto pilot, gyrocompass, sonar doppler, velocimeter, and computer.

In marine seismic work, various navigation systems, such as Loran C, are used. In many surveys, a satellite positioning system is used as the primary positioning system. Advantages of the satellite system are 24 hr all-weather operation, real-time statistical filtering to provide desired track capabilities, computer-controlled automatic ship steering to minimize on-line course correc-

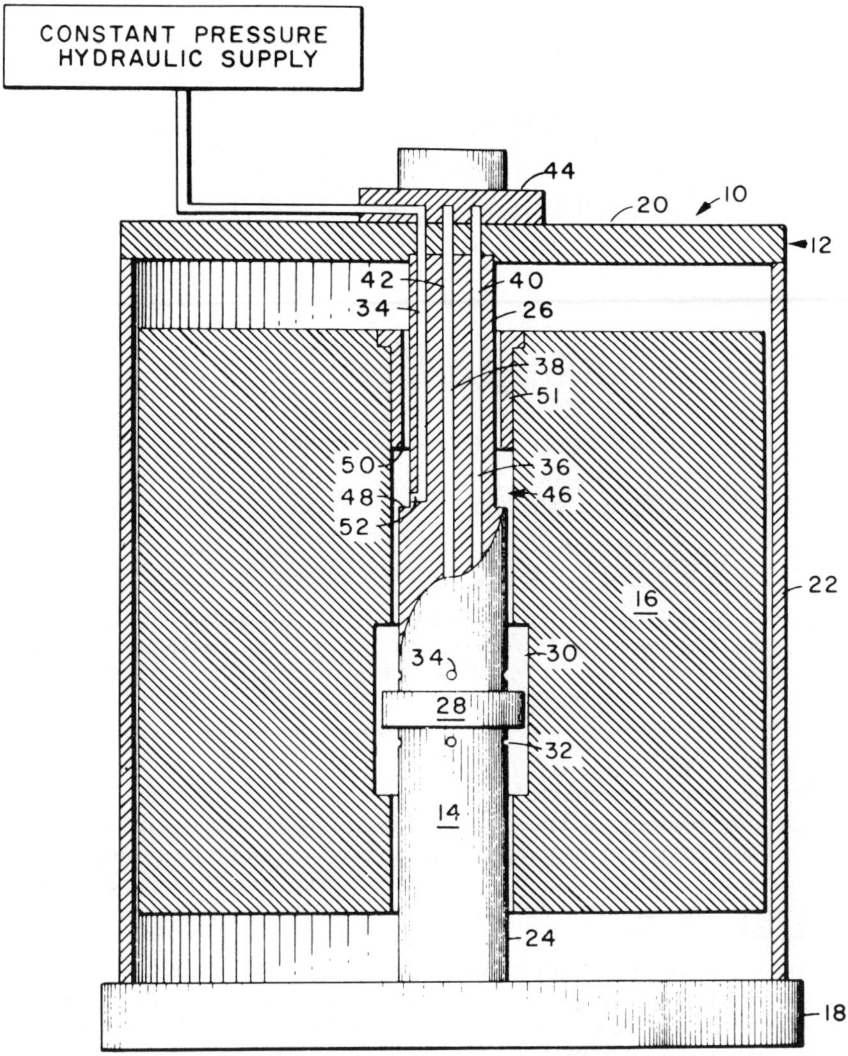

Figure 1.14. Schematic of a vertical hydraulic vibrator. Reprinted by permission of John Wiley & Sons, Inc. from Reflection Seismology by K.H. Waters 1978. Copyright © 1978 John Wiley & Sons, Inc.

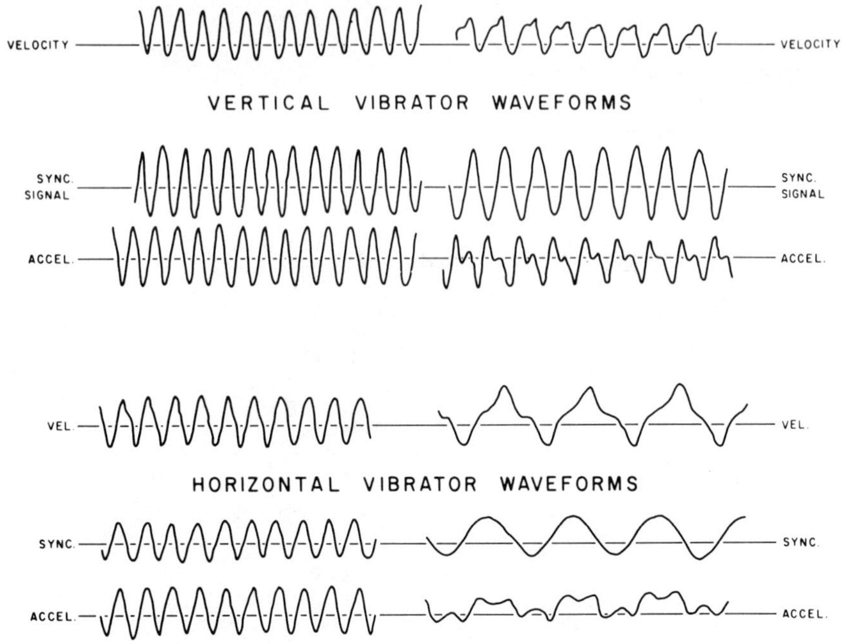

Figure 1.15. The vertical and horizontal vibrator baseplate motions corresponding to a given synchronizing electric signal. High frequencies are on the left, low frequencies on the right. Development of the double-frequency components as the frequency is lowered is evident. Reprinted by permission of John Wiley & Sons, Inc. from Reflection Seismology by K.H. Waters 1978. Copyright © 1978 John Wiley & Sons, Inc.

tions, inclinometers to provide dynamic corrections for improving satellite fix and heading accuracy, and magnetic tape recording of all sensor data for rapid analysis and computer mapping. Normal average deviation from programmed lines in water depth of 200 m or shallower is less than 300 m, and postplot accuracy of shot points is 100 m; greater accuracy can be attained at a cost premium.

An effective seismic energy source in marine exploration is the air gun. Eight air gun units are towed in an array approximately 22 m by 12 m at a depth of 6 m. Total air gun performance for this device is continuously monitored by individual gauges at the mechanics' console. The guns are fired by command of the navigation system, and a composite gun pulse from each pair of guns is displayed on monitor records and on tape. Figure 1.16 shows a typical source wavelet (i.e., gun signature pulse) at different receiver depths. Typical quality control standards would be as follows.

(1) Not more than 3 misfires in any 4 consecutive shots, 6 in any 12 shots, 8 in any 20 shots, or 16 in any 100 shots. No more than 6 percent of any line shall consist of misfires. A misfire is defined as any equipment or operator malfunction that results in an unrecoverable record, or more than two dead or excessively noisy traces, or fewer than six guns operating properly.
(2) No new line started with more than one dead trace.

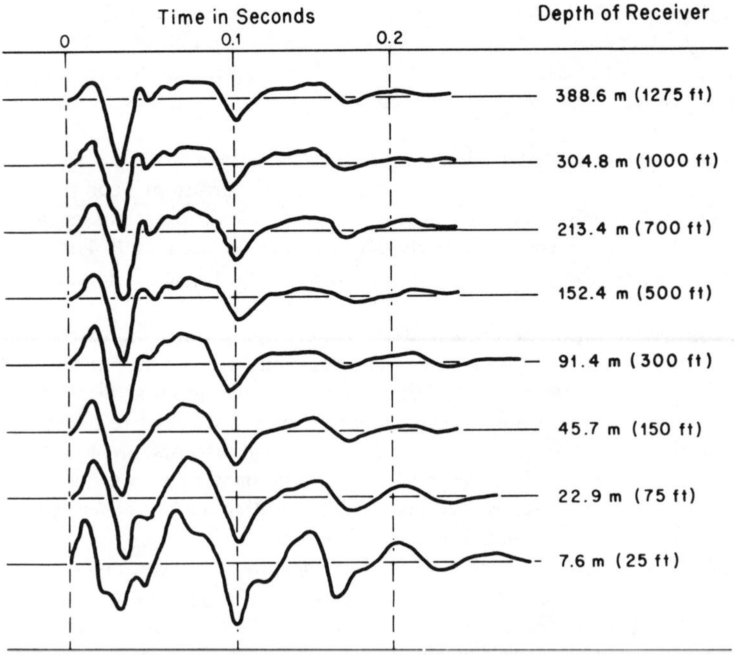

Figure 1.16. Deep-water tests of a pulse for a combination of four aquapulse guns 12.2 m (40 ft) deep. Reprinted by permission of John Wiley & Sons, Inc. from Reflection Seismology by J.H. Waters 1978. Copyright © 1978 John Wiley & Sons, Inc.

(3) Noise level on the main streamer not to exceed 5 μbar rms, except that the use of a radar target towed from the tail buoy increases the allowable noise on channel 1. (Typical streamer noise level is 3 μbar.)
(4) The angle of feathering between the main streamer and the line traverse not to exceed 12°.

The streamer trails below the water's surface behind the ship. A seismic streamer is a high-performance, operationally flexible device. Its outstanding features are field repairability, nonrotating stability, quick-coupling connectors, unbreakable bulkheads, high-sensitivity hydrophones, and flexible operational configurations to meet specific recording requirements. Downtime is minimized with quick-disconnect sections, a center-stress member, and a durable, tough, cold-resistant vinyl jacket. The center-stress cable is a wire rope (0.7 cm outside diameter) of torque-balanced galvanized aircraft steel. It has a minimum breaking strength of 5000 kg. A 0.05 cm layer of high dielectric insulation reduces leakage. A 48-channel streamer has a total length of 2400 m, divided into 48 sections (one section for each channel), each section being 50 m long. There are 40 hydrophones spaced in each of the 50 m sections. Each hydrophone is 6 cm by 1.25 cm and weighs 27 g. The hydrophones are connected in parallel in each section. The conducting wires are 24 AWG with composite stranding, with no duplication of color code. Each section is enclosed in a plastic tube (vinyl jacket) 7 cm in diameter, with bulkheads at each end so that each section is detachable and interchangeable. The plastic jacket is filled with oil to make the section neutrally buoyant. As a result, the streamer is extremely flammable. At intervals along the cable, pressure-sensitive depth controllers keep the cable at the optimum depth. Passive sections 50 m long are available for desired spacing patterns.

The 40-unit linear array of hydrophones in each section is called a group, and the signals received from all the hydrophones in a group are composited to form one signal, called a seismic trace. As a result, the entire streamer of 48 groups yields 48 seismic traces each time the source is activated. These 48 traces are recorded digitally by a 48-channel recorder on magnetic tape inside the ship.

The foregoing description pertains to deep-water exploration. Shallow-water seismic surveys are also carried out under various conditions in all parts of the world. The aim of these special investigations is to link land and marine surveys. Flexibility in the equipment and in the execution of the survey are necessary in order to survey estuaries with heavy traffic or strong currents, areas of mud flats, reefs, and other coastal features, and coastal regions with great tides. Such shallow-water areas require vessels of shallow draught as well as other special vehicles, such as pontoons, catamarans, and hovercraft.

Land exploration ranges over all areas of the world in such extremes as cities such as Los Angeles, the arctic tundra, and the tropical jungle. As an example, I will briefly describe oil exploration in the Arabian Desert. Air-conditioned trailers and helicopters have replaced the tents and camel caravans of the early oil explorers in the desert. Arrays of light charges of explosives can be used as a seismic source, but now the comparatively gentle mechanical jolts of a thumper are often used. The thumper, which is on a special vehicle, is a gas-driven chamber that can deliver a sharp thump on the ground. Several such units normally work together and pop simultaneously. Seismic field parties may consist of more than 100 men. Convoys of large, air-conditioned trailers for living and office work are towed to camp locations by heavy-duty trucks equipped for desert travel. Much of the desert region is covered with sand. Both heavy and light vehicles can be used if they are equipped with proper sand tires, but it is often necessary to follow circuitous routes to avoid the worst sand conditions. Helicopters are also used, and often personnel and some supplies are flown to the remote areas.

In summary, general exploration objectives include the mapping of subsurface geologic structures, the detection of hydrocarbon accumulations, and the estimation of total energy reserves in an area. To achieve these objectives, large amounts of seismic data are recorded and processed routinely by the oil industry. Seismic surveys are carried out on a surface grid in order to build up a three-dimensional (3-D) picture of the subsurface geology in a region. Each survey mile contains about 50 million bits of information, making modern data processing impossible without high-speed digital computers. Seismic prospecting for oil and gas has undergone a digital revolution during the past decade. Most stages of the exploration process have been affected: the acquisition of data, the reduction of these data in preparation for signal processing, the design of digital filters to detect primary echoes (reflections) from buried interfaces, and the development of technology to extract from these detected signals information on the geometry and physical properties of the subsurface. The seismic reflection is generally weak and must be strengthened by the use of signal-summing (stacking) procedures. The determination of depths to a target horizon requires knowledge of the propagational velocities of seismic waves, and a wealth of technology has evolved for this purpose. Recently, it has been possible to relate signal amplitude to the physical properties of the medium traversed and, in particular, to make inferences about the oil and gas content of the buried rocks.

Wave propagation, reflection, and refraction

The study of wave motion plays a leading role in almost all physical sciences. Some of the most important wave definitions are given in figure 1.17. Three main types of elastic waves propagate through the earth:

(1) P-waves, also called longitudinal or compressional waves.
(2) S-waves, also called transverse or shear waves.
(3) Surface waves.

Usually, only the P-waves are considered useful in seismic exploration. We try to avoid recording S-waves and surface waves; when they are recorded, they are relegated to the category of noise. Exceptions to this rule are now being made, however, and in the future S-waves may provide invaluable lithologic information that we cannot obtain from P-waves alone. Both the P-waves and the S-waves are categorized as body waves, since they propagate through a material body as a whole, as opposed to surface waves, which travel along the surface of the body. The P-waves are characterized by particle motion back and forth along the direction of propagation, whereas the S-waves are characterized by particle motion at right angles to the direction of propagation.

In a homogeneous, isotropic medium, the P-wave and the S-wave are elastic waves, each with a fixed velocity. These two velocities are characteristic of the material that makes up the medium. Such a velocity is called the *wave velocity* or the *material velocity,* and the distinction must be always made between this type of velocity and the particle velocity.

In the case of a homogeneous, isotropic medium, a few constants, known as elastic moduli or elastic constants, describe the stress–strain linear relation representing the elastic response of the medium. For historical reasons, more elastic constants are in use than are actually needed, so the various constants are not independent of each other. The most common elastic moduli are the following:

(1) Lamé's modulus λ
(2) Rigidity (shear) modulus μ
(3) Young's modulus Y
(4) Bulk modulus B
(5) Poisson's ratio σ

Any one of these five constants may be expressed in terms of any two of the remaining ones. Thus, any two of the constants can be used to define the elastic properties of a homogeneous, isotropic body.

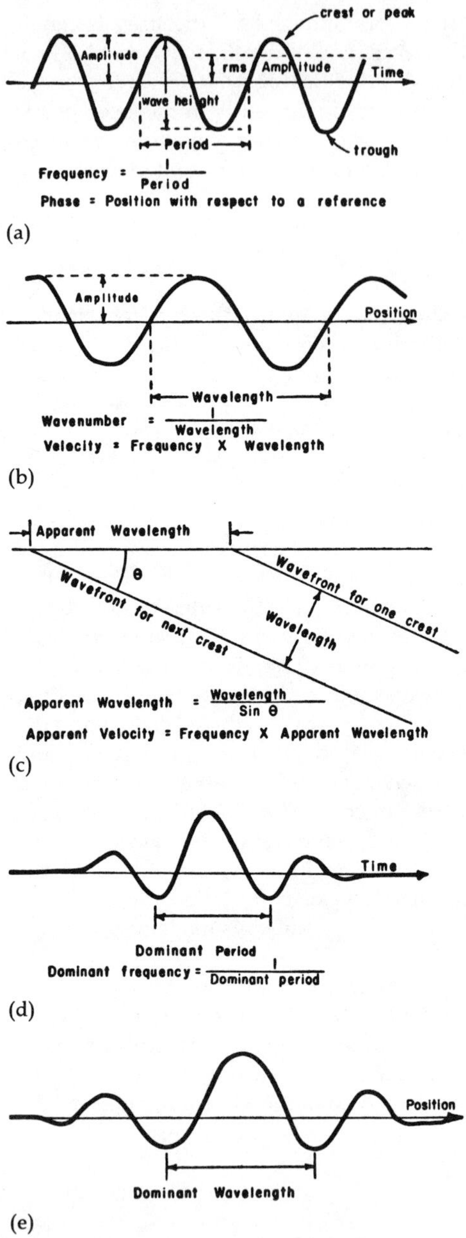

Figure 1.17. Wave definitions. For sinusoids: *(a) how displacement at one point varies with time; (b) how wave looks at different places at a given instant; (c) if wavefront approaches at an angle, the apparent wavelength will differ from the true wavelength.* For nonperiodic waves: *(d) dominant period between principal adjacent troughs (or peaks); (e) dominant wavelength is similarly measured. Reprinted by permission of the Society of Exploration Geophysicists from Sheriff 1973.*

The P-waves represent a traveling disturbance in the form of a cubical dilation or volume deformation. Thus, the direction of travel (i.e., propagation direction) is made up of a sequence of alternating zones of compressions and rarefactions. Figure 1.18 depicts a P-wave. Four wavefronts are shown, with three particles on each wavefront. The particles do not propagate but oscillate about their zero-positions in a direction coincident with the propagation direction. The velocity of the particles in this oscillatory motion is called the *particle velocity.* As mentioned earlier, the concept of particle velocity is different from that of wave or material velocity. Wave velocity is the velocity with which each wavefront propagates. The distance between two wavefronts of the same type (or phase), such as the distance between two successive rarefactions, is called the wavelength. The P-wave is the fastest wave for a given medium; therefore, for a given propagation path, it arrives before any other type of wave. This is why it is called the P-wave, as P stands for primary. The velocity of the P-wave has several expressions. If ρ stands for the density of the medium, then the more important expressions are

$$v_p = \sqrt{\frac{\lambda + 2\mu}{\rho}} = \sqrt{\frac{Y(1 - \sigma)}{\rho(1 + \sigma)(1 - 2\sigma)}}.$$

The S-wave represents a traveling disturbance in the form of a shearing strain or shape deformation. As the medium is transversed by an S-wave, the medium experiences no volume changes. The shearing strain produces a transverse displacement of the particles relative to the propagation direction; that is, the particles move in a direction at right angles to the direction of the traveling wave. An S-wave moving in the horizontal direction can be polarized so that the particle motion is confined to the horizontal plane; such a wave is called an SH-wave. When the plane of polarization is vertical, we have an SV-wave. The most common expressions for the S-wave velocity are

$$v_s = \sqrt{\frac{\mu}{\rho}} = \sqrt{\frac{Y}{2\rho(1 + \sigma)}}.$$

We immediately see that in liquids, which have a rigidity $\mu = 0$, S-waves do not propagate. Thus, in offshore exploration, where the detectors are hydrophones located within a cable in the water, we do not record S-waves.

A typical value of σ is 0.25. For such a medium the ratio

$$\frac{v_p}{v_s} = \sqrt{\frac{2(1 - \sigma)}{1 - 2\sigma}}$$

becomes $\sqrt{3}$, so that the P-wave is about 1.7 times as fast as the S-wave. Examination of this ratio shows that the P-wave velocity of a medium is always greater than its S-wave velocity. This explains the name S-wave, which stands for secondary wave. In accordance with the earlier remark that S-waves are used only rarely in seismic exploration, in what follows the terms *wave velocity, material velocity,* or simply *velocity* will always refer to the P-wave velocity.

Among the various factors that influence seismic wave velocity are lithology, structural position, depth of burial, and geologic age. Some typical values are

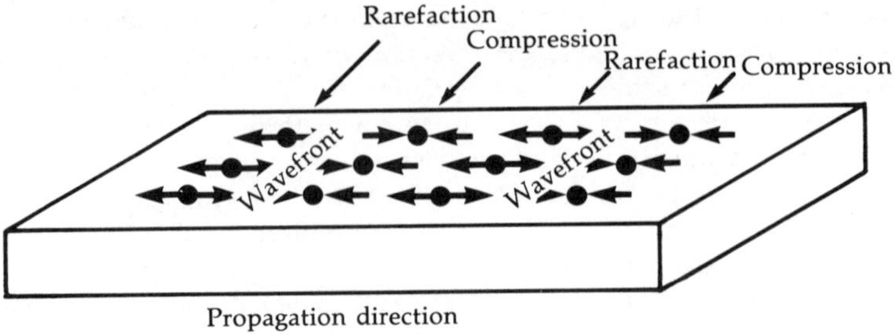

Figure 1.18. Depiction of a P*-wave.*

given in table 1.1. Usually, the greater the geologic age or the greater the depth of burial, the higher the velocity. In view of the expression for the P-wave velocity, we might think that the increase in density that accompanies greater compaction would decrease rather than increase velocity. Generally, however, the elastic constants λ and μ increase to a greater degree than does ρ, so that the velocity does increase with depth in such cases. Lithology is the most important factor, and its effect must take into account mineral composition, porosity, degree of cementation, and the nature of the interstitial fluid. Thus, the type of rock has the dominant effect on velocity. Among the sediments, limestones and evaporates generally have higher velocities than sandstones and shales. The effect that interstitial fluids have on velocities has produced much research in recent years, leading to methods for the detection of hydrocarbons.

Petroleum is found in traps associated with the sedimentary layers in the earth's crust. As a seismic wave travels in the layered earth, it encounters the interfaces between the various layers. Whenever a wave meets such an interface between media of different elastic constants and densities, the wave divides into several parts. A line drawn to represent a particular path along which a wave propagates is called a ray (see fig. 1.19). In an isotropic medium, the raypaths are perpendicular to the wavefronts. In the theory of seismic exploration, the geometry of raypaths is very important (see fig. 1.20). The raypath geometry can be quite complicated because of the new raypaths that occur each time an incident wave is divided into new waves at an interface. Usually, we assume

Table 1.1. P*-wave velocities*

Material	Depth (ft)	Velocity (ft/sec)
Sand	0	600–6500
Alluvium	0	1600–6500
Water	0	5000
Shale and sandstone		
Pleistocene	3000–4000	8000
Eocene	3000–4000	9000
Cretaceous	3000–4000	11000
Limestone		
Cretaceous	3000–4000	13500
Permian	4000	15500

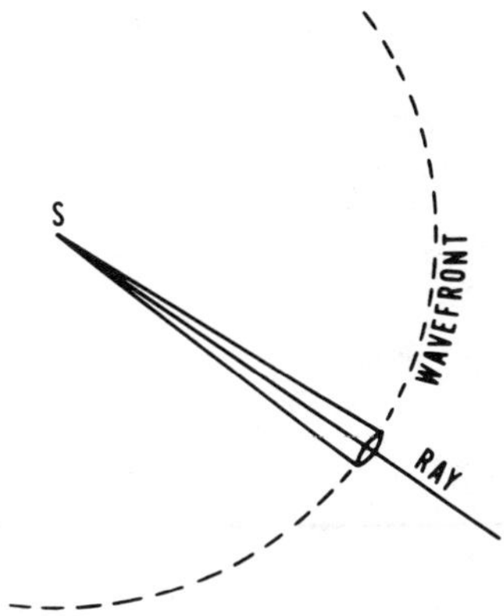

Figure 1.19. The relation between the ray and the wavefront.

that each individual layer is homogeneous and isotropic, so that the velocity is constant and the raypaths are straight within each such layer. In many cases, the use of these straight raypaths is justified, because the results obtained fall within the limits of error that we can tolerate for the particular prospect.

Figure 1.21 depicts the raypath of an incident P-wave and an incident S-wave that meet a plane interface at angles θ_i and ϕ_i, respectively. Four different waves result. One P-wave and one S-wave are reflected with angles θ_r and ϕ_r, respectively. Also, one P-wave and one S-wave are refracted (i.e., transmitted) with angles θ_t and ϕ_t, respectively. Snell's law states:

$$\frac{\sin \theta_i}{\alpha_1} = \frac{\sin \phi_i}{\beta_1} = \frac{\sin \theta_r}{\alpha_1} = \frac{\sin \phi_r}{\beta_1} = \frac{\sin \theta_t}{\alpha_2} = \frac{\sin \phi_t}{\beta_2}.$$

Here, α_1 and β_1 are the P-wave and S-wave velocities, respectively, in medium 1, and α_2 and β_2 are the velocities in medium 2. We immediately see that the law of reflection holds:

$$\theta_r = \theta_i, \qquad \phi_r = \phi_i.$$

Two limiting conditions are of interest. One is when the path of the incident wave is normal to the interface (i.e., when the angle of incidence is zero). In such a case, only two waves are formed, one reflected and one transmitted, both of the same type as the parent. Thus, for normal incidence, there is no conversion from one type of wave to the other (i.e., no conversion from P to S for an incident P-wave and no conversion from S to P for an incident S-wave). The other limiting condition occurs when the path of the transmitted P-wave is parallel to the interface. We speak of the incident angle θ_i in this case as the critical angle (see fig. 1.22).

Seismic events and their time-distance curves

The basic physical experiment in seismic exploration consists of a single source point S and many receiver points R, which lie along a straight line through the

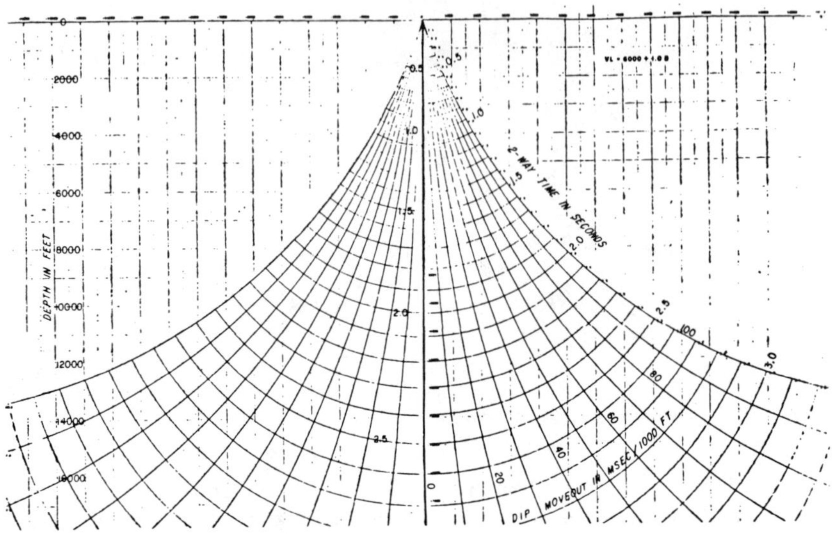

(a)

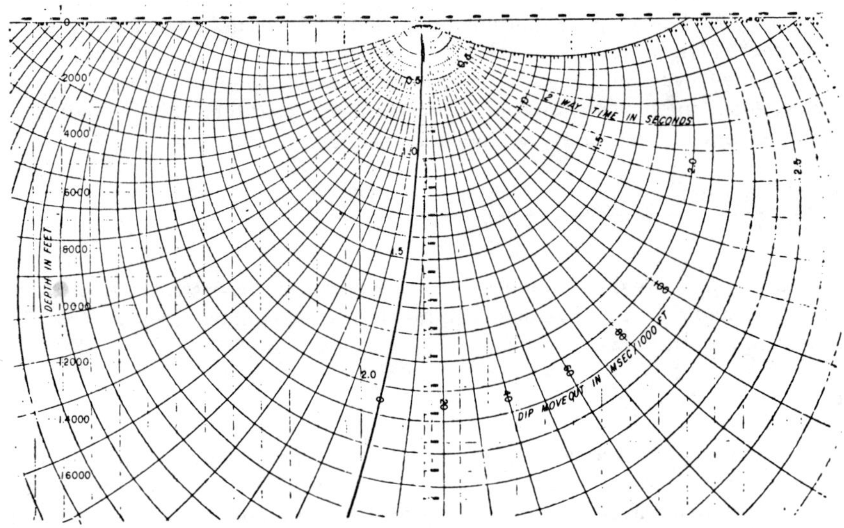

(b)

Figure 1.20. Wavefront charts: (a) wavefronts for a particular velocity function becoming farther apart with depth because of the increase of velocity with depth, with raypaths orthogonal to wavefronts; (b) wavefronts where velocity increases from right to left as well as with depth. Courtesy Chevron Oil Company. Reprinted by permission of the Society of Exploration Geophysicists from Sheriff 1973.

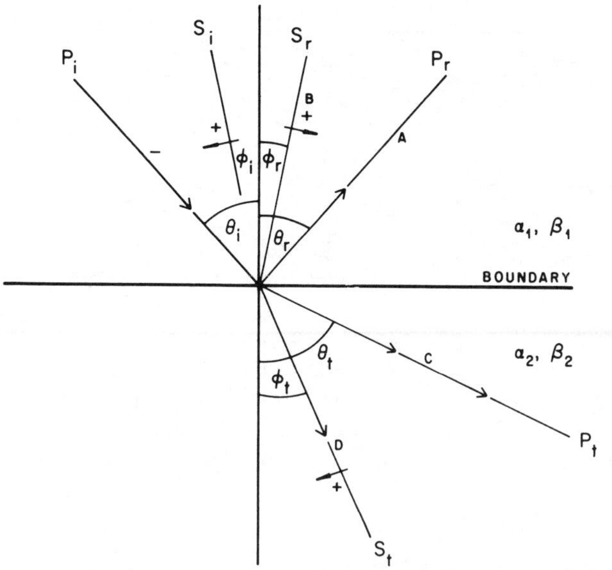

Figure 1.21 Angular relations between incident, reflected, and refracted rays when a wave encounters an interface between two different solid layers. Reprinted by permission of John Wiley & Sons, Inc. from Reflection Seismology by K.H. Waters 1978. Copyright © 1978 John Wiley & Sons, Inc.

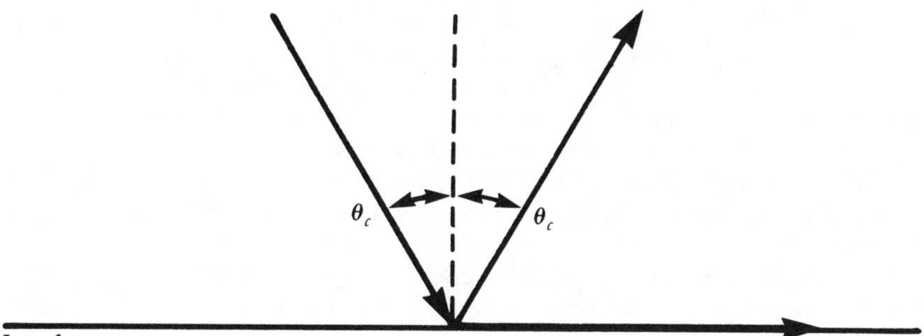

Interface

Figure 1.22. The critical angle.

source point. For the purposes of discussion, we assume that the source point and the receiver points all lie on a straight line on the surface of a flat earth. The line is called the line of exploration (see Fig. 1.23).

At the source point, seismic energy is generated at seismic reference time zero. Several types of seismic source are in common use: explosion of dynamite in a shot hole, explosion of gases in a confined chamber (Dynoseis), vibration-inducing machines (Vibroseis), release of compressed air into the water (air gun used in offshore exploration), release of high-pressure steam into the water (Vaporchoc), and discharge of electricity into the water (sparker). The seismic energy leaves the source and travels through the earth, where it is reflected and refracted at the interfaces between sedimentary layers. Some of the original

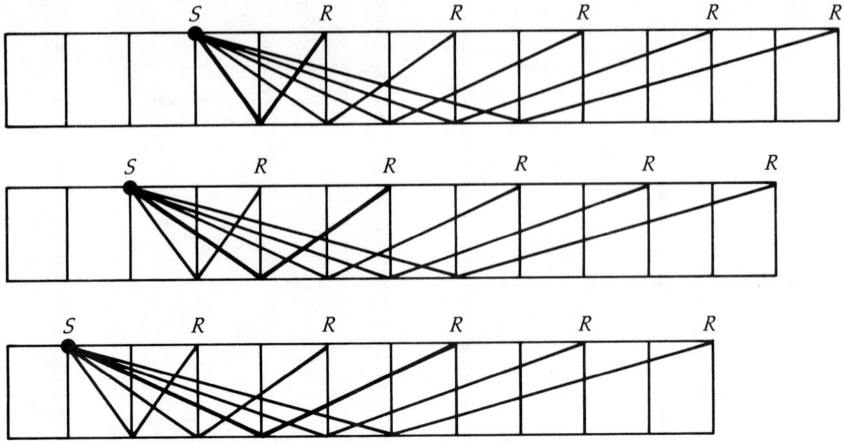

Figure 1.23. Three seismic experiments, performed one after the other. Each has one source point S *and five receiver points* R.

energy finally makes its way back up again to the surface of the earth, where it is picked up by seismic detectors. The detected energy is then transmitted by cables or by radio to the instruments, where it is recorded on magnetic tape. One signal is recorded for each receiver position R. In seismic terminology, such a recorded signal is called a *seismic trace.* A seismic trace is a function of time, with zero time the time that the source was initiated. Thus, a single physical experiment consists of one source excitation that results in many seismic traces—one seismic trace for each receiver position. Each seismic trace is labeled by the horizontal position s of the source point S and the horizontal position r of the receiver point R. Since a seismic survey consists of many such physical experiments in which the source and receiver points are moved systematically, it is absolutely necessary to identify each trace by its position coordinates (s, r) (see fig. 1.24).

We have described the seismic trace as the signal received for a particular source-receiver pair (s, r). Actually, only part of the seismic trace consists of useful signal; the remaining components must be regarded as noise. Some of this recorded noise may be associated with the source; for example, the source might be delayed or might misfire, bubbles of gas from the source may oscillate in the water, or the source pulse may be distorted in some other way. Other noise may be associated with the detectors. Energy from extraneous sources, such as animals or traffic, may be picked up by the detectors. Noise can be introduced because of poor coupling of the detectors to the ground. The cables that transfer the received signal from detectors to instruments may pick up interference from nearby electric power lines. The instruments themselves can introduce noise, especially if poor instrument settings were selected or if the instruments are malfunctioning. Recording on magnetic tape can involve loss of information. Source noise, receiver noise, and instrument noise are all important, but they can be handled and controlled by careful field work. If we disregard such noise, we are left with a seismic trace that represents a signal whose properties are due to transmission through the earth.

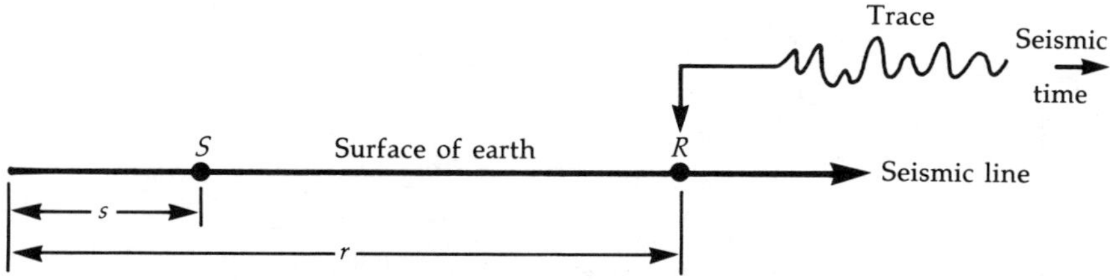

Figure 1.24. The trace is identified by position coordinates (s, r)

Because we are interested in the structure of the earth, we may regard the entire seismic profile as potentially useful information (see fig. 1.25). One component of particular interest, which represents the desired or wanted component, is made up of the primary reflections. One primary reflection is associated with each reflecting interface and represents the energy that traveled directly down from the shot to the interface, where it was reflected and then returned directly to the receiver in question. In addition to the primary reflections, there are many multiple reflections. These multiples followed paths that suffered one or more reflections at other interfaces, so they followed a circuitous route instead of the direct one of the primary reflection (see fig. 1.26).

Thus, in reflection seismic exploration, the useful signal consists of primary reflection events. The received trace also consists of multiple reflections, as well as energy that traveled horizontally in a direct path from the source to the detectors, or in shallow, high-velocity layers. It also consists of received surface waves (ground roll) and shear waves, as well as diffractions that represent energy that has been scattered by small features in the earth.

Except for the primary reflections, all other energy represented on the seismic trace is usually regarded as various kinds of noise or unwanted signal. The characteristics of the types of noise differ. Many are coherent; that is, they arrive at different detectors in a systematic fashion. Others are incoherent; that is, they appear to be random. Noise caused by wind, for example, would be random. In some cases, the source of the noise and its physical characteristics are clearly understood, and the process of separating it from the useful signal is straightforward. In other cases, the separation is statistical. Sometimes we have a model of the cause of the noise, as in the variations in arrival times that occur because the earth is not flat and the elevations of the detectors differ, or because the depths of the sources differ. At other times, we have to make use of a theoretical model of wave propagation.

In simple terms, we can think of the seismic source as producing a very sharp pulse at time zero. This pulse divides itself into wavelets, and each wavelet travels along a particular raypath from the source to the receiver. The time at which the wavelet arrives at the receiver is called the time of the event associated with that raypath. Thus, the received seismic trace is made up of many superimposed wavelets, each wavelet associated with its particular raypath and each wavelet appearing on the seismic trace at its respective arrival time.

Let us choose a given event—for example, the primary reflection from a horizontal interface (see fig. 1.27). For the source S, each receiver R picks up a wavelet corresponding to this event. Each receiver R has a given offset distance x (i.e., the distance from the source to that receiver). Each wavelet in the

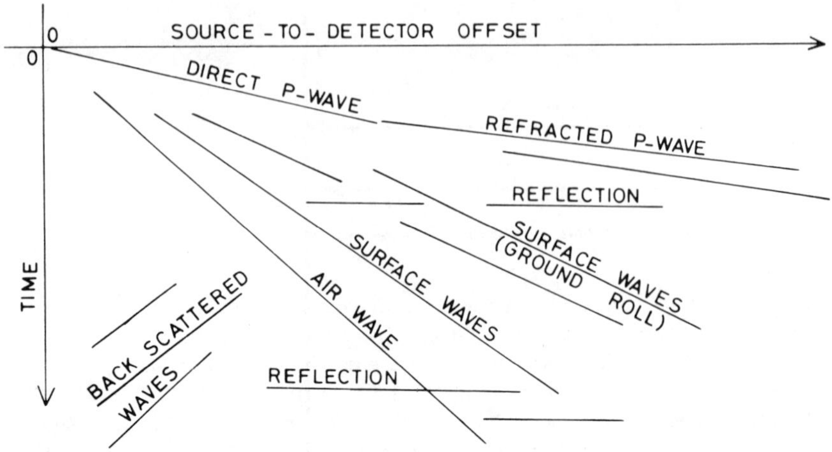

Figure 1.25. A schematic diagram showing what can appear in a typical section. The lines represent the arrival times of the different phases as a function of the horizontal source-to-detector distance. From the slopes of these lines, the apparent velocities of these phases can be measured. Reprinted by permission of Birkhauser-Verlag from Al-Sadi 1980.

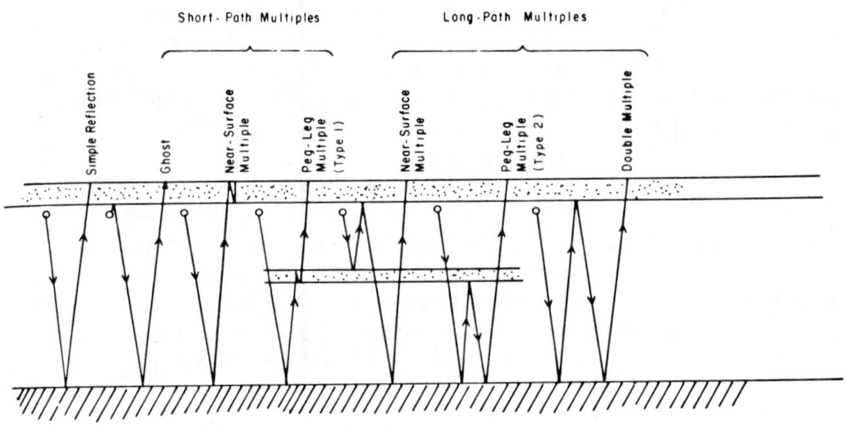

Figure 1.26. The simple or primary reflection together with some different types of multiple reflections. Reprinted by permission of the Society of Exploration Geophysicists from Sheriff 1973.

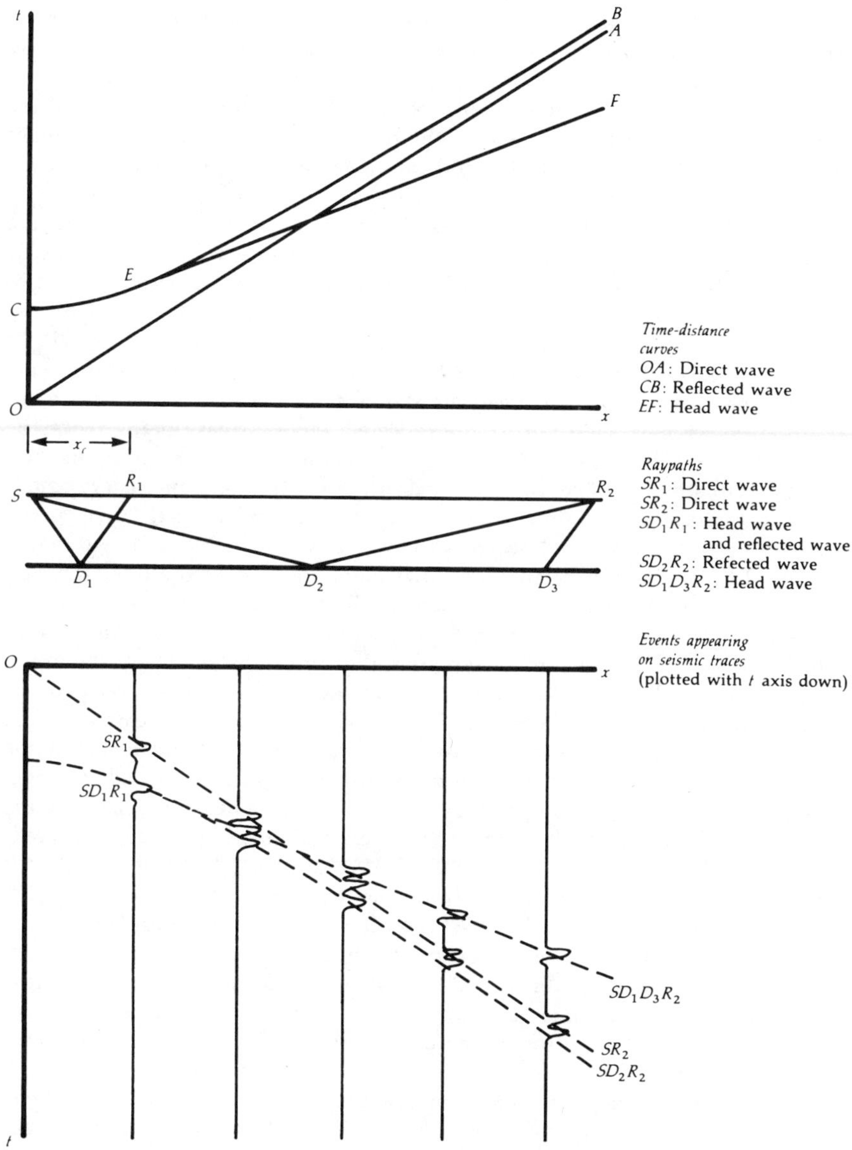

Figure 1.27. The time-distance curve is a curve drawn through an event that we can see on the seismic profile. It is thus the theoretical manifestation of an event that we can see and measure on the received traces.

reflected event is associated with a different arrival time t, depending on x. The curve of t versus x is called the *time-distance curve* for this event, denoted by $t(x)$. A time-distance curve is thus a plot of arrival time of an event versus the horizontal distance from source to receiver. Through analysis of traveltime curves, the geophysicist can identify the type of the returned wavelet, obtain information concerning velocity, and be aided in the processing of seismic data.

Figure 1.28 shows the relationships of the raypath diagram and the traveltime curves of the seismic events for the case of two flat layers. The upper layer is layer 1, with velocity v_1; the lower layer is layer 2, with velocity v_2. Several observations can be made from this figure. When the horizontal distance from source to receiver is zero (i.e., the case of a coincident source-receiver pair, or offset $x = 0$), the direct-wave arrival time is zero. (The direct wave is the wave with a horizontal raypath from source to receiver.) Also, for the coincident source-receiver pair, we see that the reflected wave arrival time is twice the depth of the interface divided by v_1. The slope of the reflected time-distance curve is zero at $x = 0$. The slope of the direct-wave time-distance curve equals $1/v_1$. In addition to the direct raypath and the reflected raypath, there is the critically refracted raypath, known as the *head wave.* This refracted raypath hits the interface at the critical angle, follows along just below the interface, and then leaves the interface at the critical angle. The slanted segments correspond to velocity v_1, whereas the horizontal segment corresponds to v_2. The case in which the horizontal segment is zero gives the critical distance x_c of the initial head wave. At distances less than x_c, only the direct-wave arrival and the reflected-wave arrival are present. The head-wave arrival first appears at x_c, and at this distance the arrival time of the head wave and the arrival time of the reflected wave are equal. At x_c, the direct wave is also present and it arrives sooner; that is, at x_c, the quickest way to go from source to receiver is along the horizontal path (direct path). At x_c, the slopes of the head-wave and reflected-wave time-distance curves are the same and equal to $1/v_2$. When the offset is less than the crossover distance x_d, the first arrivals (often called first breaks) are from the direct wave. When the offset is greater than x_d, the first breaks are from the head wave. Thus, for refraction prospecting, the receivers are spread at a distance greater than x_d. As the offset distance becomes large, the reflection time-distance curve becomes asymptotic to the direct time-distance curve. Figure 1.29 depicts the wavefronts for the case given in figure 1.28. Another case is shown in figure 1.30.

On the seismic profile, we pick out events and draw curves through them. Such empirical curves are the theoretical manifestations of the theoretical time-distance curves. A smooth curve through the reflection wavelets would be the reflection time-distance curve, and likewise for the other time-distance curves. Because we can often see the events on the display of the seismic traces, we can draw such smooth curves and thus physically obtain the mathematical concept of time-distance curves. Thus, a time-distance curve is not an abstract concept but has physical manifestation.

The set of traces from a single physical experiment—namely, one source point and many receivers—has a special name: the common source point (CSP) gathers or CSP profile. The physical experiment is repeated for many different source locations by moving the source and receivers in the direction of exploration. The

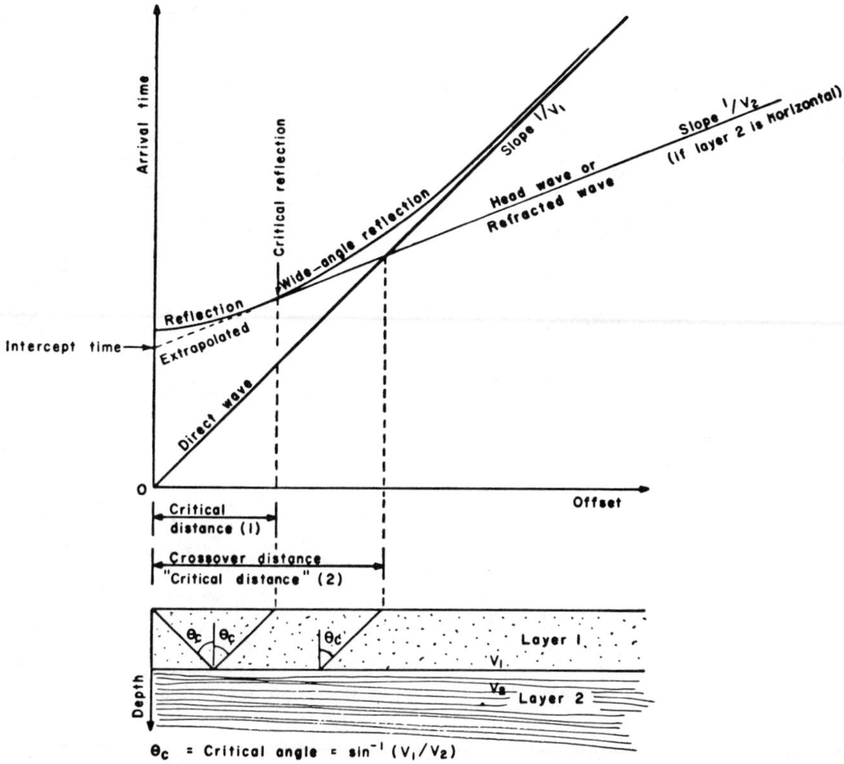

Figure 1.28. Critical distance. The curvature of the reflection is hyperbolic. The reflection amplitude may be exceptionally large in the vicinity of critical reflection. Critical distance is used in either of two different meanings: (1) the preferred usage—the offset x_c *at which the reflection time equals the refraction time; that is, the offset for which the reflection occurs at the critical angle; (2) the offset* x_d *at which a refracted event becomes the first break; the intersection point for the traveltime curves for two refractors. The preferred term is* crossover distance. *Reprinted by permission of the Society of Exploration Geophysicists from Sheriff 1973.*

data are then processed by a computer and displayed in final form. Figure 1.31 exhibits a stacked seismic section, whereas figure 1.32 shows the same data after it has undergone wave-equation migration. If we compare the two sections, we see that the effect of migration has been to give a better picture of the actual cross section through the earth.

Seismic data collection

The purpose of the seismic reflection method is to determine the distance and directions of remote and inaccessible bodies within the earth. The essential features of the process of seismic data collection are as follows:

(1) Active sources of energy at the surface of the earth, such as dynamite, air guns, and chirp signal generators.
(2) Propagation of waves outward from the sources into the earth.

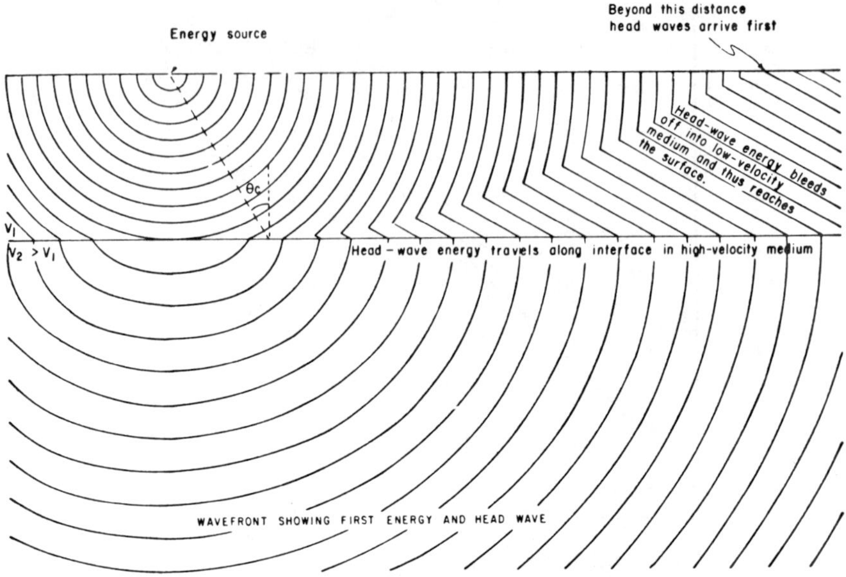

Figure 1.29. Head waves and wavefronts showing first energy. Reprinted by permission of the Society of Exploration Geophysicists from Sheriff 1973.

(3) Reflection of waves from the interfaces between geological layers in the earth's crust.
(4) Propagation of waves upward from the interfaces.
(5) Detection of the reflected waves at the surface of the earth by a spread of receivers located at various distances from the source.

The recorded traces of the received waves make up the seismograms, or seismic record section, and represent the basic data that must be analyzed and interpreted. Because each receiver is usually made up of a group or an array of geophones or hydrophones, the term *group* is often used for receiver.

Each recorded trace is a time series made up of reflected events together with various interfering waves and noise. The desired reflected events are the primary reflections—that is, waves that travel down to a given reflector and then back up to the surface, where they are recorded. An important type of undesirable interfering wave is the multiple reflection. A multiple reflection has a raypath that goes down to a given reflector, then up to another reflector, then down to still another reflector, then up again, and so on, until the multiple reflected event is recorded on the trace (see fig. 1.33). In any layered system, there are infinitely many possible multiple reflections. The presence of multiple reflections makes the identification of primary reflections difficult, and thus it is necessary to attenuate the multiples as much as possible on the final display. Hence, in the computer processing, in addition to making the aforementioned corrections, it is necessary to attenuate the multiple reflections and other noise.

A common type of multiple is the ghost reflection (see fig. 1.34), which occurs when a wave travels upward from a source located at a certain depth and is reflected by the surface or by the base of a low-velocity surface layer (the

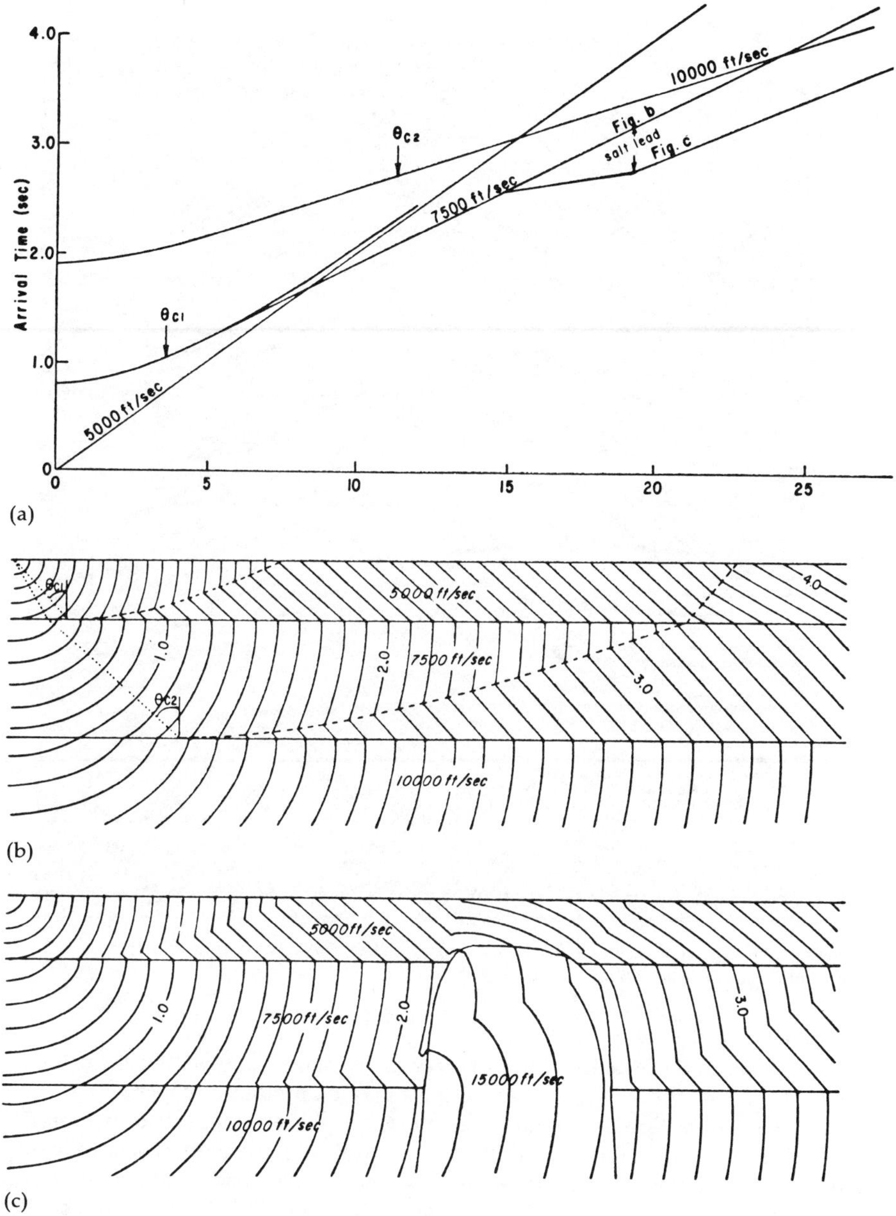

Figure 1.30. (a) Time-distance curve. (b) Wavefront diagram showing first arrivals. (c) Wavefront diagram showing an idealized salt-dome. Reprinted by permission of the Society of Exploration Geophysicists from Sheriff 1973.

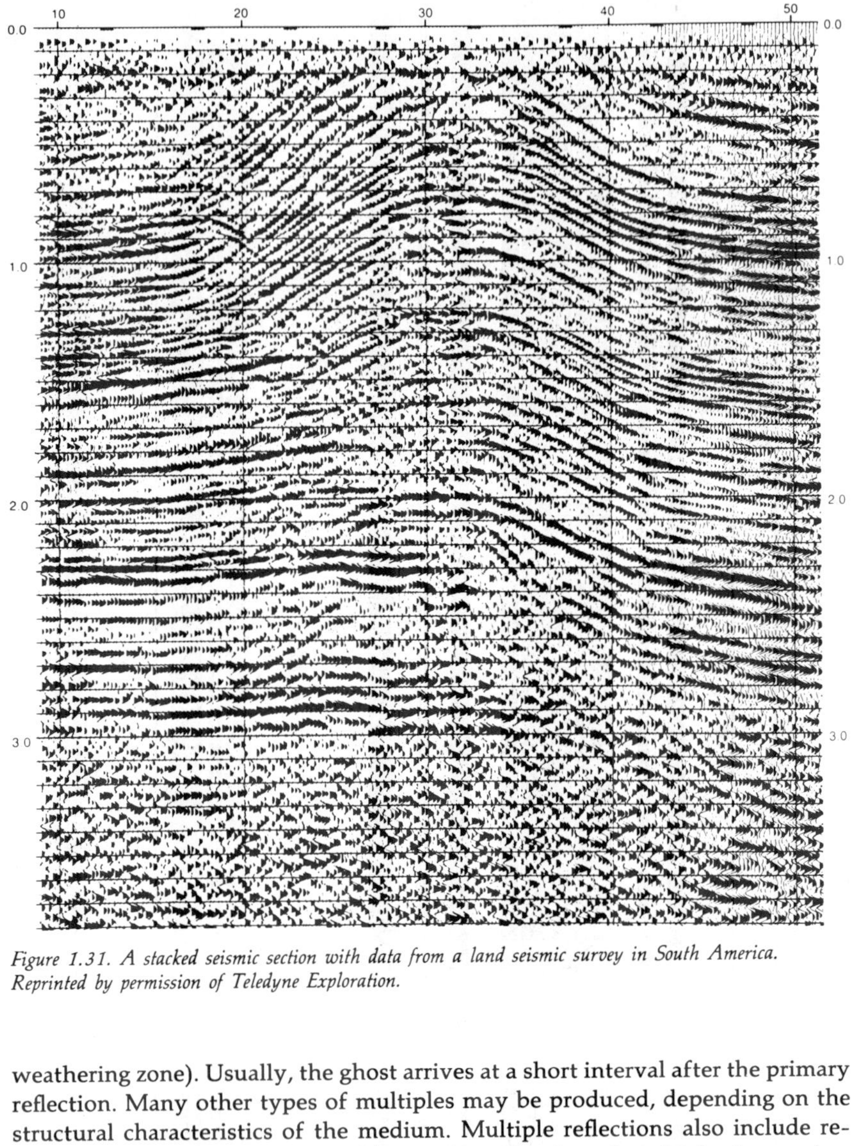

Figure 1.31. A stacked seismic section with data from a land seismic survey in South America. Reprinted by permission of Teledyne Exploration.

weathering zone). Usually, the ghost arrives at a short interval after the primary reflection. Many other types of multiples may be produced, depending on the structural characteristics of the medium. Multiple reflections also include reflected refraction or refracted reflection. Identification of multiples is very important. Mistaking such an arrival for a primary reflection introduces a serious error into the interpretation of reflection data. Multiple reflections are usually of less energy than the singly reflected (or primary) waves. An exception would be a multiple that has reached the surface after being reflected several times from a number of strong interfaces that have large reflection coefficients.

To achieve these goals, the seismic data are collected in a special way. For each source, a spread (or array) of detectors is laid out. The source is activated (the shot), and the traces are recorded. Then the entire configuration is moved laterally and the process is repeated. By moving the configuration in small

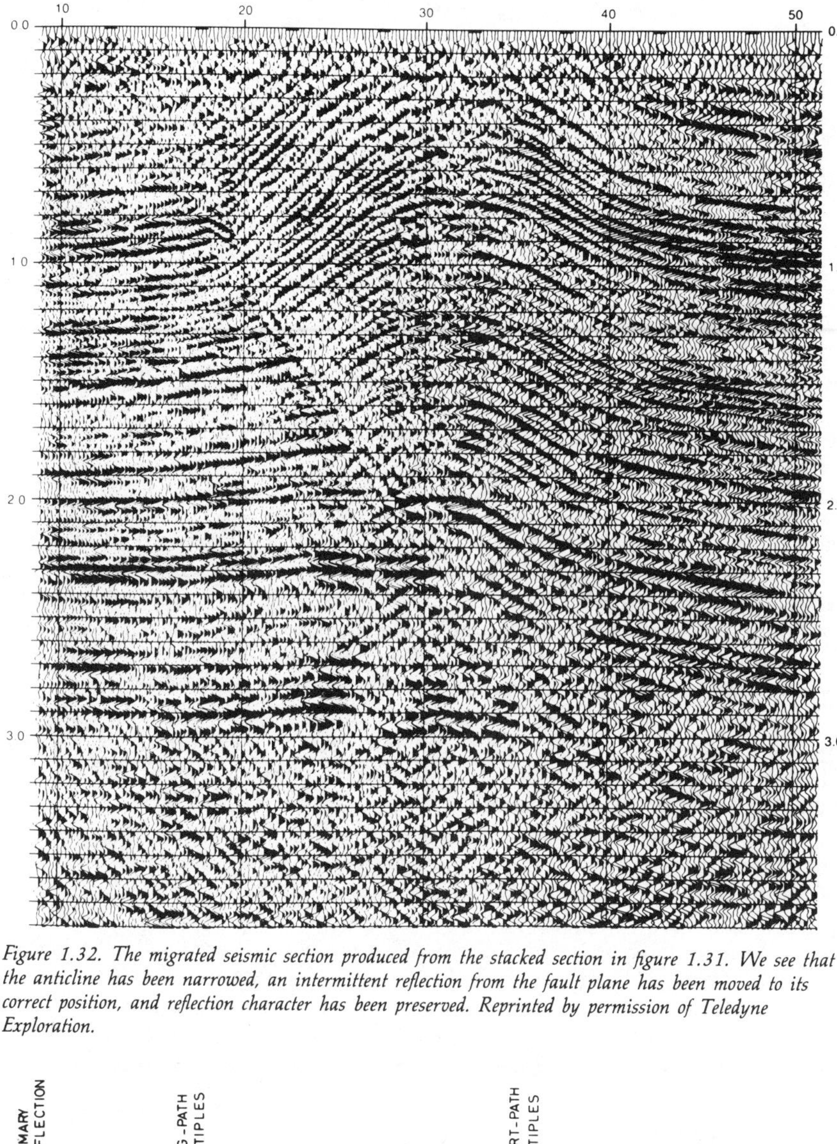

Figure 1.32. The migrated seismic section produced from the stacked section in figure 1.31. We see that the anticline has been narrowed, an intermittent reflection from the fault plane has been moved to its correct position, and reflection character has been preserved. Reprinted by permission of Teledyne Exploration.

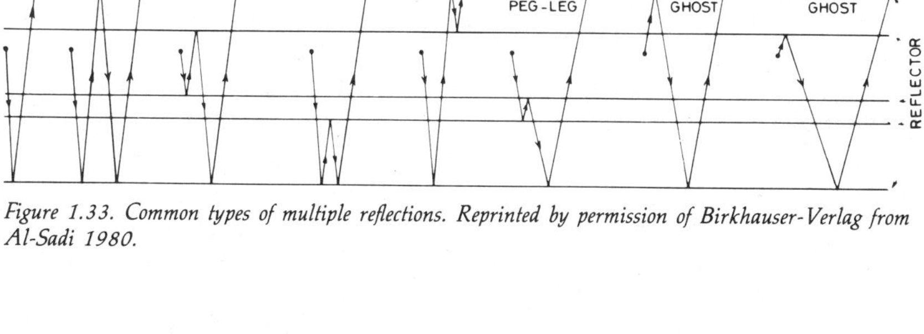

Figure 1.33. Common types of multiple reflections. Reprinted by permission of Birkhauser-Verlag from Al-Sadi 1980.

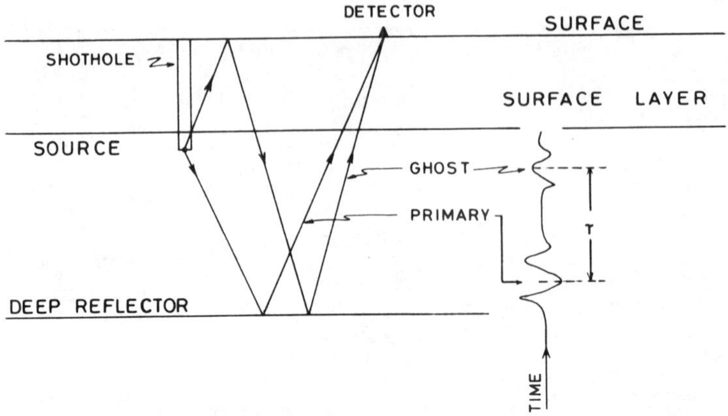

Figure 1.34. Ghost reflection. Reprinted by permission of Birkhauser-Verlag from Al-Sadi 1980.

enough increments, each depth point is covered several times (multiple coverage). In sixfold coverage, six traces will be recorded for each depth point, from the near trace, which has the shortest offset distance between source and receiver, to the far trace, which has the longest. Thus, the recording of seismic data by a multiple coverage scheme introduces a considerable amount of overlap or redundancy. The redundancies allow us to make the necessary corrections and to attenuate severely the unwanted interference and noise. The word *multiplicity* is used to refer to the number of common midpoint traces that sample essentially the same portion of reflector but with different offsets. A 12-fold CMP recording, for example, involves recording each subsurface point 12 times, once from each of 12 different offset distances.

A basic diagram in the study of seismic data collection is shown in figure 1.35. The figure shows a raypath from surface source S to horizontal interface, where reflection occurs at depth point D. The raypath then returns to the receiver R at the surface. The midpoint between source and receiver is labeled M. Because the interface is horizontal, the reflection point D occurs directly below the midpoint M. (This geometry follows from the fact that angle of incidence equals angle of reflection.) The horizontal axis is the x-axis, and the vertical axis is the z-axis. Depth is measured positively in the downward direction on the z-axis. Four important coordinates are represented on the horizontal axis:

(1) The source coordinate s
(2) The receiver coordinate r
(3) The midpoint coordinate m
(4) The half-offset coordinate h (the distance from either S to M or M to R)

The coordinates s and r are the primary coordinates, and m and h are derived coordinates. The relationships are

$$m = \frac{r + s}{2}, \qquad h = \frac{r - s}{2}.$$

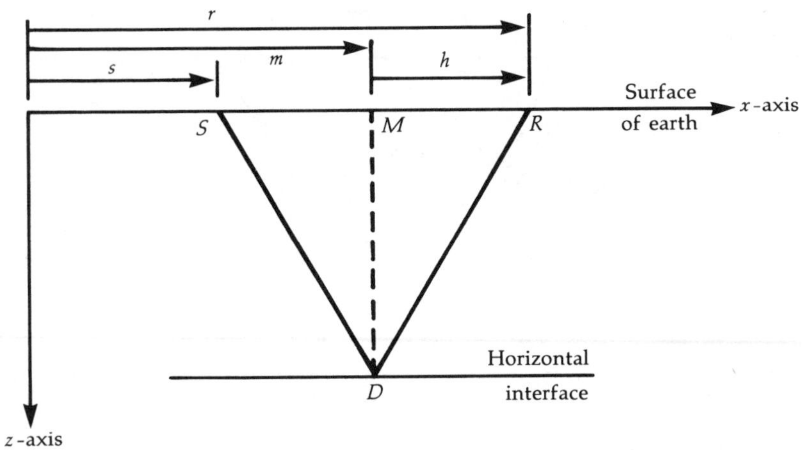

Figure 1.35. Definition of coordinates.

The definition of half-offset is taken to be $(r - s)/2$ instead of $(s - r)/2$, so that positive half-offset corresponds to waves traveling in the positive x-direction. Half-offset is used instead of total offset $(r - s)$ for reasons of mathematical symmetry. The inverses of the foregoing equations are

$$s = m - h, \qquad r = m + h.$$

A seismic trace is a time series associated with a source-receiver pair designated by the point (s, r). A seismic survey consists of collecting data associated with as many of these points as required. We cannot make a continuous survey of each and every point but must pick relatively few points that are representative of the entire area. Such a procedure represents sampling in the space domain.

Over the years, the seismic industry has developed some very powerful methods of sampling. To illustrate the multiple coverage method, let us envisage a single flat reflecting plane at depth. Supposing that our exploration efforts are restricted to a straight line along the surface, we focus our attention on a single point D on the interface plane under the surface line. In order to record a reflection from point D, the source s and the receiver r must be placed in symmetric positions with respect to the surface point M directly above the subsurface point D. Figure 1.36 shows three possible sources and the corresponding receivers required to record a reflection from point D. The surface point M is the midpoint between s_1 and r_1, between s_2 and r_2, and between s_3 and r_3. If we use this setup, we see that we record three separate reflections from the given depth point D. We have, in effect, sampled the depth point D three times; that is, the diagram represents a threefold coverage of the depth point D.

Why do we want to do this? Why is one sample of the depth point D not enough? There are several important reasons for multiple coverage with regard to noise attenuation, especially with regard to the suppression of the so-called

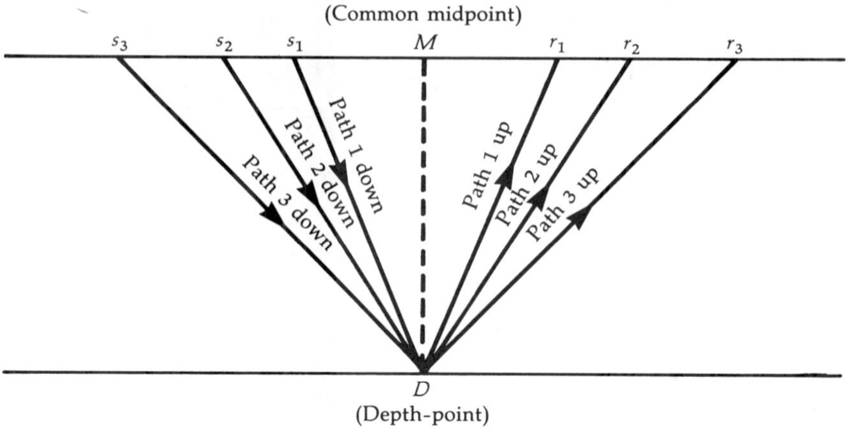

Figure 1.36. Threefold coverage.

multiple reflections. We will have much to say about these matters in the following discussions.

Figure 1.37 shows a typical geometry for the recording of seismic reflection data. A single impulsive source and four receivers are evenly spaced along a line on the surface of the ground. A single source position involves the recording of a seismic trace at each of the four receivers. The field procedure is shown in figure 1.38. The traces from source s_1 are recorded at receiver positions r_1, r_3, r_5, r_7. The distance between r_1, r_3, between r_3, r_5, or between r_5, r_7 is called the group interval. The entire experiment is then moved to the right by one-half group interval. Thus, the traces from source s_2 are recorded at receiver positions r_2, r_4, r_6, r_8. The field procedure is repeated in this manner, resulting in this case in seismic traces for four different offsets for each midpoint. A collection of such traces for the given midpoint makes up the CMP gather, which in this case is a fourfold gather. If we had moved the experiment to the right by one group interval, we would have obtained only a twofold gather.

A convenient way to display this field procedure is shown in figure 1.39. Each point represents a seismic trace for a typical marine recording geometry. The boat moves in the negative direction along a straight line and tows a source and a streamer cable of receivers. As it moves ahead each one-half group interval (where the group interval is defined as the physical distance between two adjacent receivers on the cable), it sets off a source impulse and records the resulting traces.

The entire data set from a single seismic line is a function of three coordinates, s, r, t, where t is time. Because the data set fills a three-dimensional volume, it is usually displayed by means of cross sections. These cross sections are called sections, gathers, or profiles. The four basic gathers are as follows (see fig. 1.40):

(s = constant, r, t): common source point (CSP) gather
(s, r = constant, t): common receiver point (CRP) gather
(m = constant, h, t): common midpoint (CMP) gather
(m, h = constant, t): common half-offset point (CHP) gather

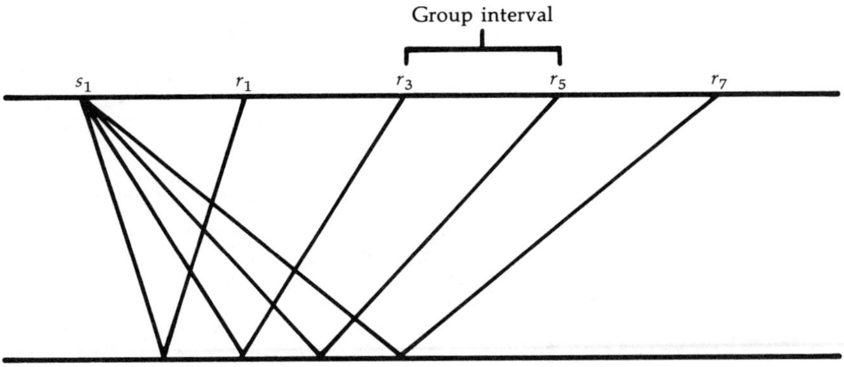

Figure 1.37. One-sided spread of receivers.

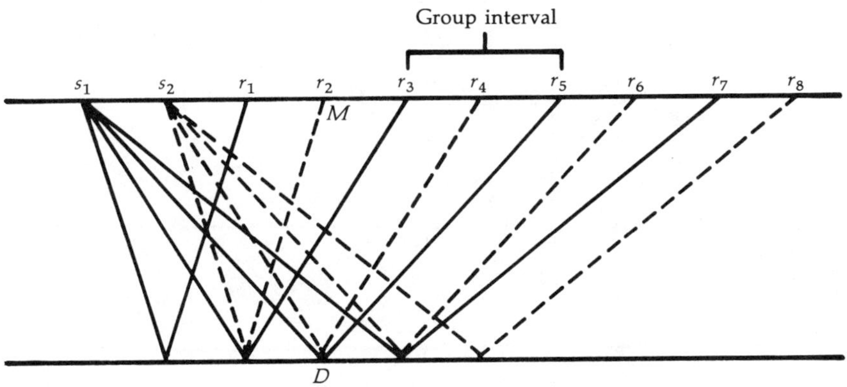

Figure 1.38. Geometry for two successive shots, s_1 *and* s_2*, showing data redundancy that allows us to make up common midpoint (CMP) gathers. The CMP gather for a midpoint* M *at location* r_2*, for example, would include trace* s_1r_5 *and trace* s_2r_4*.*

Special cases of the CHP gather (for $h = 0$, h_{min}, and h_{max} respectively) are:

$(m, h = 0, t)$: zero-offset section
$(m, h = h_{min}, t)$: near-trace section
$(m, h = h_{max}, t)$: far-trace section

Some typical seismic sections are shown in figures 1.41 and 1.42.

In seismic processing, the word *mute* is often used to refer to a change in the relative contribution of the components of a seismic section with record time. In the early part of the record, the long offset traces may be muted or excluded from the section because they are dominated by refraction arrivals or because their frequency content after normal moveout (NMO) correction is appreciably lower than other traces. The transition where they begin to contribute may be either abrupt or gradual. A large, abrupt transition introduces frequencies that might not be otherwise present and may distort the design criteria for a decon-

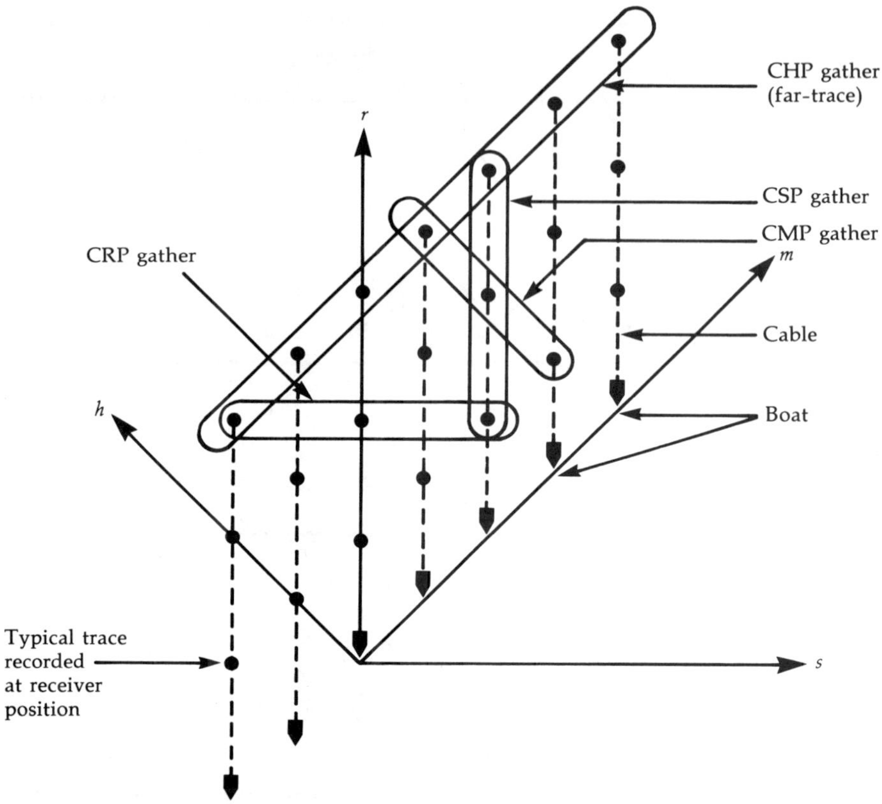

Figure 1.39. Relationships among (horizontal) x*-coordinates* s, r, m, h. *All axes represent distances on the path of the boat that moves in the negative* x*-direction. Each point represents a seismic trace. The source is at the boat, and the seismic trace is associated with the receiver* r *position.*

volution operator. Muting may be done over certain time intervals to keep ground roll, air waves, or noise bursts out of the section.

Digital processing of seismic data

Seismic data are usually processed and interpreted in a computer center in a central location, such as Houston, London, or Singapore. With the advent of minicomputers, however, more and more processing is being done in the field. In the acquisition of seismic data, countless confusing signals are recorded simultaneously with the desired primary signal. The unwanted signals are referred to broadly as noise. In a processing center, use is made of a software system that enhances the signal while attenuating the noise. A significant and continuing research and development program exists to further improve the signal content of the seismograms. Figure 1.43 shows reels of tape that store a multitude of types of data. Such magnetic tapes are processed according to the specifics dictated by the oil company and then are filed and held as inventory or returned to the oil company.

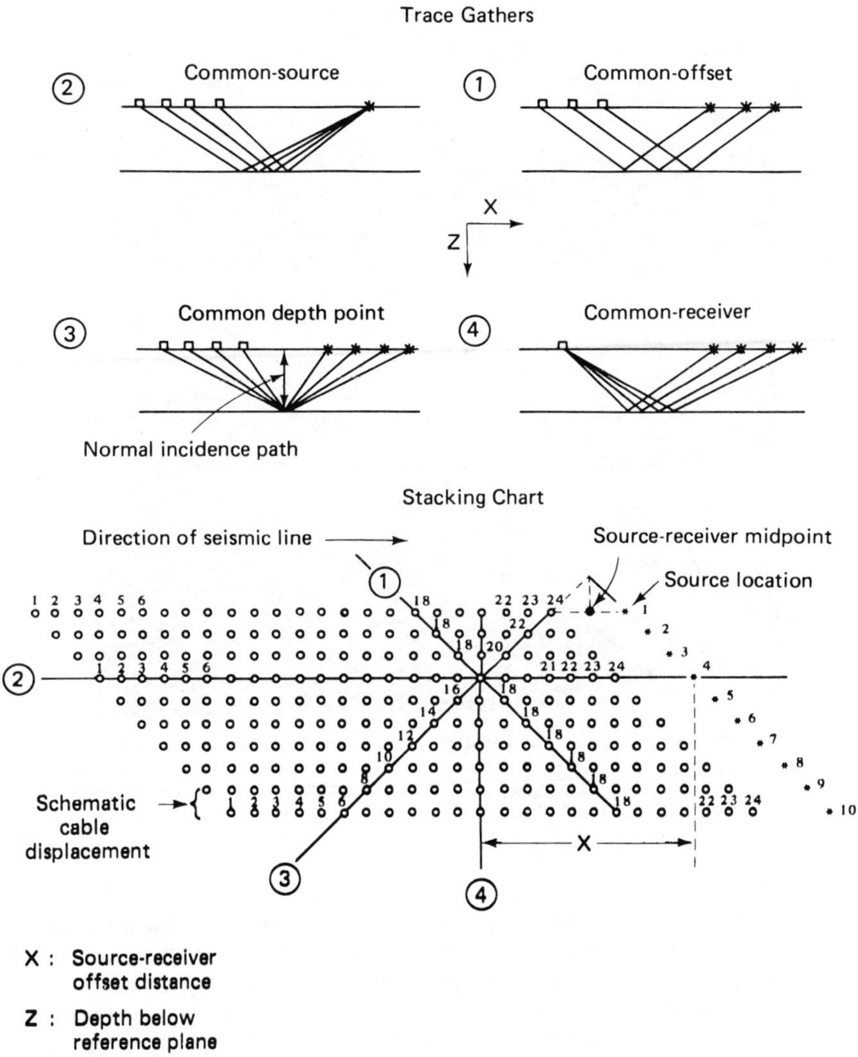

Figure 1.40. Source and receiver positions corresponding to four principal planes used for sorting seismic traces. Adapted by permission of Geophysics from Taner et al. 1974 and Prentice-Hall, Inc. from Robinson and Treitel 1980.

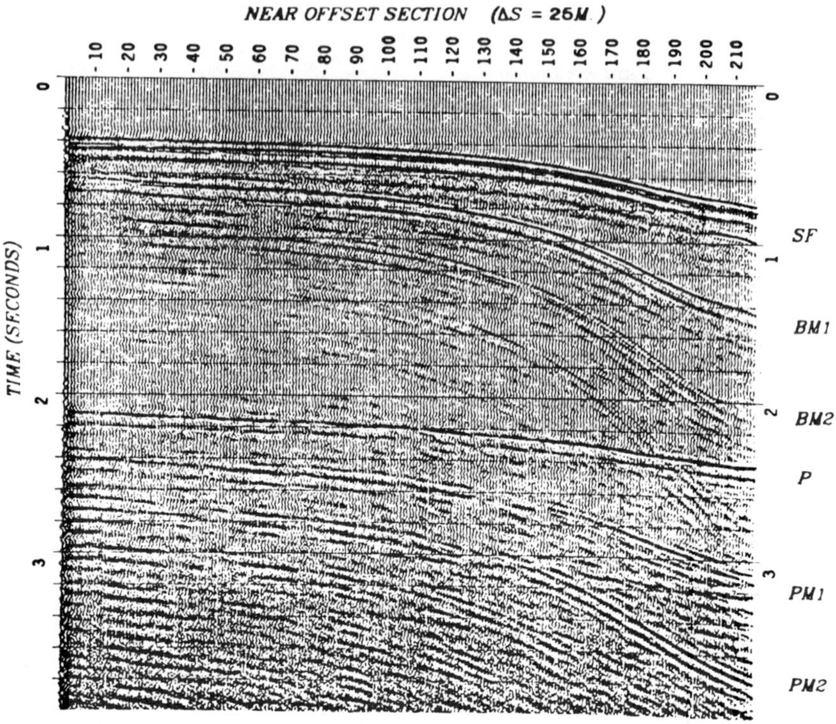

Figure 1.41. Near-offset section, offshore Labrador (Flemish Cap). Shotpoint spacing is 25 m. Offset distance is about 9 shotpoints. Labeled events are SF (seafloor or bottom), BM1 (i.e., first bottom multiple), BM2 (i.e. second bottom multiple), P (i.e., primary), PM1 (i.e., first peg-leg multiple), and PM2 (i.e., second peg-leg multiple). Reprinted by permission of Geophysics from Morley and Claerbont 1983.

Figure 1.44 shows the final product in seismic exploration—the final seismic depth section after all the processing methods have been applied. This depth section represents a portion of the geologic structure indicating the possibility of oil accumulation. On the basis of such depth sections, oil companies submit bids for acreage in government lease sales.

It may be said that geophysical exploration is pushing more and more into the remote areas of the world that have difficult accesses and operating conditions. Many offshore areas are also coming under active interest. In a seismic exploration program conducted by a seismic contractor or an oil company, literally millions of seismic traces are collected. These seismic traces must be analyzed so as to yield contour maps of the subsurface sedimentary structure. The maps are used to make decisions regarding where to drill exploratory oil and gas wells. Because of the great costs of drilling and the much greater costs of leasing potential oil-producing land or obtaining oil concessions from foreign governments, the geophysical results are expected to be accurate. They must stand up to the test of the drill. The usefulness of the seismic method rests on the fact that, relatively, it is extremely accurate. A seismic survey can delineate a geologic structure 4000 meters under the ground, for example, to an accuracy

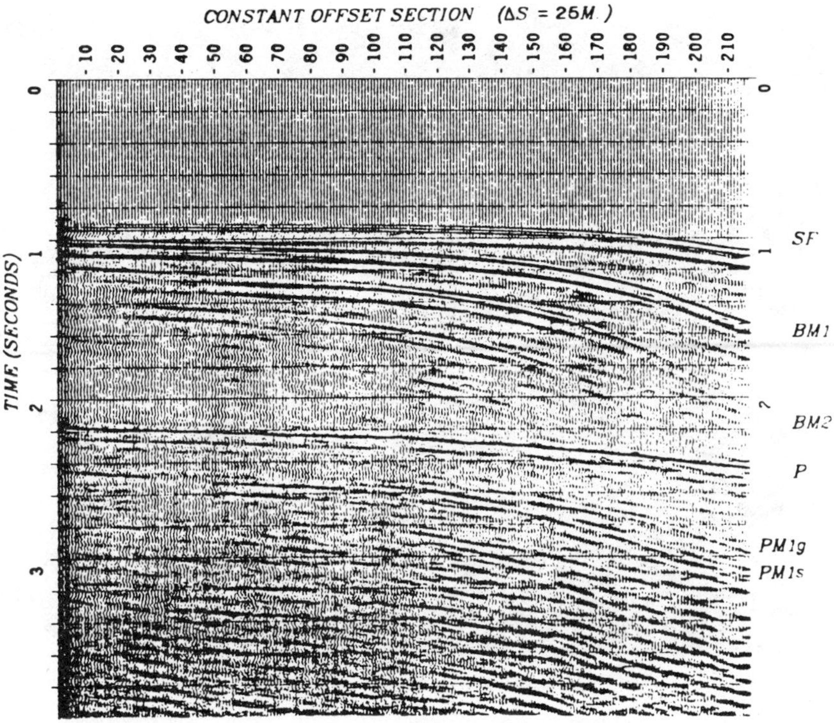

Figure 1.42. Constant offset section from same section from same line as in figure 1.41. Offset distance is about 46 shotpoints (i.e., an offset half-way down the cable). Note that the first-order peg-leg multiple is now split into two distinct arrivals: PM1s (i.e., the first peg-leg multiple at the shot location) and PM1g (i.e. the first peg-leg multiple at the geophone location). The maximum split is some 200 msec around shot points 180 to 200. This occurs, as one might expect, where the seafloor has maximum dip, i.e., where the difference between seafloor depths at the shot and geophone positions is greatest. Reprinted by permission of Geophysics from Morley and Claerbout 1983.

of the order of a few tens of meters. Such accuracy is beyond what reasonably may be expected in view of the extreme conditions under which the data are often collected. Much arctic exploration, for example, has to be done in the winter while the tundra is frozen; temperatures of 50° below zero, storms, and winter darkness make this work dangerous. The accuracy of a seismic survey depends on solving difficult interpretation problems caused by the intricate wave patterns that result from the complex configurations of the strata in the upper 6 km of the earth's crust. The purpose of digital processing of seismic data is to convert the complex wave patterns recorded on the seismic records into meaningful information that can be used to determine the underground structure.

Ideally, it would be nice to be able to put all the data into a computer, fit the necessary wave equations, and come out with the final solution. We are far from such an ideal method, however. Instead, we must process the data through a sequence of operations, each of which involves serious approximations and shortcuts. Seismic analysis is so complex that even an approximate understanding of what parameters are important and what are not in a given situation is

Figure 1.43. Storage of seismic data. Courtesy of Digicon Geophysical Corp.

worth having. Each of the operations is based on the physical point of view acquired over the years by the field geophysicists who analyze and interpret the reflection seismic records. One of the things learned is that the analysis requires an understanding of the statistical makeup of the seismic record section. We do not want to know where every seismic wave is actually moving; rather, we want to know how many move here and there on the average and the mutual buildup and cancellation of effects. This statistical interpretation forms the basis of all the important digital processing techniques used in seismic data analysis: velocity analysis, static and dynamic corrections, stacking, source signature analysis, deconvolution, wavelet processing, and migration. To put each technique into context with the other techniques in the processing sequence, we will briefly describe these operations.

In the petroleum industry today, few major decisions are made without the aid of a computer and applications software systems. For many years, data services, skilled people, and applications software have helped petroleum companies tackle problems that never could have been solved without computer technology. Today, an oil or geophysical company would make use of a complete seismic processing system supported on a large-scale computer that is dedicated to oil exploration and production applications. Access to the system, documentation, and consulting services are available to the company's geophysicists. Such a system provides large-scale computing power with industry-recognized applications software from leading consultants and petroleum service companies. A typical software system serves as a geophysical language for controlling geophysical data processing. The seismic processing system provides

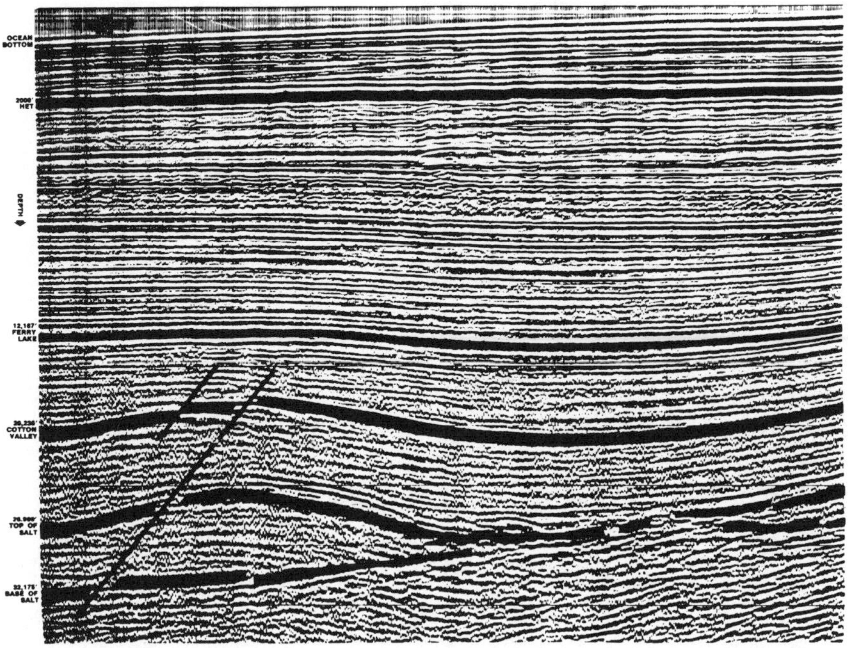

Figure 1.44. Final seismic depth section. This final seismic section represents a portion of a geologic structure indicating the possibility of oil accumulation and provides the basis on which oil companies submit bids for this acreage in a government offshore lease sale. Courtesy of Digicon Geophysical Corp.

complete seismic processing capabilities, such as basic and advanced seismic processing algorithms, a comprehensive set of standard seismic processing analyses, and methods of arranging seismic system job processing. Modularity gives the geophysicist the ability to control the seismic processing sequence.

All of the currently used field tape formats can be demultiplexed and sorted. These formats include SEG-A, SEG-B, and SEG-C, both gapped and gapless. Various conventional seismic acquisition geometries can be accommodated, as well as the preprocessing of areal array data. Output from the preprocessing function can vary, but it is normally output in the SEG-Y format. A functional diagram of the preprocessing system is shown in figure 1.45. As an aid in setting up subsequent processing steps, so-called special sorts can be generated on the preprocessor. The special sorts include near-trace sorts, field record sorts, single coverage sorts, and selected CMP sorts. Several utility programs available on the preprocessor also serve as analytical aids; these are coverage diagrams and formatted dump programs. Formatted dump programs aid in examining both field tapes and preprocessed tapes for troubleshooting purposes.

The software system is a problem-oriented symbol language as well as a seismic processing system. It consists of a monitor, a number of modules or elements of seismic processes, and a set of system jobs for performing selected standard or special seismic processes. Using the system language, geophysicists can build their own jobs using existing modules. New modules can be added and used along with existing modules.

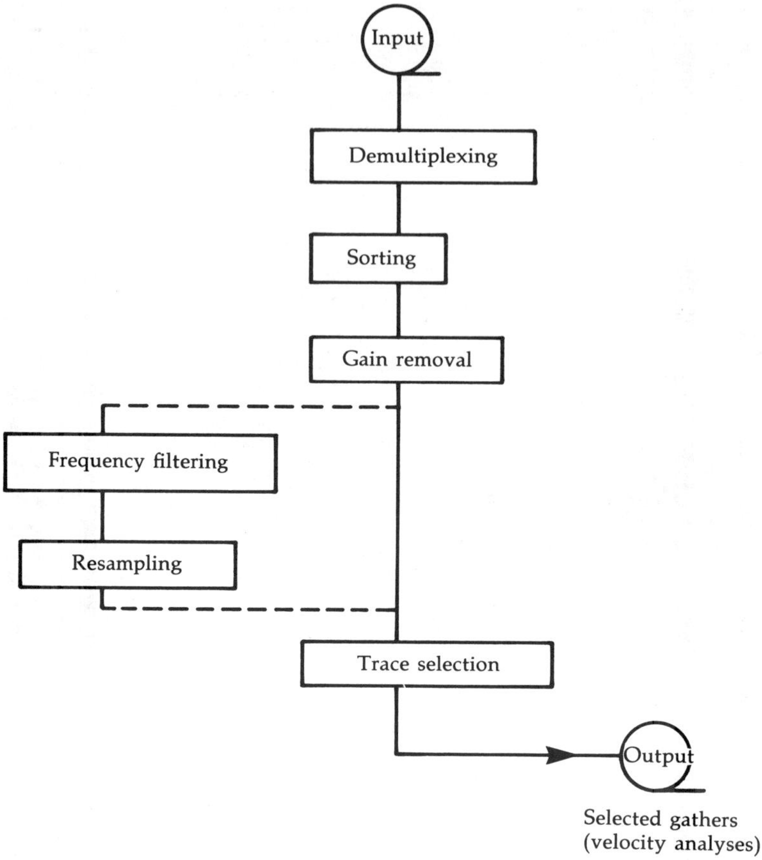

Figure 1.45. Preprocessing system.

Separate paths are available for processing marine and land seismic data. Each type of data processing has special requirements, and the systems are structured accordingly.

In the next chapter, we will discuss velocity analysis. Velocity determination performed by the system can allow input of a brute stack velocity or use of predetermined velocity functions provided by the system itself. For determining the optimum stacking velocity functions, the selected velocity function can be varied over a velocity range in percentage increments. A selected number of CMPs are then stacked at desired intervals. The best stacking results determine the velocity function to be used at each tested CMP. Velocity analysis processing is shown in figure 1.46.

The main processing steps are deconvolution, stacking, and migration. Stacking will be treated in the following chapter. Attendant amplitude corrections, time corrections, automatic gain control (AGC), and frequency filtering are also required at various stages. The main processing sequence is shown in figures 1.47 and 1.48.

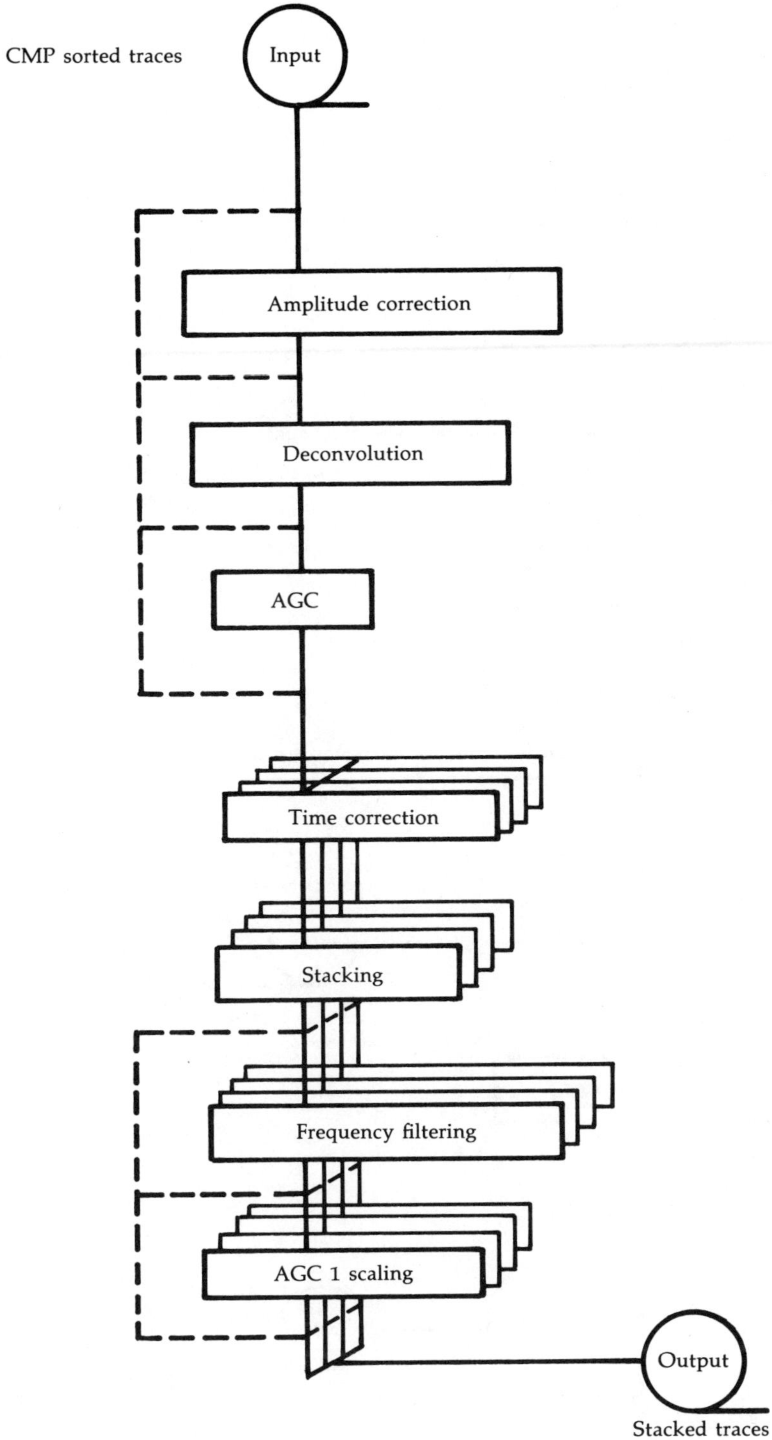

Figure 1.46. Velocity analysis.

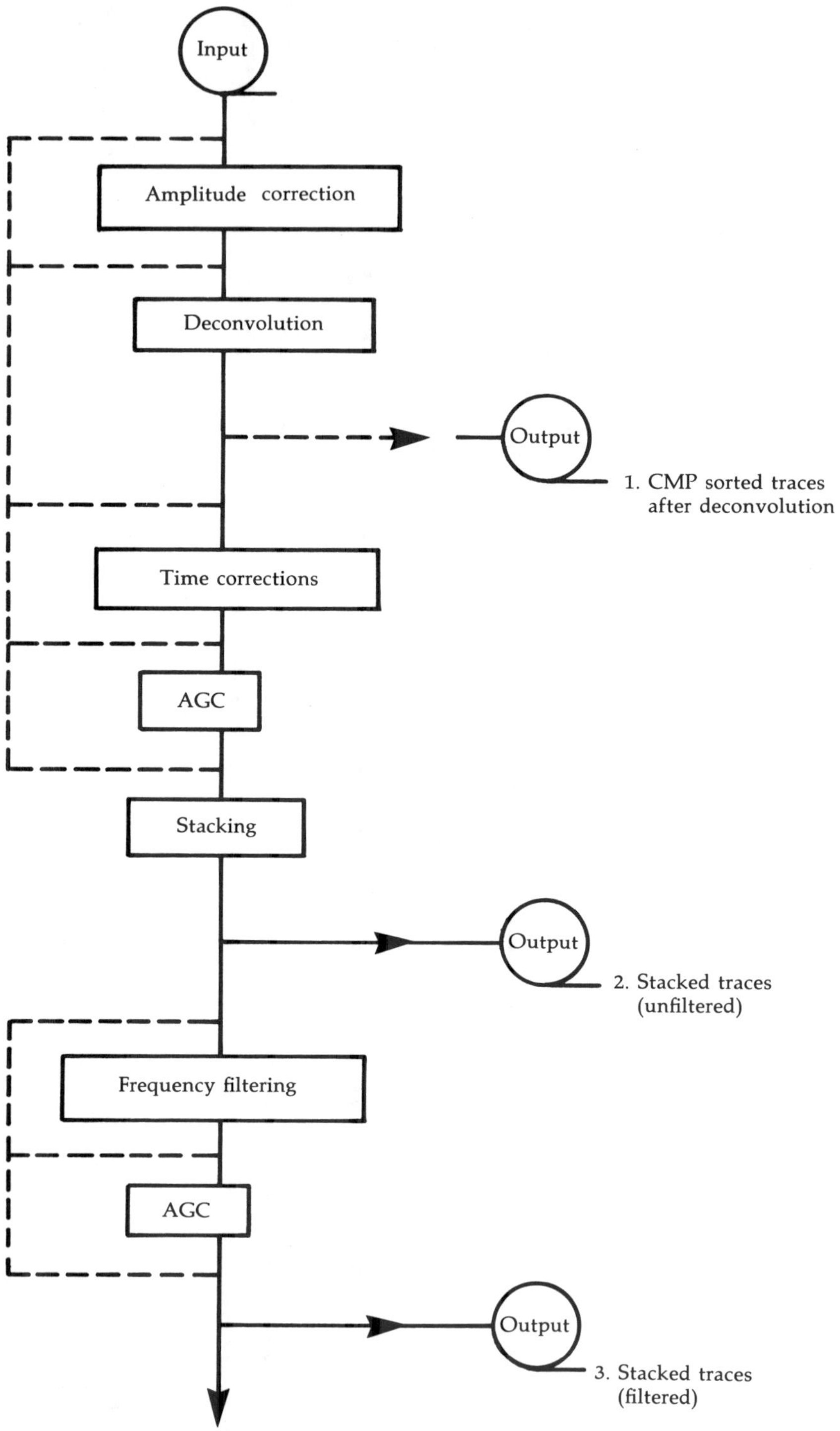

Figure 1.47. Main processing sequence.

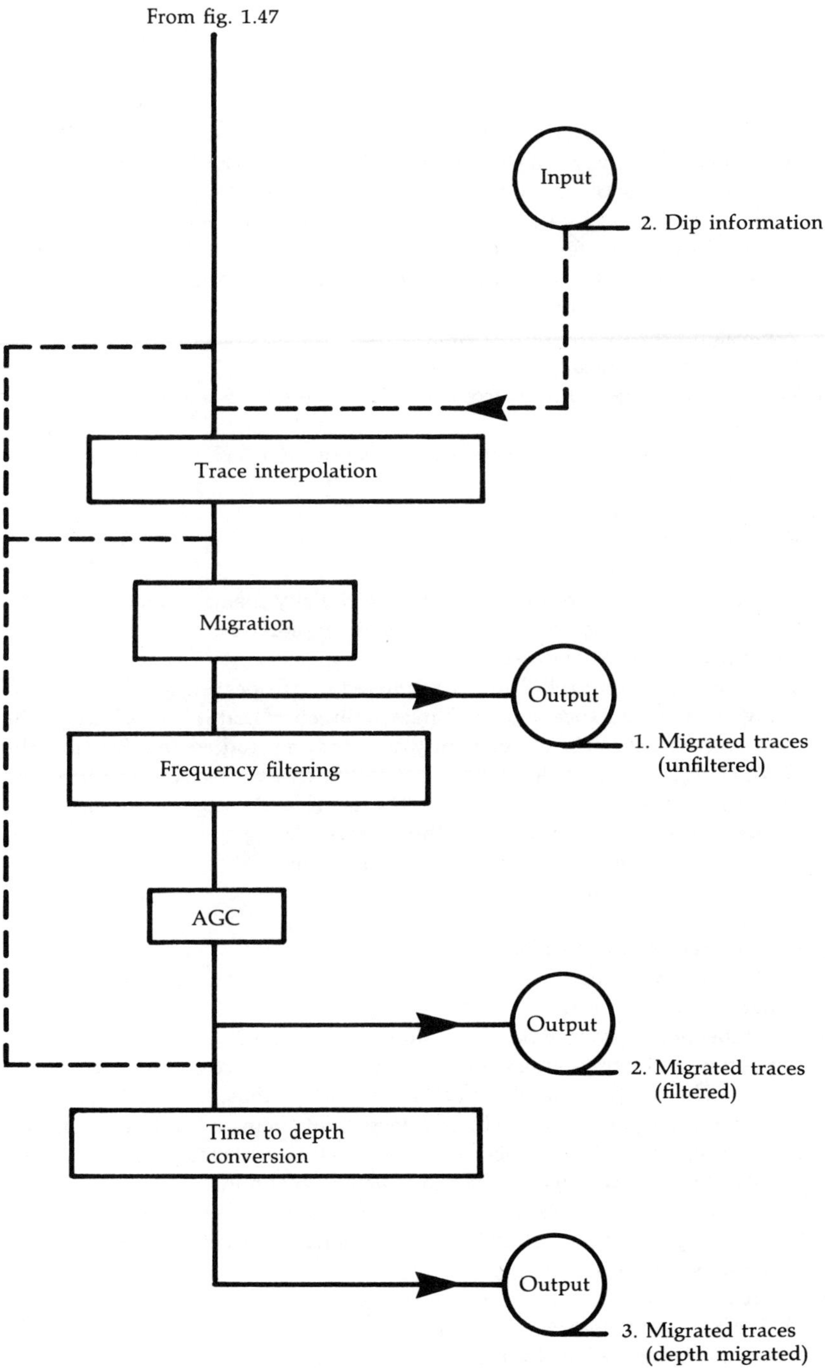

Figure 1.48. Main processing sequence continued.

Exercises

1. Briefly define the discipline of geophysics and name its main branches.
2. A geophysical survey is a program to make geophysical measurements over an area, with the objective of learning about the geology. The objective of many surveys is to locate accumulations of hydrocarbons, minerals, water, and so forth. In a few words, describe each of the following: electrical survey, geothermal prospecting, gravity survey, magnetic survey, reflection survey, refraction survey.
3. What is meant by the term *electric log?*
4. Explain what is meant by *anisotropy?*
5. What is a homogeneous, isotropic medium in terms of seismic wave propagation?
6. The term *ground roll* is used in seismic exploration. What does it mean?
7. What is hole fatigue?
8. Explain hole noise.
9. Why are shot holes plugged after use?
10. What is an implosion?
11. The term *horizon* refers to the line where the sky seems to meet the earth. In what sense is the word used in seismic exploration?
12. Explain wavefronts and rays.
13. The phenomenon of reflection is employed in prospecting since the downgoing wave suffers such a great change in direction that it comes back to the surface, where it can be conveniently detected and studied. In refraction, the wave changes at an interface, but not to the extent observed in reflection. Describe a type of refraction that gives rise to a phase that, under normal stratification, can travel back to the surface. Assume that the velocity of each layer increases with the depth of the layer.
14. What is a diving wave?
15. What is an isotime, or isochron, map?
16. Explain Fermat's principle.
17. What is a marker bed?
18. What is mode conversion?
19. In a reflection survey, what does multiplicity mean?
20. Explain why the following is usually the case in an actual reflection seismic survey: The multiplicity of a stack increases by steps, the shallowest data often being a twofold stack, slightly deeper data being a fourfold stack, and so on, until the full multiplicity of the stack is achieved after the muted events have passed beyond the most distant geophones.
21. A diversity stack is a stack in which the components are weighted inversely as their mean power over certain intervals. When and why is such a stack used?
22. What is wind noise?
23. What is an acoustic log?
24. Define the word *slowness* as used in seismology.
25. In seismic usage, what does *vertical exaggeration* mean?
26. Explain the divergence of seismic energy due to the geometrical spreading of a propagating wave.

27. Explain, in simple terms, the word *frequency* (as measured in cycles per second).
28. In recording a seismic trace, what is meant by gain control?
29. Together with the normal reflection shooting, special surveys are conducted for following up the variation of the depth and velocity of the surface layer along the seismic line. Why is this done, and what are the most commonly used methods?
30. What is the weathering layer?
31. A CSP gather is realized from a single source, whereas a CMP gather requires several sources. Is there any way, however, that the data in a CMP gather can be interpreted in terms of a single physical experiment? Can the data in a stacked section in some way be regarded as the result of a single physical experiment?
32. The group spacing on the surface is double the depth point spacing on the subsurface interface. In other words, if a certain spacing of depth points on the interface is required to resolve the geology, then the group spacing must be double this. Explain this.
33. Figure 1.49 depicts a uniformly dipping reflector at angle θ. Can you find an expression for the offset d of the actual reflection point from the midpoint?
34. Figure 1.50 shows reflections from two flat reflectors and one dipping reflector. Which is the dipping reflector, and can we tell in what direction it dips?
35. Draw a diagram illustrating the basic field arrangement of a typical fivefold CMP spread. The recording track has a switch that allows selection of the correct sequence of receivers as the source point is moved forward. As convenient, portions of the cable with the receivers are moved from behind and connected to the front end of the cable.
36. Given a single dipping horizontal interface of dip angle θ, find the equation for the time-distance curve for the following:
    (a) CSP gather
    (b) CMP gather
    (c) CRP gather
    (d) CHP gather
37. Given two dipping horizons of dips θ_1 and θ_2, find parametric equations for the time-distance curves for (a) CSP gather and (b) CMP gather.
38. The normal raypath from a point on the surface of the earth to a point on a subsurface interface is defined as that raypath which strikes the interface at right angles. Show that, if the source and receiver are at the given surface point, the least-time path for a reflection from the interface is the normal raypath.
39. The image raypath from a subsurface diffracting point to a point on the surface of the earth is defined as that raypath which strikes the surface at right angles. Show that, if the source is at a subsurface point, the least-time path from that point to the surface of the earth is the image raypath.
40. Given a dipping horizon of dip θ, draw, for a given point on the surface, the normal raypath and the image raypath. For a given point on the horizon, draw the image raypath and the normal raypath. Repeat the exercise for two dipping interfaces of dips θ_1 and θ_2. Can equations be found for these raypaths?

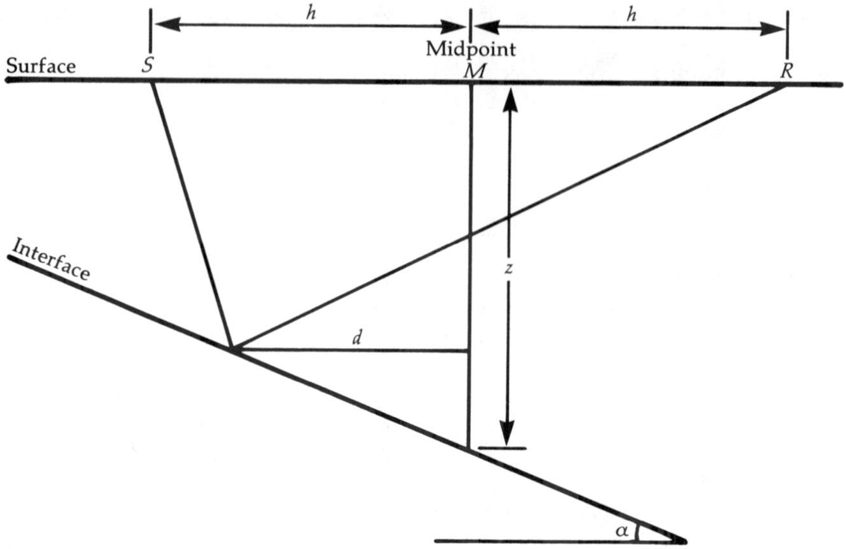

Figure 1.49. Uniformly dipping reflector.

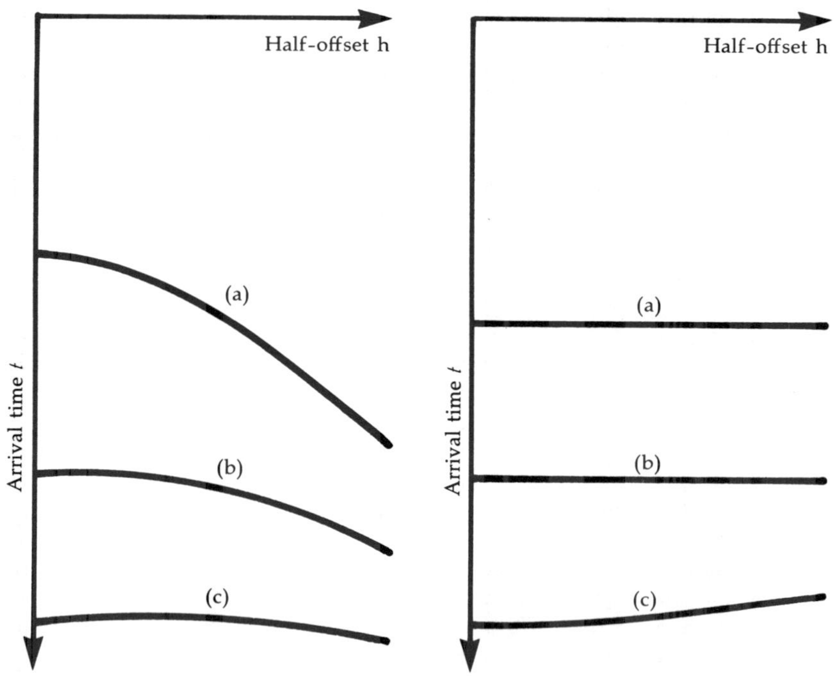

*Figure 1.50. Reflections from straight interfaces: (*left*) before NMO correction; (*right*) after NMO correction.*

41. Let an incident wave in layer 1, of velocity v_1, strike an interface at angle of incidence θ_1. Let the refracted wave in layer 2, of velocity v_2, have angle of refraction θ_2. By the use of the principle of least-time, establish Snell's law:

$$\frac{\sin \theta_1}{v_1} = \frac{\sin \theta_2}{v_2}.$$

42. Given a wavefront with a certain curvature, draw a circle with its center on the straight-line extension of the raypath and with radius R, such that the arc of the circle best fits the wavefront. The center is thus the center of a hypothetical spherical wave, and the radius R of the circle is called the radius of curvature of the wavefront. Let an incident wave in layer 1, of velocity v_1, have radius of curvature R_1. Let the angle of incidence be θ_1. Let the refracted wave in layer 2, of velocity v_2, have radius of curvature R_2. Let the angle of refraction be θ_2. Show that

$$\frac{\cos^2\theta_1}{R_1 v_1} = \frac{\cos^2\theta_2}{R_2 v_2}.$$

This equation might be called Snell's second law.

43. The amplitude A of a seismic wavelet is inversely proportional to the radius of the energy sphere propagating outward from the source. The conventional spherical decay equation is

$$\frac{A_2}{A_1} = \frac{r_1}{r_2},$$

where r_1 and r_2 are the distances traveled. The radius of curvature also changes, however, because of refraction at reflecting interfaces. This change in radius of curvature is such that

$$\frac{\cos^2\theta_1}{R_1 v_1} = \frac{\cos^2\theta_2}{R_2 v_2},$$

where θ_1 is the angle of incidence, θ_2 is the angle of refraction, and R_1 and v_1 are the values for the radius of curvature and velocity, respectively, in the incident medium, and R_2 and v_2 are those values in the refraction medium. Assume normal incidence ($\theta_1 = \theta_2 = 0$), and compute a correction factor to the conventional spherical decay that compensates for wavefront refraction across discontinuities.

44. Figure 1.51 shows a 12-fold line whose profile is a downdip to the right. Sketch the time responses for shot points 12, 24, and 36 that simulate continuous singlefold records of a 24-trace split spread. Also sketch the records for the six CDP locations: SP 12, 18, 24, 30, 36, and 42. Finally, sketch the NMO-corrected versions.

45. A crew operating in Michigan is shooting for nominal 12-fold coverage with a 24-trace split spread with a two-group offset (five-group drop). The group interval is 220 ft. There are indications of a reef on an east-west line, which

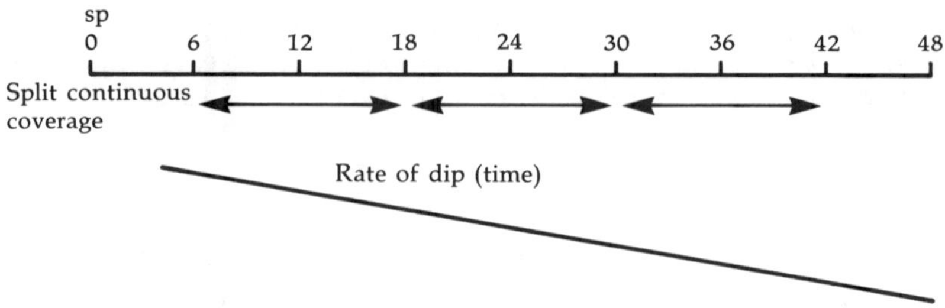

Figure 1.51. A 12-fold line with interface downdip to the right.

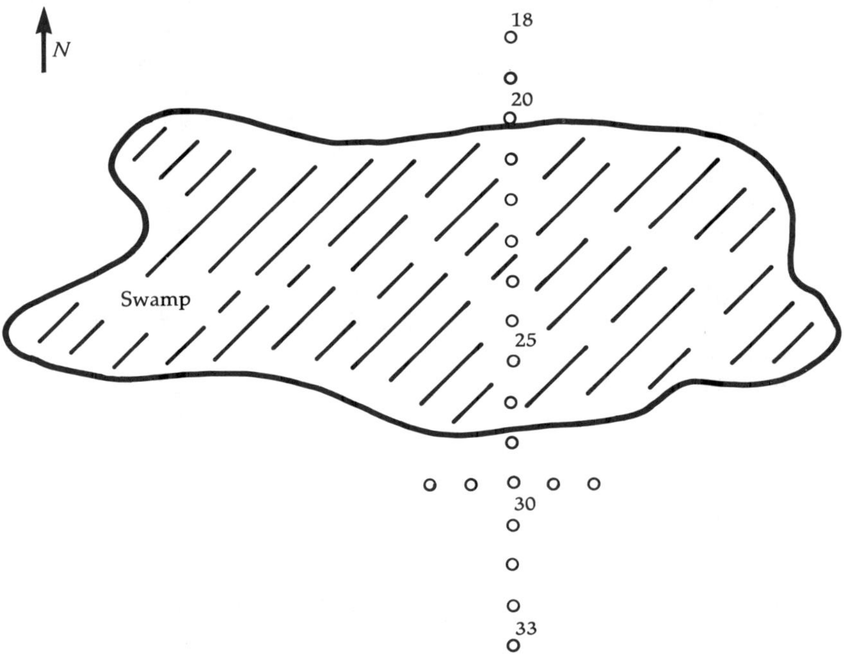

Figure 1.52. Map view.

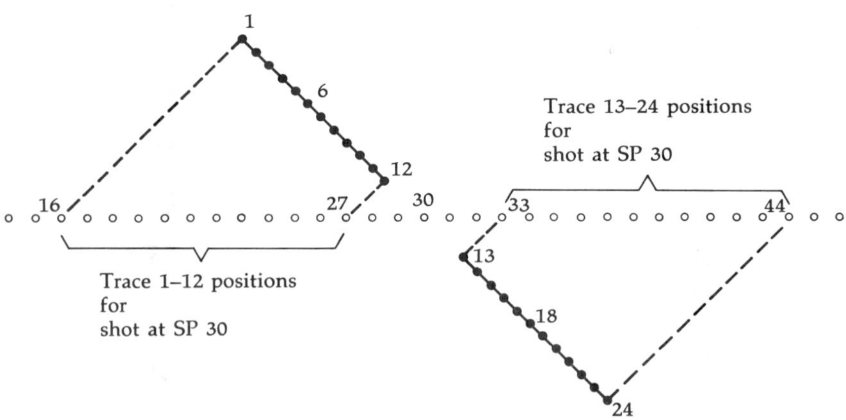

Figure 1.53. Construction of the stacking diagram for SP 30.

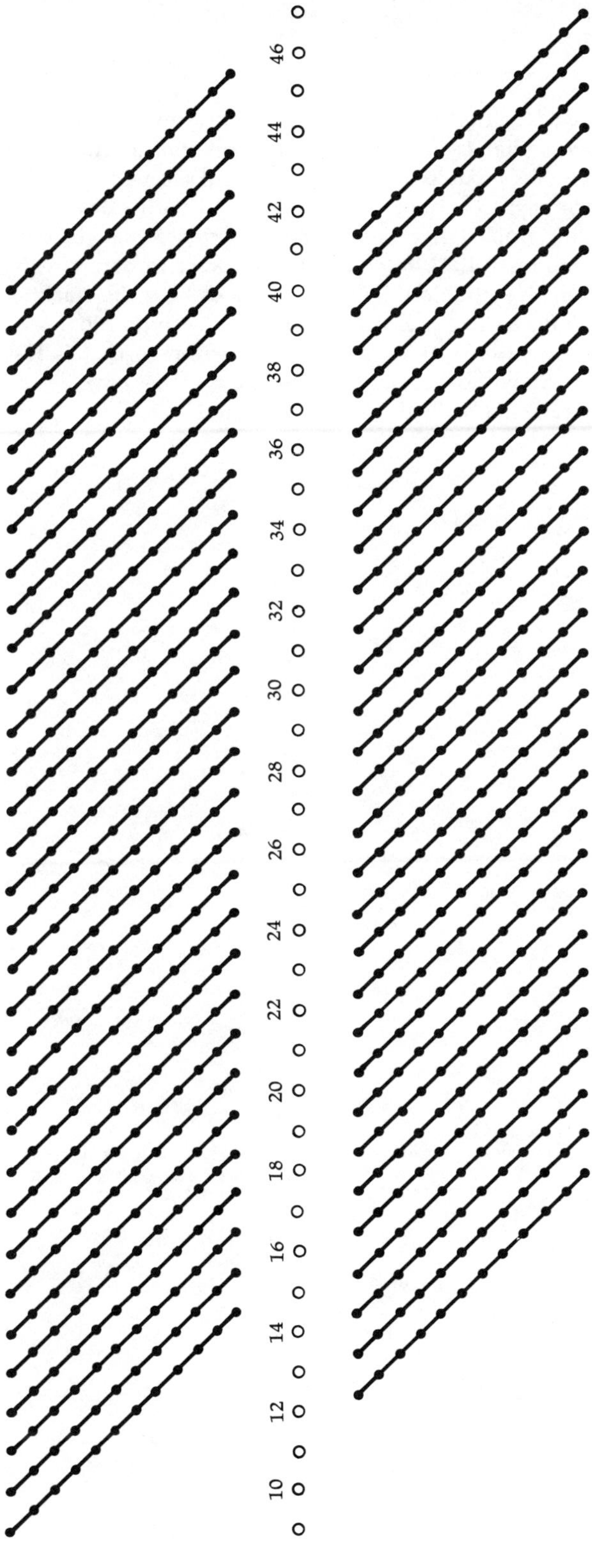

Figure 1.54. Stacking diagram.

they need to delineate in the north-south direction. Unfortunately, there is a swamp 1700 ft across that is just north of the reef indication on the east-west line. They cannot shoot in the swamp or lay out seismometers, but they can lay cable across it to connect seismometers on both sides of the swamp to the recording truck. They know from experience that the data quality in the area requires full 12-fold coverage in order to detect reefs. Figure 1.52 is a map of the situation, figure 1.53 shows how the stacking diagram is drawn, and figure 1.54 is a stacking diagram with no provisions for the swamp. Show on the stacking diagram the coverage that is lost because of the swamp. Show how to shoot to replace that lost coverage so that full 12-fold coverage is obtained across the swamp.

2. *Velocity analysis and stacking*

Velocity analysis

The earth's sedimentary layers are approximately horizontal, but they have features such as anticlines, unconformities, and faults that can serve as traps for petroleum. In order to map the subsurface, the geophysicist must convert the received seismic traces that record the events as a function of time into a function of depth; that is, a time function recorded at the surface must be transformed into a depth function. Unlike radio waves, seismic waves have a velocity that is very much dependent on the medium. Thus, the velocity changes as the waves travel into the earth. Generally, the velocity increases with depth, although occasionally there may be layers in which a decrease in velocity occurs. For a given surface point, the velocity plotted as a function of depth is called the velocity function. Thus, in reflection seismology, there are two equally important variables: time of reflected events and velocity. With a knowledge of these variables, the depth to the reflecting horizons can be determined. Because there are important lateral changes in velocity—that is, because the velocity function varies from one location to another—a given velocity function cannot be assumed to be valid for an entire prospect. As a result, the velocity function must be continually corrected from place to place over the area of exploration.

One method of measuring the velocity function is to drill a deep hole, such as an oil well, and determine the velocity by placing seismic detectors in the hole at various depths. The velocity function is computed from the observed traveltimes to the detectors from sources at the surface. This technique is known as shooting a well (see fig. 2.1).

A more common method of determining velocity from an oil well is by use of a continuous velocity log (CVL), an instrument made up of a seismic pulse generator with attached detectors. As the instrument is gradually pulled up the well, the changes in transit time across the fixed distance between the detectors are recorded as a continuous curve, known as the sonic log. This fixed distance is only a few feet, so the vertical resolution is good (see figs. 2.2 and 2.3).

The velocities determined by the CVLs are reasonably representative of the velocities of seismic waves through the corresponding formations, except under the following circumstances:

(1) Invasion of the porous formation by drilling fluid is great enough that the fastest time path is through the altered formation. Thus, the velocity is not representative of the true formation velocity.
(2) The hole diameter is very large, or very irregular, which either results in

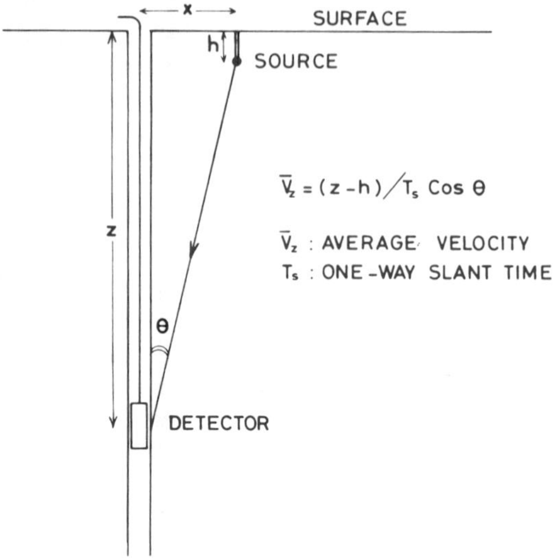

Figure 2.1. Shooting a well. Reprinted by permission of Birkhauser-Verlag from Al-Sadi 1980.

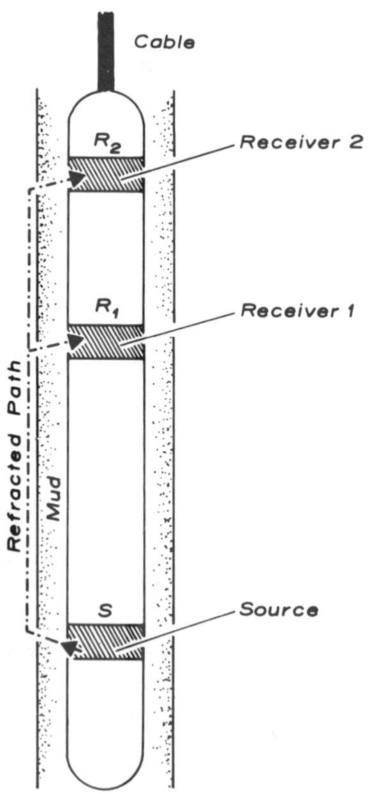

Figure 2.2 Schematic of a P-wave velocity logging tool. Reprinted by permission of John Wiley & Sons, Inc. from Reflection Seismology by K.H. Waters 1978. Copyright © 1978 John Wiley & Sons, Inc.

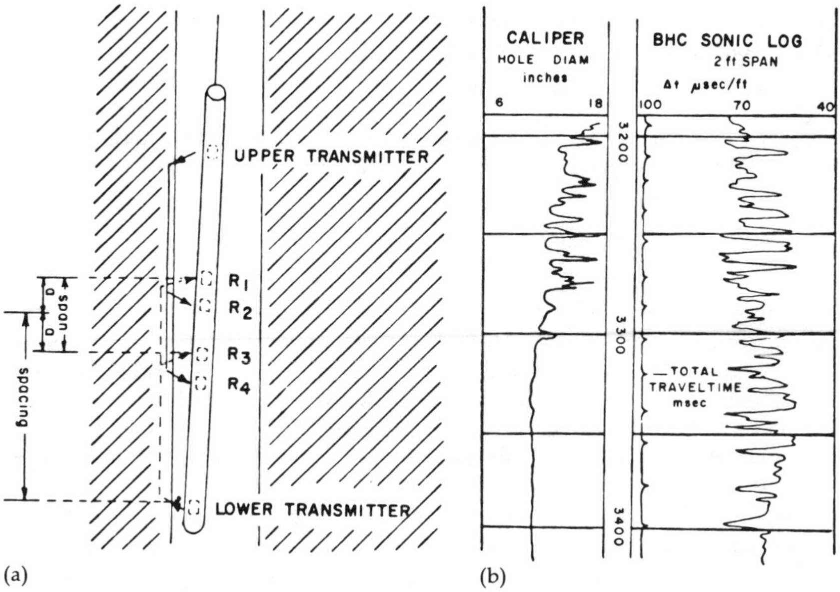

(a) (b)

Figure 2.3. Sonic log: (a) schematic of a compensated sonic logging sonde; (b) the sonic log. Reprinted by permission of Schlumberger and the Society of Exploration Geophysicists from Sheriff 1973.

unequal times through the fluid or causes the level of the signal to become too low.

(3) The formation velocity is lower than the P-wave velocity through the drilling fluid (about 1524 m/sec or 5000 ft/sec).

CVLs are available over some portion of the total drilled footage for thousands of wells and are cataloged by log service companies, along with electrical, neutron, gamma-density, and other logs.

In most cases, it is necessary to estimate the velocity by measurements confined to the surface, since oil wells are available only in old prospects. Therefore, the actual seismic data must be used to estimate velocity. Velocity can be estimated by considering the time differentials of the same event received by a lateral array of detectors. Any such estimate always depends on a *ceteris paribus* (other things being equal) assumption. Computers can determine velocities by carrying out calculations based on many intricate time-distance relationships, and the results can present empirical velocity as a function of traveltime (or depth) in a form called a *velocity spectrum.* In this chapter, we will study how this velocity estimation is done. The empirical velocity so determined is called *stacking velocity,* and the problem is to relate this empirical quantity to a mathematical expression from which we can extract the thicknesses and velocities of the subsurface layers.

Discussion of velocity determination requires knowledge of seismic trace sorting procedures, which were discussed in chapter 1. As we have seen, the data set recorded from a seismic survey line is described in terms of three coordinates: midpoint m, half-offset h, and time t. Likewise, there are three main activities in seismic data processing, one associated with each coordinate:

(1) Time coordinate t: deconvolution
(2) Half-offset coordinate h: velocity analysis and stacking
(3) Midpoint coordinate m: migration

At this point, we will not introduce deconvolution or migration, so that we can devote full effort to velocity analysis and stacking.

Static and dynamic corrections

Traveltime adjustments, or corrections, made to the seismic trace represent the first important step in the digital processing sequence. Geophysicists divide time corrections into static and dynamic categories. A static correction consists of the application of a time shift, or translation, to an entire trace. In other words, a constant time correction term is added or subtracted from all reflection times, regardless of record time or reflector depth (see fig. 2.4). Dynamic corrections, in contrast, vary with record time and therefore depend on reflector depth.

Let us consider static corrections. A region of very low velocity extends from the earth's surface to a depth of several tens to hundreds of feet; at this point, velocities change either gradually or abruptly from values near 2000 to 5000 ft/sec or more. Time delays associated with this weathered layer disrupt reflection continuity (i.e., trace-to-trace alignment) and pose a major obstacle to successful processing of data acquired on land. In addition, elevation changes and near-surface inhomogeneities severely degrade trace-to-trace continuity. Because of these lateral variations in the near surface layers, each trace is corrected by a time shift, which has the effect of placing the source and receiver on a fictitious horizontal datum plane. This time shift is additively composed of a source correction and a receiver correction, which together are referred to as static corrections. Thus, the purpose of static computations is to remove time variations caused by anomalous conditions at the earth's surface. Seismic lines recorded at sea, however, do not usually require static adjustments because of the uniform water layer of constant elevation.

An assumption underlying all automatic correction programs is that a simple translation of a trace converts it into a model trace that would have been recorded had sources and receivers been vertically displaced downward to a reference plane with no weathering material present (see fig. 2.5). This time delay is assumed to be surface-consistent—that is, to be the sum of a source-related component and a contribution characteristic of a given receiver position. The validity of these two assumptions is confirmed by the great success of modern static correction programs. A more complete analysis, however, shows that the near-surface layers behave as complicated filters whose impulse responses distort amplitude and phase characteristics of seismic wavefronts.

The dynamic corrections convert each trace to the equivalent trace that would have been received if the source and receiver were at the same lateral point—namely, the point midway between the actual source and receiver locations. In this conversion, we are referring only to traces made up of so-called primary reflections. According to ray theory, a primary reflection results from a raypath down from the source to the reflecting horizon and then directly upward to the receiver. Given the velocity function, this raypath can be computed by means

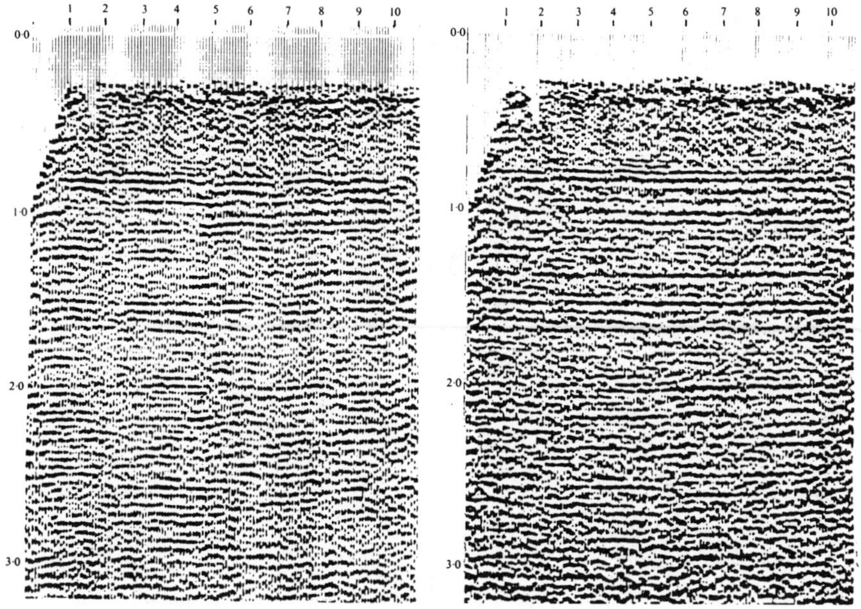

(a) (b)

Figure 2.4. Static corrections, applied to seismic data to eliminate the effects of variations in elevation, weathering thickness, or weathering velocity. The objective is to determine the reflection arrival times that would have been observed if all measurements had been made on a (usually) flat plane with no weathering or low-velocity material present. The information on which these corrections are based derives from uphole time data, refraction first breaks, and/or data-smoothing considerations. Uphole shooting is the direct measurement of near-surface traveltimes. This is so expensive that one cannot sample near-surface conditions everywhere along a spread; when surveying with a surface source, one may not have any uphole information at all. First-break ABC method and variations for more complex assumptions are used for this determination. Data-smoothing methods assume that patterns of irregularity that most events have in common result from near-surface variations; hence, the static corrections are those trace shifts that minimize such irregularities. Most automatic statics-determination programs work with this assumption and employ statistical methods to achieve the minimization. Underlying the very concept of static corrections is the assumption that a simple time shift of an entire seismic trace will yield the seismic record that would have been observed if the geophones had been displaced vertically downward to the reference datum. This assumption is not strictly true, of course, and a complete solution of the problem requires consideration of other factors, including the interplay of dynamic and static corrections and lateral as well as vertical velocity variations. The cross sections displayed here show an example of the application of automatic statics: (a) before automatic statics; (b) after automatic statics. Courtesy Regional Surveys and Geocom. Reprinted by permission of Cambridge University Press from Telford et al. 1976.

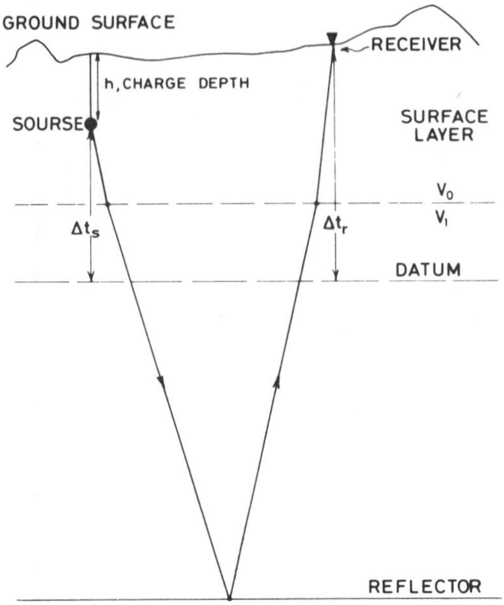

Figure 2.5. Source and receiver static corrections. Reprinted by permission of Birkhauser-Verlag from Al-Sadi 1980.

of Snell's law. If the layers are horizontal, then all the reflection points (or depth points) are always directly beneath the midpoint between source and receiver. If the layers are dipping, then the depth points are offset from the midpoint. Thus, the dynamic corrections depend on both the velocity function and the dip of the reflecting beds. The component of the correction caused by the separation of source and receiver is called the normal moveout (NMO) correction and the component caused by dip is called the dip correction.

Thus, four important corrections must be made to the recorded traces: source correction, receiver correction, NMO correction, and dip correction (see fig. 2.6).

Normal moveout

Normal moveout, denoted by NMO or by Δt, is the most important of the traveltime corrections. In this section, we will illustrate the concept of normal moveout in terms of the simplest subsurface model—a single layer bounded by a flat earth surface on top and a flat reflecting interface below. This is the single (flat) layer model. An alternative term for moveout is *stepout.*

Figure 2.7 depicts the geometry of a common midpoint (CMP) gather as well as the data display. A single horizontal reflecting plane is shown. A typical path, such as from source s_2 to receiver r_2, results in a seismic trace that is plotted as a function of time directly below the receiver position r_2. Each trace shows a single event in the form of a pulse due to the reflection at depth point D. The times of occurrence of these pulses vary from trace to trace, depending on the length of the path and the velocity function. For simplicity, we assume that

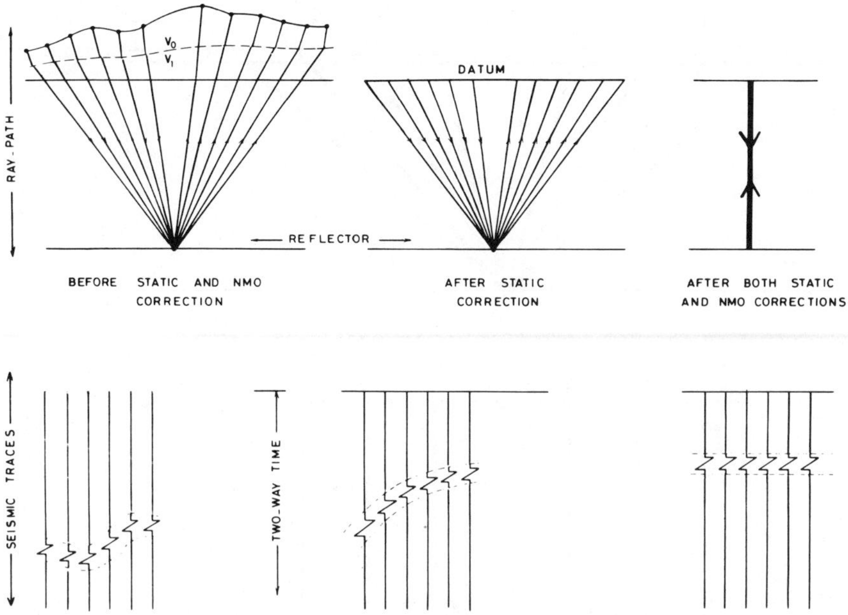

*Figure 2.6. Raypath diagram (*above*) and seismic traces (*below*), showing static and dynamic correction on a CMP gather. Reprinted by permission of Birkhauser-Verlag from Al-Sadi 1980.*

velocity is a constant v, so that we can analyze reflection time in terms of path length. The basic diagram is shown in figure 2.8 for a single path. Seismic time is two-way time—that is, time from source to depth point plus time from depth point to receiver. Let t be the traveltime from S to D to R. Then the one-way time (i.e., the time from S to D, or equivalently from D to R) is $t/2$. The zero-offset traveltime is the two-way time associated with the midpoint M; that is, t_0 is the time from M to D and back to M. In the diagram, triangle DMR is a right triangle, with legs of length h and $vt_0/2$ and hypotenuse of length $vt/2$. Using the Pythagorean theorem, we obtain the equation

$$(vt/2)^2 = (vt_0/2)^2 + h^2.$$

We can also write this equation as

$$\left(\frac{t}{2}\right)^2 - \frac{h^2}{v^2} = \left(\frac{t_0}{2}\right)^2.$$

Here t_0 and v are constant, so this equation is the equation of a hyperbola in the coordinates $(h, t/2)$. Thus, we see that the events due to a horizontal reflecting plane lie upon a hyperbola on the CMP gather (see fig. 2.9). The asymptotes of the hyperbola are $t/2 = \pm h/v$. Here we have written the hyperbola in terms of one-way time $t/2$ and half-offset h. More often, this hyperbola

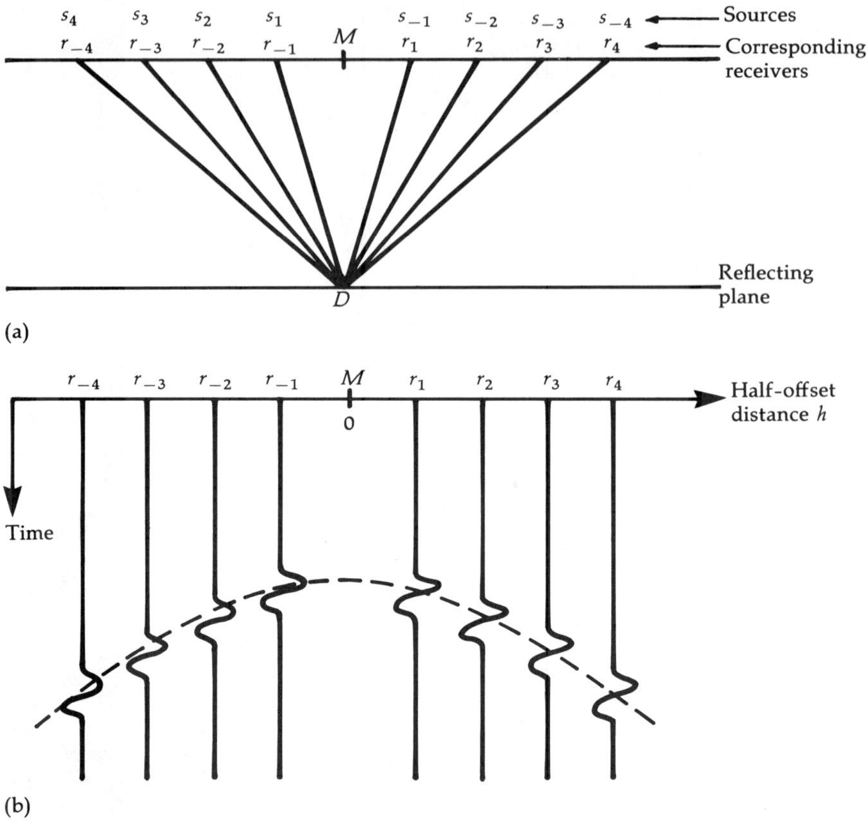

Figure 2.7. Common midpoint (CMP) gather: (a) geometry; (b) data display. Trace r_1 *results from shot* s_1*, trace* r_2 *from shot* s_2*, and so on. The dashed curve shows the locus of the reflections from the given horizontal reflecting plane.*

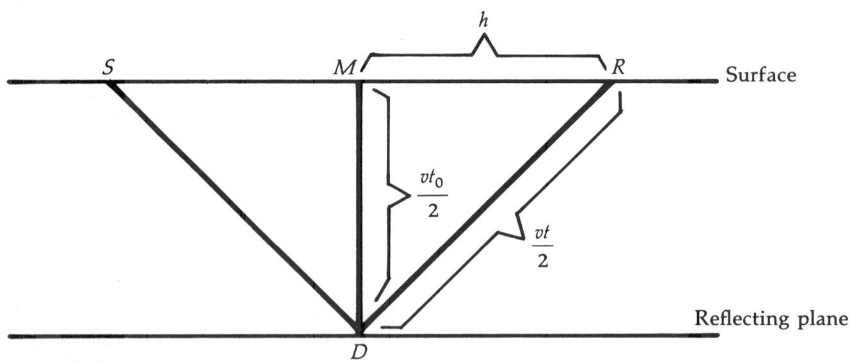

Figure 2.8. Illustration of distances in terms of reflection times.

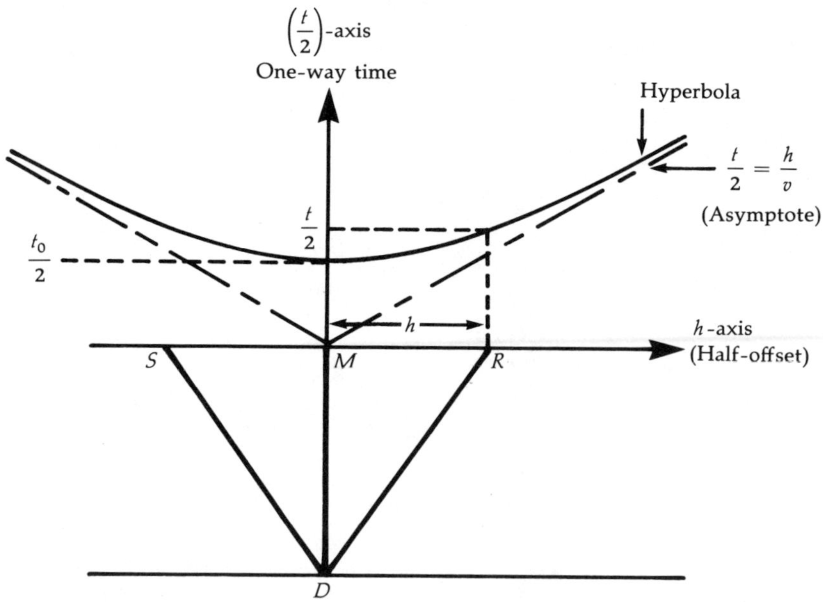

Figure 2.9. Hyperbolic event in terms of half-offset and one-way time.

is written in terms of two-way time t and full-offset $x = 2h$. We then have the equation for the hyperbola

$$t^2 - \frac{x^2}{v^2} = t_0{}^2,$$

and the asymptotes are $t = \pm x/v$.

Consider the reflected event with traveltime t for offset x between source and receiver. If the traveltime is t_0 for offset zero, then the NMO is defined as

$$\Delta t = t - t_0.$$

This definition is good for any seismic model, not just for the simple model of a single flat interface. In other words, the NMO is the extra traveltime required for the reflected pulse to travel the path for offset x over and above the traveltime required to travel the path for offset zero (see fig. 2.10).

For the model with a single horizontal interface, the NMO is

$$\Delta t = \left[t_0 + \frac{x^2}{v^2}\right]^{1/2} - t_0.$$

Using the parabolic approximation, we can approximate this equation as

$$\Delta t = \frac{x^2}{2t_0 v^2}.$$

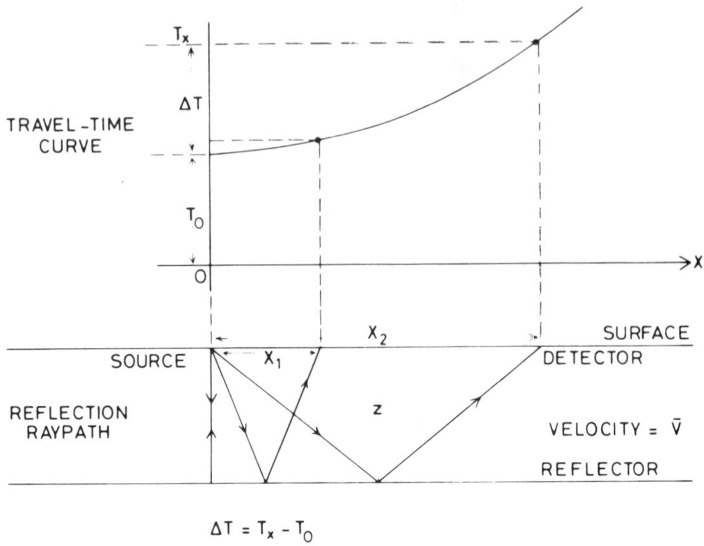

Figure 2.10. Definition of normal moveout (NMO). Reprinted by permission of Birkhauser-Verlag from Al-Sadi 1980.

The parabolic approximation is explained in the next section.

We see that Δt is a function of three variables: the offset x, the velocity v, and the two-way vertical time t_0. For a given reflector (i.e., a given t_0) the NMO increases as full-offset x increases. For a given full-offset x, the NMO decreases as the reflector depth (i.e., t_0) increases (see fig. 2.11).

When we process reflection data, we must compute Δt as accurately as possible. Normally, both x and t_0 are known, since both are observed quantities. Thus, the accuracy of Δt determination depends largely on how successful we are in estimating v. An error in the estimation of v incurs a corresponding error in Δt, commonly referred to as the residual NMO,

$$\Delta^2 t = \Delta t_e - \Delta t,$$

where Δt corresponds to the velocity v and Δt_e to the actually applied (estimated) velocity v_e. Using the approximation formula for Δt, we have

$$\Delta^2 t = \frac{x^2}{2t_0}\left(\frac{1}{{v_e}^2} - \frac{1}{v^2}\right).$$

Thus, there is positive $\Delta^2 t$ (overcorrection) when $v_e < v$ and negative $\Delta^2 t$ (undercorrection) when $v_e > v$ (see figs. 2.12, 2.13, and 2.14).

Let us summarize the main facts up to this point. As we know, all the traces in the prospect can be segregated (or gathered) into sets, called gathers, such that all traces within each gather have a common midpoint between source and receiver. Under a *ceteris paribus* assumption, let us suppress for the time being everything except the NMO correction. The appropriate NMO correction would convert each trace in this gather into the same equivalent trace—namely, the primary reflection trace that would have been received if source and receiver

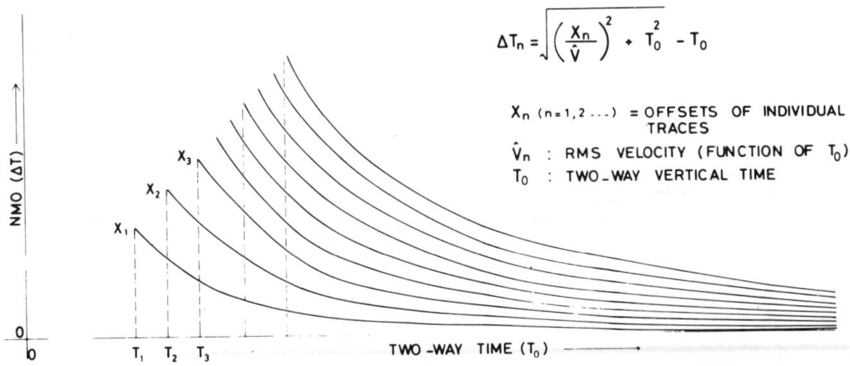

Figure 2.11. Schematic diagram of typical NMO curves, where each curve is computed for a given source-to-detector offset. The times T_1, T_2, T_3, . . ., *define the lower limit of the applicability of the NMO correction. Below this limit, the signal is so distorted (by the NMO correction) that it is better to discard. Reprinted by permission of Birkhauser-Verlag from Al-Sadi 1980.*

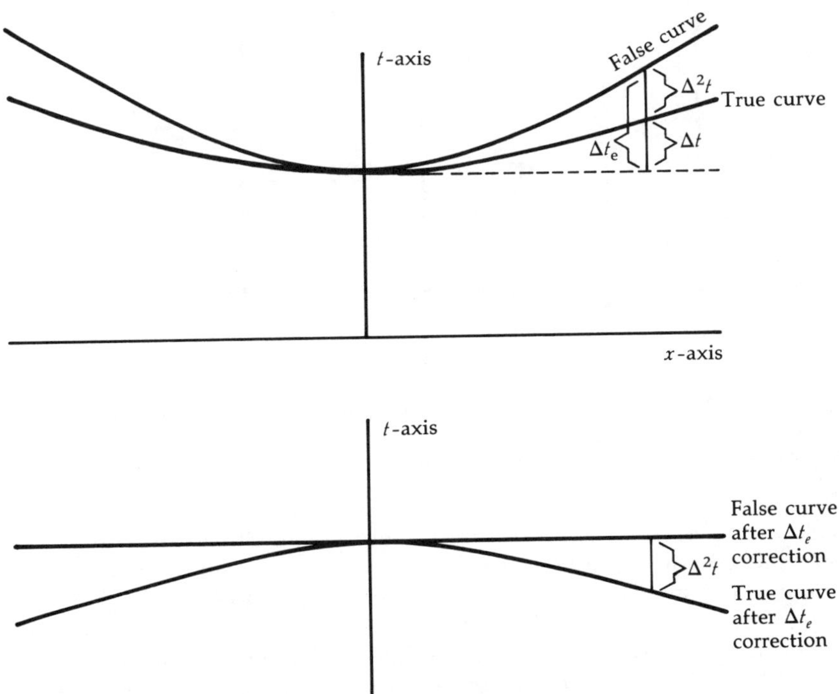

Figure 2.12. When $v_e < v$, *the NMO correction* Δt_e *overshoots (overcorrects).* (top) *True time-distance curve with actual velocity* v *and estimated time-distance curve with false velocity* v_e. (bottom) *The NMO correction* Δt_e *would make the false curve horizontal, but it makes the true curve (on which the actual seismic event lies) bend downward.*

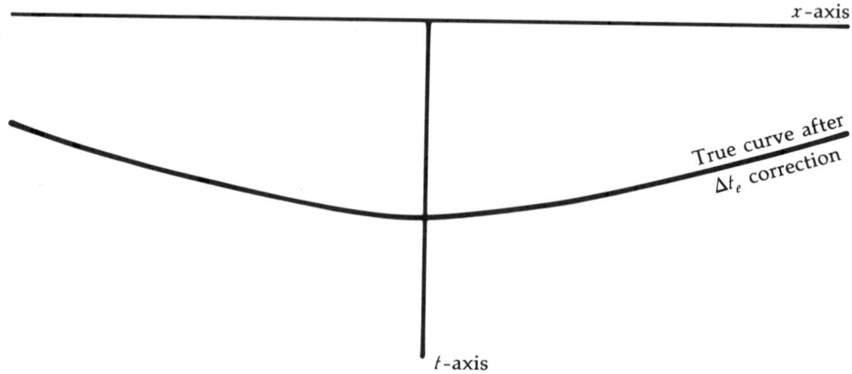

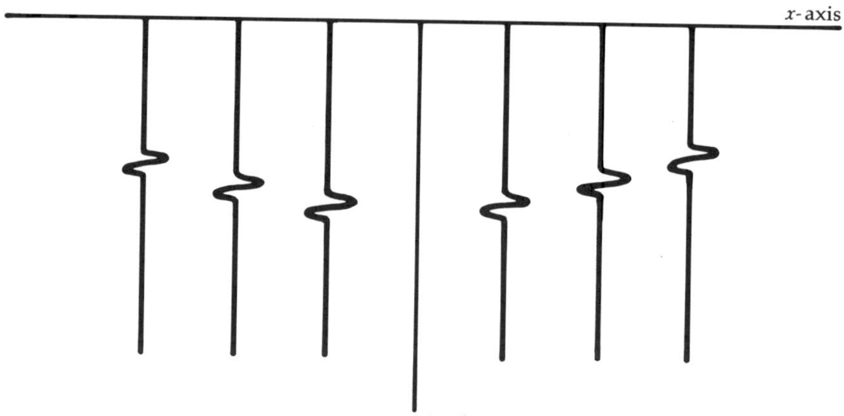

Figure 2.13. When $v_e < v$, *the NMO correction* Δt_e *overshoots.* (top) *The same curve as the bottom curve of figure 2.12, but with the* t-*axis pointing down.* (bottom) *The appearance of the overcorrected seismic event. It bends in the wrong direction.*

were directly at the common midpoint in question. In other words, under the appropriate NMO correction, the primary reflections of all the traces in the gather will be in phase, thus making the corrected traces coherent.

In this section, we have illustrated the NMO correction in the case of a single layer with flat boundaries (i.e., the single flat interface model). After the application of the static corrections, the reduced record time represents the two-way slant time, where both the source and receiver are effectively on the same datum. If we now examine a reflection event on all the contributers of one CMP, we find that these events fall on a hyperbola (in the single flat interface case). This is because of the dependence of traveltime on the trace offset. The NMO equation

$$\Delta t = \left[t_0^{\,2} + \frac{x}{v^2} \right]^{1/2} - t_0$$

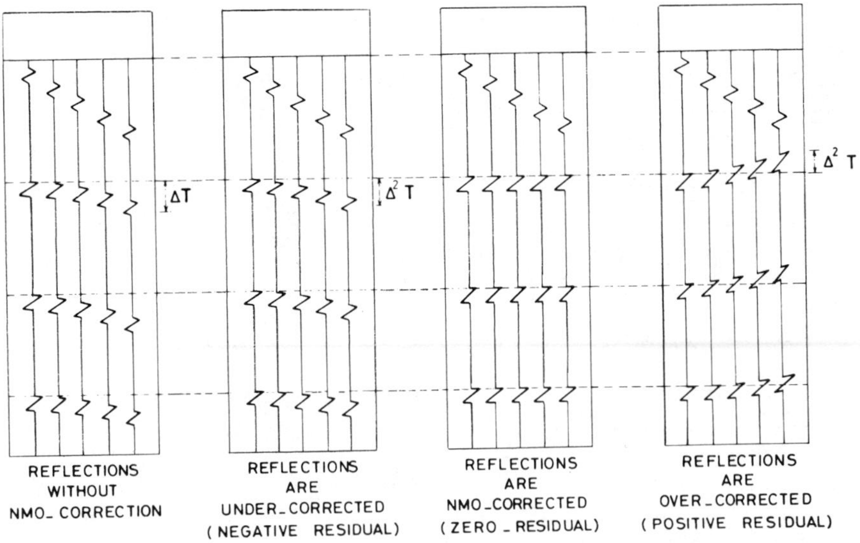

Figure 2.14. One side of NMO-corrected gather. Left to right: *No correction, undercorrection, proper correction, overcorrection. Reprinted by permission of Birkhauser-Verlag from Al-Sadi 1980.*

defines the NMO correction for our single horizontal interface model. The NMO correction is carried out by computing Δt for each trace. Then the event whose record time before the correction is $t(x)$ will be at time $t(x) - \Delta t$ after the correction. Since Δt is a function of record time $t(x)$, different parts of the trace will be time-shifted differently; hence, the correction is called dynamic (or time-dependent) to distinguish it from the time-independent or static correction. When a proper NMO correction Δt is applied, the reflection traveltime is reduced to two-way vertical time. Because all the traces have the same CMP, the reflection events will fall along a straight line, producing a cophasal alignment (see fig. 2.15). Excellent alignment—that is, zero residual ($\Delta^2 t = 0$)—is achieved by proper choice of velocity function, which can be derived by a velocity analysis. We will develop the concept of a velocity analysis as the chapter proceeds.

In the case of dipping interfaces or other complex structure, moveout (or stepout) is defined as $\Delta t = t - t_0$ where now t_0 is the zero-offset, normal-incident raypath to the interface in question (see fig. 2.16). Except in simple cases there is no easy formula, so either ray-tracing methods or wave-equation methods are required in order to compute Δt (see exercises 5, 6, 7, and 8).

The parabolic and circular approximations

As we know, the simplest subsurface model we can imagine is a single horizontal layer underlain by a half-space. A given point on the surface is designated as the common midpoint M, and source-receiver pairs symmetrically placed about M give us a set of traces called the common midpoint (CMP) gather. The reflection event from the single horizontal interface has the hyperbolic time-distance curve

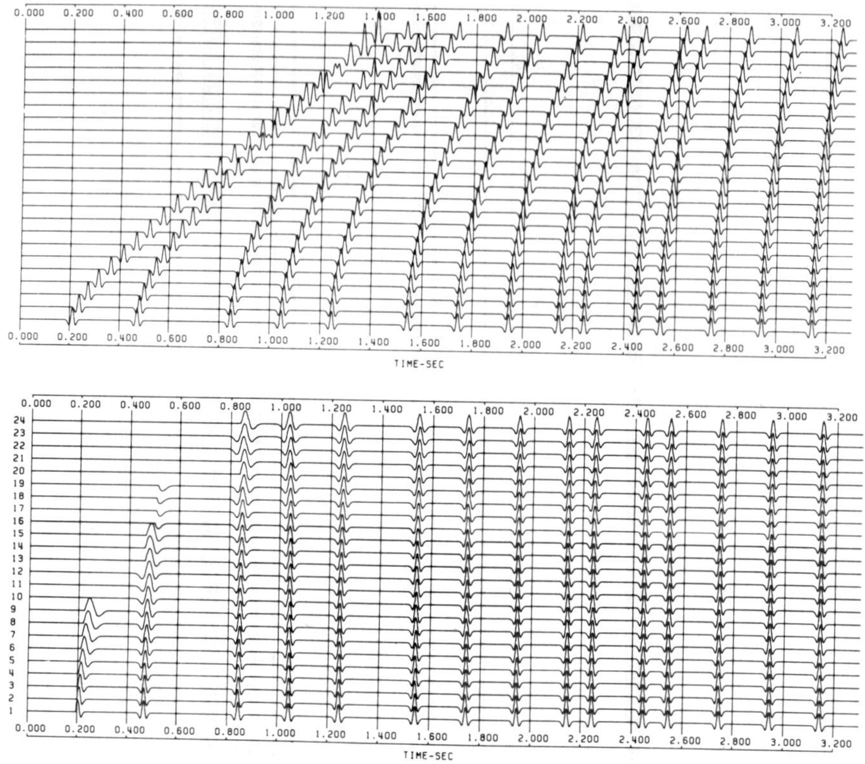

Figure 2.15. (top) *Synthetic seismogram before NMO correction.* (bottom) *Synthetic seismogram after NMO correction. Reprinted by permission of Geophysics from Dunkin and Levin 1973.*

$$t^2 - \frac{x^2}{v^2} = t_0{}^2,$$

where v is the velocity of the single layer, x is the offset between source and receiver, t is the two-way reflection time for offset x (i.e., time from source down to reflecting horizon and up to receiver), and t_0 is two-way vertical reflection time (i.e., time for offset x equal to zero). We derived this hyperbolic curve in the previous section. It is simple, but simple things are important in seismic processing and require constant study.

In the early days, seismic prospecting was done with modest offset distances. The greatest offset distance x would be small compared to the depth of the target horizon. This situation has changed in present-day prospecting, where great offset distances are in common use. Let us go back, however, to the old days, and assume that the offset distance x is small in magnitude. We can then approximate the hyperbola by either a parabola or a circle in this region of small offset.

Let us look first at the parabolic approximation. We write the hyperbola as

$$t^2 = t_0{}^2 + \frac{x^2}{v^2} = t_0{}^2\left(1 + \frac{x^2}{t_0{}^2 v^2}\right),$$

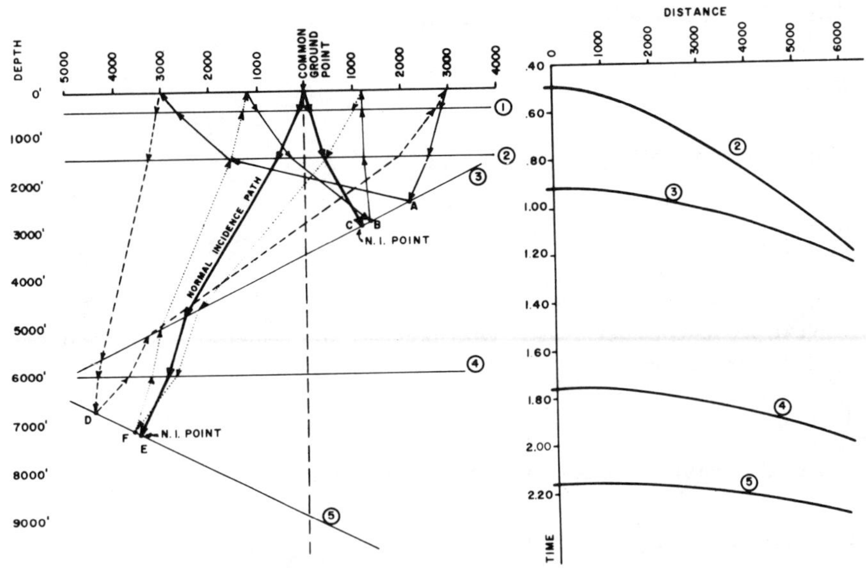

Figure 2.16. The common midpoint (or equivalently common ground point) model with dipping interfaces. The curves in the right hand diagram are the curves of traveltime versus offset distance for each of the interfaces. Because of the principle of reciprocity, the travel time from shot −3000 to receiver 3000 is the same as the traveltime from shot 3000 to receiver −3000, and so on. Thus the curves are symmetric about the zero-offset point, which explains why only their right sides are displayed. Reprinted by permission of Geophysics from Taner and Koehler 1969.

which, upon taking the positive square root, gives

$$t = t_0 \left(1 + \frac{x^2}{t_0{}^2 v^2}\right)^{1/2}.$$

The parabolic approximation says that for R small in magnitude, the quantity

$$(1 + R^2)^{1/2}$$

can be approximated by

$$1 + \frac{R^2}{2}.$$

This approximation is, in fact, the first three terms of the Taylor series expansion. Thus, the time-distance curve can be approximated as

$$t \cong t_0 \left(1 + \frac{x^2}{2t_0{}^2 v^2}\right)$$

or

$$t \cong t_0 + \frac{x^2}{2t_0 v^2}.$$

The right-hand side is a parabola of the form

$$t_0 + \frac{\alpha x^2}{2},$$

where the constant

$$\alpha = \frac{1}{t_0 v^2}$$

is called the curvature, and t_0 is the t-intercept.

Next, let us consider a circle with radius ρ and center at $x = 0$, and $t = t_0 + \rho$; that is, the circle

$$(t - t_0 - \rho)^2 + x^2 = \rho^2.$$

If we take the negative square root, we can write

$$t - t_0 - \rho = -\rho\left(1 - \frac{x^2}{\rho^2}\right)^{1/2}.$$

The parabolic approximation gives

$$t - t_0 - \rho \cong -\rho + \frac{x^2}{2\rho}$$

or

$$t \cong t_0 + \frac{x^2}{2\rho}.$$

Again the right-hand side is a parabola. If we let the radius ρ be $t_0 v^2$, this parabola derived from the circle is the same as the parabola derived from the hyperbola. The curvature α of the parabola is equal to the reciprocal of the radius of the circle; that is, $\alpha = \rho^{-1}$. It is for this reason that α^{-1} is called the radius of curvature of the parabola.

We have shown that the time-distance hyperbola can be approximated either by the parabola

$$t = t_0 + \frac{x^2}{2t_0 v^2}$$

or by the circle

$$(t - t_0 - t_0 v^2)^2 + x^2 = (t_0 v^2)^2,$$

where both approximations are valid for offset x small in magnitude (see fig. 2.17). The radius of both curves is $t_0 v^2$. Often the normal moveout (NMO) is expressed in terms of the approximating parabola as

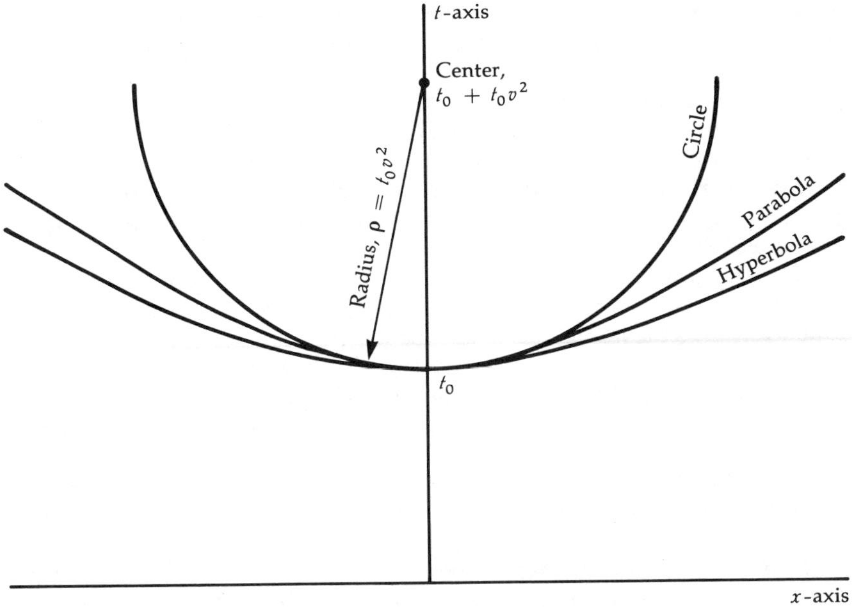

Figure 2.17. The single-interface time-distance hyperbola and the approximating parabola and circle.

$$\Delta t = \frac{x^2}{2t_0 v^2}.$$

The multiple-flat-interface (layercake) model

Well surveys are the ultimate way to obtain the velocity function of the subsurface strata. Such surveys do not justify drilling a well on their own account (except in exceptional circumstances) and thus are available only in old prospects where there are existing oil wells. As a result, the velocity function must be estimated from the data taken in a seismic survey. In such a survey, only surface sources and surface receivers can be used, so the problem is estimating subsurface information from surface data. (Of course, shot holes are drilled to place the explosive charge below the weathered layer, but we still consider these relatively shallow explosions as a surface experiment. The corrections required to bring all sources and receivers to a common datum level are called static corrections.)

Each seismic trace is a function of time, and each trace is associated with the horizontal coordinates of its source point and its receiver point. The distance between source and receiver is called the (full) offset x. The time it takes a wave to go from the source down to a deep horizon and then by reflection back up to the receiver is called the (two-way) traveltime t. A display of the seismic data that exhibits traveltime t versus offset x will show an event for each reflecting horizon. The locus of such an event is a curve that we can consider as a function $t(x)$, which gives traveltime t as a function of offset x. Such a curve is called a *time-distance curve.* The important point is that we can see these curves as the reflected events appearing on the seismic data and that such events are caused by the stratiform structure of the sedimentary rocks existing in the earth's crust (see fig. 2.18).

Our approach is the following. We set up a mathematical model of the stratified earth and derive the resulting time-distance curves from this model. These theoretical time-distance curves depend on the parameters of the model. We then fit the theoretical time-distance curves to the empirical time-distance curves appearing on the seismic data. In this way, we can make inferences about the parameters. Thus, we are able to obtain statistical estimates of the parameters from the seismic data. The most important parameters are the thicknesses and wave velocities of the layers in the stratified model; the estimates of these parameters give the desired velocity function.

We will now discuss in some detail a typical and well-known model used in geophysics—the familiar model of a flat-layered earth. A horizontal line represents the surface, and below the surface are media whose interfaces are parallel to the surface. The media have various thicknesses and compressional velocities. The path of a wave originating at the shot point and arriving at an offset point after being reflected at depth satisfies certain parametric equations. The first equation gives an expression for the offset distance, the second equation gives the traveltime, and the third equation represents Snell's law of refraction of the wave at each interface. From these equations we obtain the equation for the time-distance curve of the waves originating at the shot point and reflected back to the surface from the reflecting interface. This earth model, which is made up of horizontal layers, is a simplification of the true situation in the earth; however, it provides us with the equations for the time-distance curve that is used daily in the interpretation of seismic records.

In petroleum exploration, we are usually working with more-or-less flat-lying layers. The changes in the wave velocity as one moves in the horizontal direction are usually the result of slow changes in the density and elastic properties of the strata, and thus the changes in the wave velocity in this direction are mostly small. As one moves in the vertical direction, however, the changes in the wave velocity result from lithological changes of the layering as well as increasing pressure due to increasing depth. As a result, the vertical changes in velocity are generally much more rapid than the horizontal changes. Because of the gradual nature of the horizontal changes, it is common practice to divide the prospect into smaller subareas, within each of which the horizontal velocity distribution is assumed to be constant, so that the same vertical velocity distribution can be used over the entire subarea. The interpreter must then merge these separate subareas together in constructing the final map of the prospect.

A model of the subsurface structure that is often used to take into account the velocity variations with depth consists of a number of horizontal (i.e., flat-lying) layers of different wave velocities, in which the velocity within each layer is constant. If we follow a seismic raypath in such a model, we find that the ray bends at each layer in accordance with Snell's law. Let us now derive the equations for the raypath.

The so-called *layercake model* is illustrated in figure 2.19. The x-axis represents the surface, and the z-axis represents depth measured positively down. Under the surface are n media whose interfaces are parallel to the surface. The thicknesses of the media are $\Delta z_1, \Delta z_2, \ldots, \Delta z_n$, and the compressional wave velocities

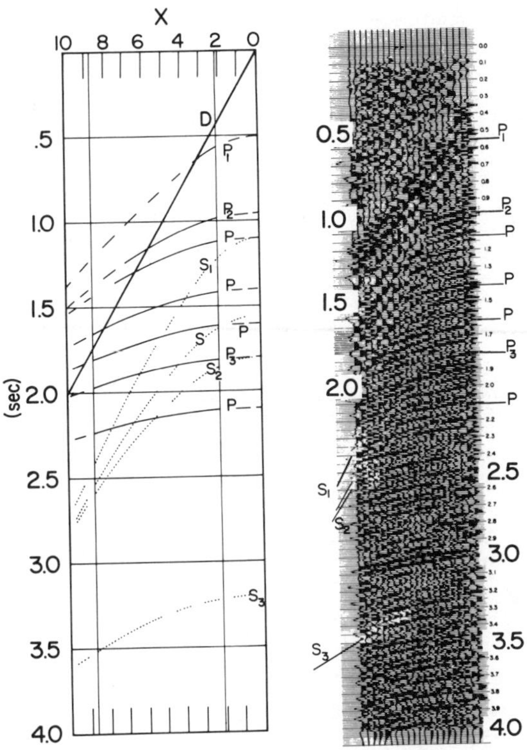

(a) (b)

Figure 2.18. (a) Time-distance plot showing positions of computed reflection hyperbolas corresponding to interpreted normal moveout velocities and observed arrival times. Offset distance (X) is in thousands of feet. Solid hyperbolas indicate interpreted P*-wave reflections and dotted hyperbolas are the interpreted* S*-wave reflections. Solid sections of the hyperbolas indicate portions of the total array used in the velocity analyses. Note that all* S*-wave reflections arrive at least 0.5 sec after the computed direct water (D) arrival, and that there is no interference between wave types. (b) One CMP gather that was part of the input into the original velocity analyses. Positions of interpreted* S*- and* P*-wave reflection hyperbolas are indicated. Note the destructive effect of* S_1 *and* S_2 *upon the higher velocity compressional wave arrivals they encounter. Reprinted by permission of Geophysics from Tatham and Staffa 1976.*

within the media are $v_1, v_2, \ldots, v_n$, respectively. In standard geophysical terminology, the velocities $v_1, v_2, \ldots, v_n$ are called the *interval velocities*, as they are the velocities within the intervals defined by the interfaces. In seismic work, x denotes the full offset between source and receiver, and t denotes the two-way traveltime (i.e., downward time plus upward time). Because the layers are horizontal, the downward path and the upward path are completely symmetric. Therefore, instead of considering both, we may consider either the downward path or the upward path. Thus, we will base our analysis on one of the paths, either the downward path from S to D or the upward path from D to R. The results for the two-way path S to D to R would be double the results for the one-way path. Hence, the reader should note that h now refers to half-offset (i.e., half the full-offset) and τ refers to one-way time (i.e., half the two-way time).

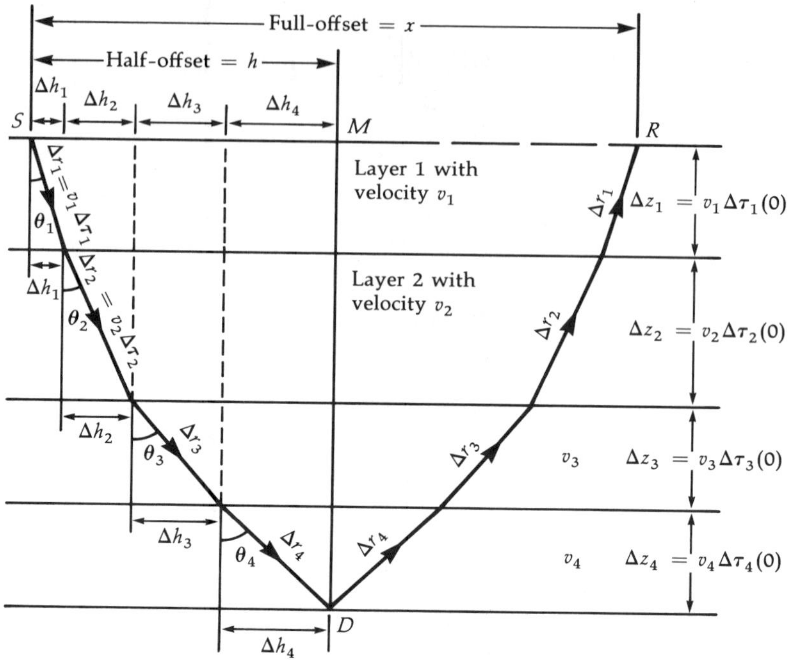

Figure 2.19. Layercake model with n = *4 layers.*

The downward path from S to D is made up of n segments, which are indexed by a subscript k. The length of the kth segment is Δr_k, and the angle the segment makes with the vertical is θ_k. The traveltime of the wave along this segment is $\Delta\tau_k$, and since the velocity for that segment is v_k, we have $\Delta r_k = v_k \, \Delta\tau_k$. The vertical projection of the segment is Δz_k, and the horizontal projection is Δh_k:

$$\Delta z_k = \Delta r_k \cos\theta_k, \qquad \Delta h_k = \Delta r_k \sin\theta_k.$$

If we let $\Delta\tau_k(0)$ denote the traveltime for a vertically traveling wave to cover the vertical distance Δz_k, then $\Delta z_k = v_k \, \Delta\tau_k(0)$.

Either the downgoing path or the upgoing path can be characterized by parametric equations as follows. From the diagram, we have

$$\Delta h_k = \Delta z_k \tan\theta_k, \qquad \Delta\tau_k = \frac{\Delta r_k}{v_k} = \frac{\Delta z_k}{v_k \cos\theta_k},$$

where $\Delta\tau_k$ is the traveltime for the slant path Δr_k. We recall that Snell's law is

$$p = \text{constant Snell parameter} = \frac{\sin\theta_k}{v_k} \qquad \text{for } k = 1, 2, \ldots, n.$$

Thus, we have

$$\sin \theta_k = pv_k,$$

$$\cos \theta_k = (1 - \sin^2 \theta_k)^{1/2} = (1 - p^2 v_k^2)^{1/2}.$$

Let M be the surface point directly above the depth point D. The distance between M and R is called the half-offset and is given by

$$h = \sum_{k=1}^{n} (\Delta h_k).$$

The traveltime from D to R is given by

$$\tau = \sum_{k=1}^{n} (\Delta \tau_k).$$

From the foregoing relations, we find, using $x = 2h$ and $t = 2\tau$, that

$$x = \sum_{k=1}^{n} \frac{2pv_k \Delta z_k}{(1 - p^2 v_k^2)^{1/2}},$$

$$t = \sum_{k=1}^{n} \frac{2\Delta z_k}{v_k (1 - p^2 v_k^2)^{1/2}}.$$

These two equations relate full-offset x and two-way traveltime t to the depth z. However, these equations cannot be solved explicitly for t in terms of x; that is, the Snell parameter p cannot be eliminated by algebraic manipulations. The curve of t as a function of x—that is, $t(x)$—is called the *time-distance curve.* Thus, these two equations give the time-distance curve (as a function of a parameter p) of the waves originating at the source point and arriving at the receiving point R. Sometimes we will write the time-distance curve as $t(x)$ to indicate that t is a function of x. The significance of the Snell parameter p is realized by taking $k = 1$ in the equation $p = (\sin \theta_k)/v_k$. We obtain

$$p = \frac{\sin \theta_1}{v_1},$$

which shows that p is proportional to the sine of the angle of issue of a downgoing ray (or to the sine of the angle of emergence of an upgoing ray). Thus, when the source point S and the receiver point R coincide directly above the depth point D (i.e., when $x = 0$), we see that $\theta_1 = 0$, so $p = 0$.

The time-distance curve is symmetric; that is,

$$t(x) = t(-x).$$

Thus, we may confine ourselves to nonnegative values of the Snell parameter p. For $p = 0$, as we have seen, we obtain the case of a vertical travelpath given by

$$x = 0, \qquad t(0) = \sum_{k=0}^{n} \frac{2\Delta z_k}{v_k}.$$

The two-way time $t(0)$ is called the *vertical traveltime* and is denoted by $t_0 = t(0)$. The traveltime t_0 is seen to be the sum of the two-way vertical traveltimes in each of the layers; that is,

$$t_0 = t(0) = \sum_{k=0}^{n} 2\Delta\tau_k(0), \qquad \text{where } \Delta\tau_k(0) = \Delta z_k / v_k.$$

Suppose that we know the angle of issue θ_1 of a downgoing ray and that we know the thicknesses and the velocities of each layer. We first compute the parameter $p = \sin\theta_1/v_1$. Then we can compute all of the remaining angles θ_k by Snell's law $\theta_k = \sin^{-1}(pv_k)$. Finally, we compute the time-distance curve by the expressions

$$x = \sum_{k=1}^{n} 2(\tan\theta_k)\Delta z_k,$$

$$t = \sum_{k=1}^{n} 2(v_k \cos\theta_k)^{-1} \Delta z_k.$$

Alternatively, we may compute the vertical time increments $\Delta\tau_k(0) = \Delta z_k / v_k$ and then compute the slant traveltime t by the expression

$$t = \sum_{k=1}^{n} 2(\cos\theta_k)^{-1} \Delta\tau_k(0).$$

Let us consider an example. The earth model consists of two horizontal layers. The first layer is 1.0 km thick and has an interval velocity of 2 km/sec. The second layer is 1.5 km thick and has an interval velocity of 3 km/sec. We consider three rays, with Snell parameter p equal to 0, 1/12, and 1/6, respectively. These three rays are shown in figure 2.20. The angles shown in the figure are obtained from the equations

$$\text{Ray } A\text{:} \quad p = 0 = \frac{\sin\theta_1}{2} = \frac{\sin\theta_2}{3}, \qquad \text{so } \theta_1 = 0, \quad \theta_2 = 0;$$

$$\text{Ray } B\text{:} \quad p = \frac{1}{12} = \frac{\sin\theta_1}{2} = \frac{\sin\theta_2}{3}, \qquad \text{so } \theta_1 = 9.59°, \quad \theta_2 = 14.48°;$$

$$\text{Ray } C\text{:} \quad p = \frac{1}{6} = \frac{\sin\theta_1}{2} = \frac{\sin\theta_2}{3}, \qquad \text{so } \theta_1 = 19.47°, \quad \theta_2 = 30.00°.$$

The half-offsets for each of these rays are

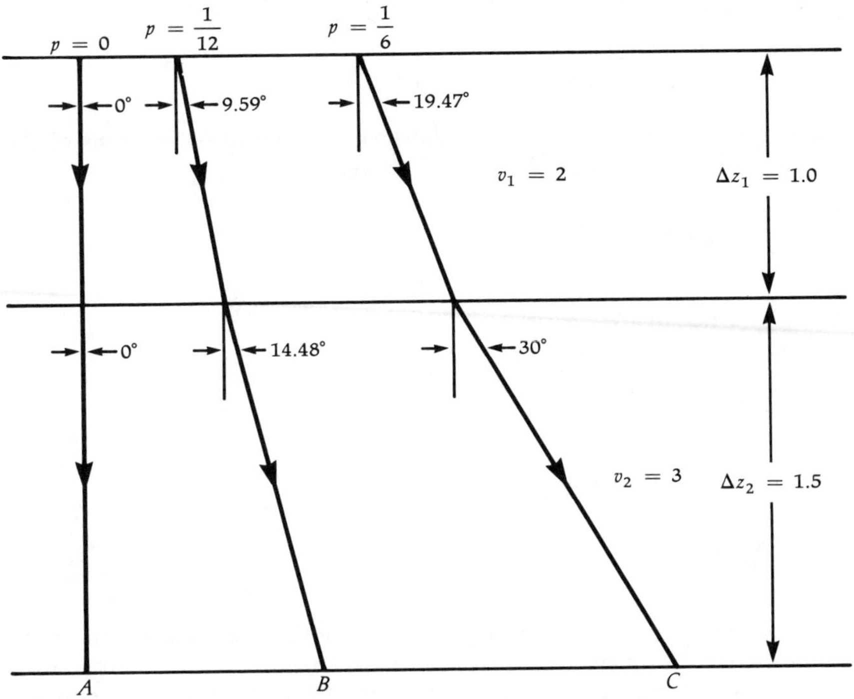

Figure 2.20. Rays A, B, *and* C *(downgoing legs only).*

Ray A:
$$\begin{aligned} \Delta h_1 &= (\Delta z_1) \tan 0° = 0 \\ \Delta h_2 &= (\Delta z_2) \tan 0° = \underline{0} \\ \text{Half-offset} &= 0; \end{aligned}$$

Ray B:
$$\begin{aligned} \Delta h_1 &= (1.0) \tan 9.59° = 0.169 \\ \Delta h_2 &= (1.5) \tan 14.48° = \underline{0.387} \\ \text{Half-offset} &= 0.556; \end{aligned}$$

Ray C:
$$\begin{aligned} \Delta h_1 &= (1.0) \tan 19.47° = 0.353 \\ \Delta h_2 &= (1.5) \tan 30.00° = \underline{0.866} \\ \text{Half-offset} &= 1.219. \end{aligned}$$

The one-way traveltimes are

Ray A:
$$\begin{aligned} \Delta\tau_1(0) &= \Delta z_1/v_1 = 1.0/2 = 0.500 \\ \Delta\tau_2(0) &= \Delta z_2/v_2 = 1.5/3 = \underline{0.500} \\ \text{Vertical one-way traveltime} &= 1.000; \end{aligned}$$

Ray B:
$$\begin{aligned} \Delta\tau_1 &= \Delta\tau_1(0)/\cos 9.59° = 0.507 \\ \Delta\tau_2 &= \Delta\tau_2(0)/\cos 14.48° = \underline{0.516} \\ \text{Slant one-way traveltime} &= 1.023; \end{aligned}$$

Ray C: $\Delta\tau_1 = \Delta\tau_1(0)/\cos 19.47° = 0.530$

$\Delta\tau_2 = \Delta\tau_2(0)/\cos 30.00° = 0.577$

Slant one-way traveltime $= 1.107$.

There is an interesting equation that gives the full-offset in terms of the incremental slant two-way times Δt_k. It is derived as

$$\begin{aligned} x &= \sum_{k=1}^{n} 2 \tan \theta_k \, \Delta z_k \\ &= \sum_{k=1}^{n} 2 (\sin \theta_k)(\Delta z_k / \cos \theta_k) \\ &= \sum_{k=1}^{n} (pv_k)(v_k \, \Delta t_k), \end{aligned}$$

so that the interesting equation is

$$x = p \sum_{k=1}^{n} v_k^2 \, \Delta t_k.$$

This equation states that the full-offset x is equal to the parameter p times the summation of the squared velocities weighted by the two-way slant-time increments. If we normalize this summation by the total slant time t, we obtain a quantity that is called the slant mean-square velocity. Its square root is called the slant root-mean-square (rms) velocity, given by

$$v_{\text{srms}} = \left[\frac{1}{t} \sum_{k=1}^{n} v_k^2 \, \Delta t_k \right]^{1/2}, \qquad \text{where } t = \sum_{k=1}^{n} \Delta t_k.$$

This velocity is different for each raypath; that is, the slant rms velocity depends on the Snell parameter p. In terms of this velocity, the time-distance curve is

$$x = ptv_{\text{srms}}^2,$$

$$t = \sum_{k=1}^{n} \Delta t_k = \sum_{k=1}^{n} (\cos \theta_k)^{-1} \Delta t_k(0).$$

In the case of normal incidence (i.e., a vertical raypath), Snell's parameter p is zero, and the slant rms velocity is the conventional rms velocity, defined as

$$v_{\text{rms}} = \left[\frac{1}{t(0)} \sum_{k=1}^{n} v_k^2 \, \Delta t_k(0) \right]^{1/2}, \qquad \text{where } t(0) = \sum_{k=1}^{n} \Delta t_k(0).$$

Continuing our example, we have

Ray A: $v_{\text{rms}}^2 = [2^2(0.5) + 3^2(0.5)]/[0.5 + 0.5] = 6.5$

$= 2.550$ (conventional rms velocity);

Ray B: $v_{srms} = [x/pt]^{1/2} = [h/p\tau]^{1/2} = [0.556/(1/12)(1.023)]^{1/2}$
$= 2.554$ (slant rms velocity for $p = 1/12$);

Ray C: $v_{srms} = [1.219/(1/6)(1.107)]^{1/2}$
$= 2.570$ (slant rms velocity for $p = 1/6$).

We see that as the angle of issue increases, the slant rms velocity increases, because the ray spends an increased proportion of its journey in the lower (high-velocity) layer.

Let us now summarize the results of our example in tabular form of a table. We convert half-offset to offset x and one-way time to time t, each by multiplication by 2.

Path	P	θ_1	x	t	v_{srms}
A	0	0	0	2.000	2.550
B	1/12	9.59°	1.112	2.046	2.554
C	1/6	19.47°	2.438	2.214	2.570

Before we proceed with the hyperbolic approximation problem, it would be helpful to find out more about the time-distance curve for the layercake model. One important property is that the slope of the time-distance curve at any point is equal to the value of the parameter p corresponding to that point (fig. 2.21). To establish this result, let the upcoming ray have angle of emergence θ_1, with the normal to the surface of the earth (see fig. 2.22). Hence, the wavefront makes an angle θ_1 with the surface of the earth. A pulse coming up the ray meets the surface at time t. Because of the slant of the raypath, we see that a pulse coming up the adjacent ray will meet the surface at time $t + dt$. Thus, the extra path length is $v_1\, dt$. The extra offset is dx. We thus have

$$\sin\theta_1 = \frac{v_1\, dt}{dx}.$$

The Snell parameter p, however, is defined as $(\sin\theta_1)/v_1$. Thus, the foregoing equation gives the required result:

$$p = \frac{\sin\theta_1}{v_1} = \frac{dt}{dx}.$$

The (x^2, t^2) method

Of the known physical phenomena in our physical world, 99 percent are nonlinear, but 99 percent of the known mathematical techniques and methods are linear. The only recourse we have is to convert nonlinear things into linear things in order to deal with them mathematically. We have to be content that such a transformatoin cannot hold in the large but is valid only in a small region surrounding the point of greatest interest.

The single-flat-interface time-distance curve is the hyperbola

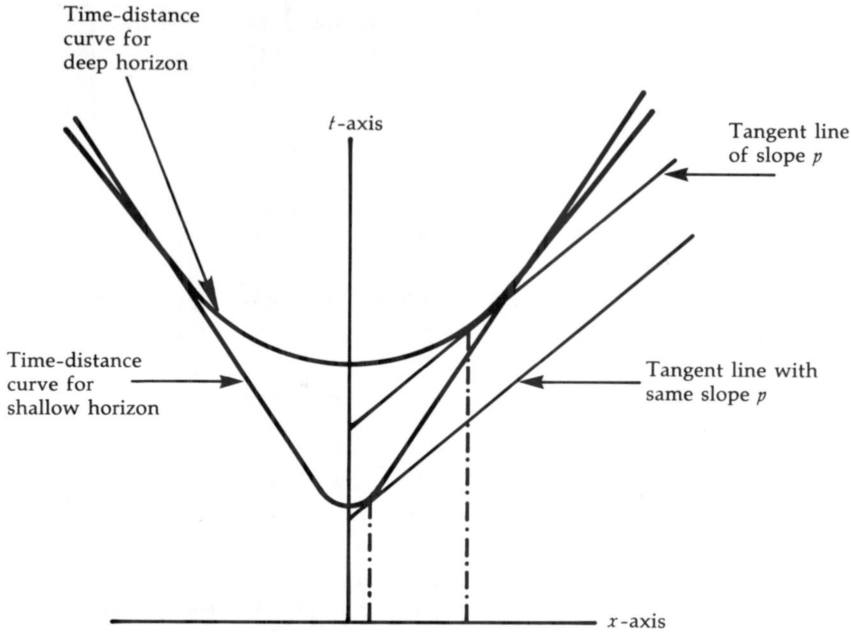

Figure 2.21. Slope dt/dx *of the time-distance curve is equal to Snell parameter* p.

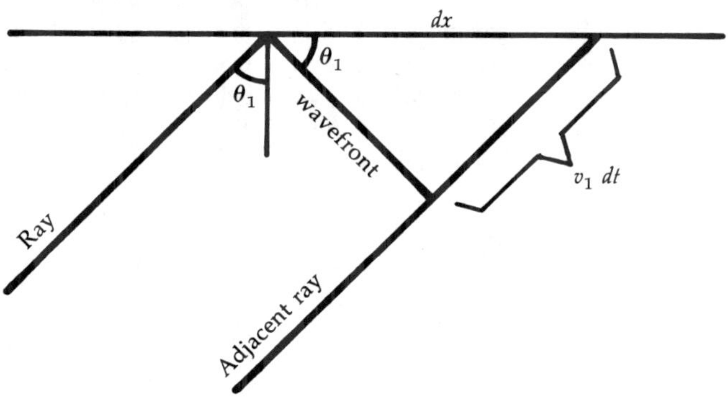

Figure 2.22. Emergent wave.

$$t^2 - \frac{x^2}{v^2} = t_0^2.$$

We can convert this curve into a straight line by the x^2, t^2 transformation

$$u = x^2, \quad w = t^2.$$

The hyperbola becomes the straight line

$$w = \frac{u}{v^2} + t_0^2,$$

with slope $1/v^2$ and w-intercept t_0^2. Graphically, straight lines are much easier to work with than hyperbolas. So much for one interface; the conversion works in the large.

When we come to two or more flat interfaces, the time-distance curve is no longer a hyperbola but is a higher-order curve; however, it is almost hyperbolic in shape. Thus, the (x^2, t^2) transformation converts it into an almost straight line, curved slightly downward as u increases. To make things simple in our geophysical analysis, we would like to approximate this slightly curved line by a straight line (see fig. 2.23). Such an approximation cannot hold in the large, but we can pick some point at which we want good approximation, and the accuracy will fall off as we get away from that point.

Let us call the required point (u_1, w_1). We will pick the straight line as the one that goes through this point and has the same slope as the slightly curved line at this point. These two conditions are enough to determine the straight line.

The n-flat-interface time-distance curve $t(x)$ is given by the parametric equations

$$x = ptv_{\text{srms}}^2,$$

$$t = \sum_{k=1}^{n} (\cos \theta_k)^{-1} \Delta t_k(0).$$

Under the (x^2, t^2) transformation, the curve $u = x^2, w = t^2(x)$ is a slightly curved line. The slope of this line is

$$\frac{dw}{du} = \frac{dt^2}{dx}\frac{dx}{du} = 2t(x)\frac{dt(x)}{dx}\frac{1}{2x},$$

but the slope of the time-distance curve is

$$p = \frac{dt(x)}{dx}.$$

Hence,

$$\frac{dw}{du} = \frac{t(x)p}{x},$$

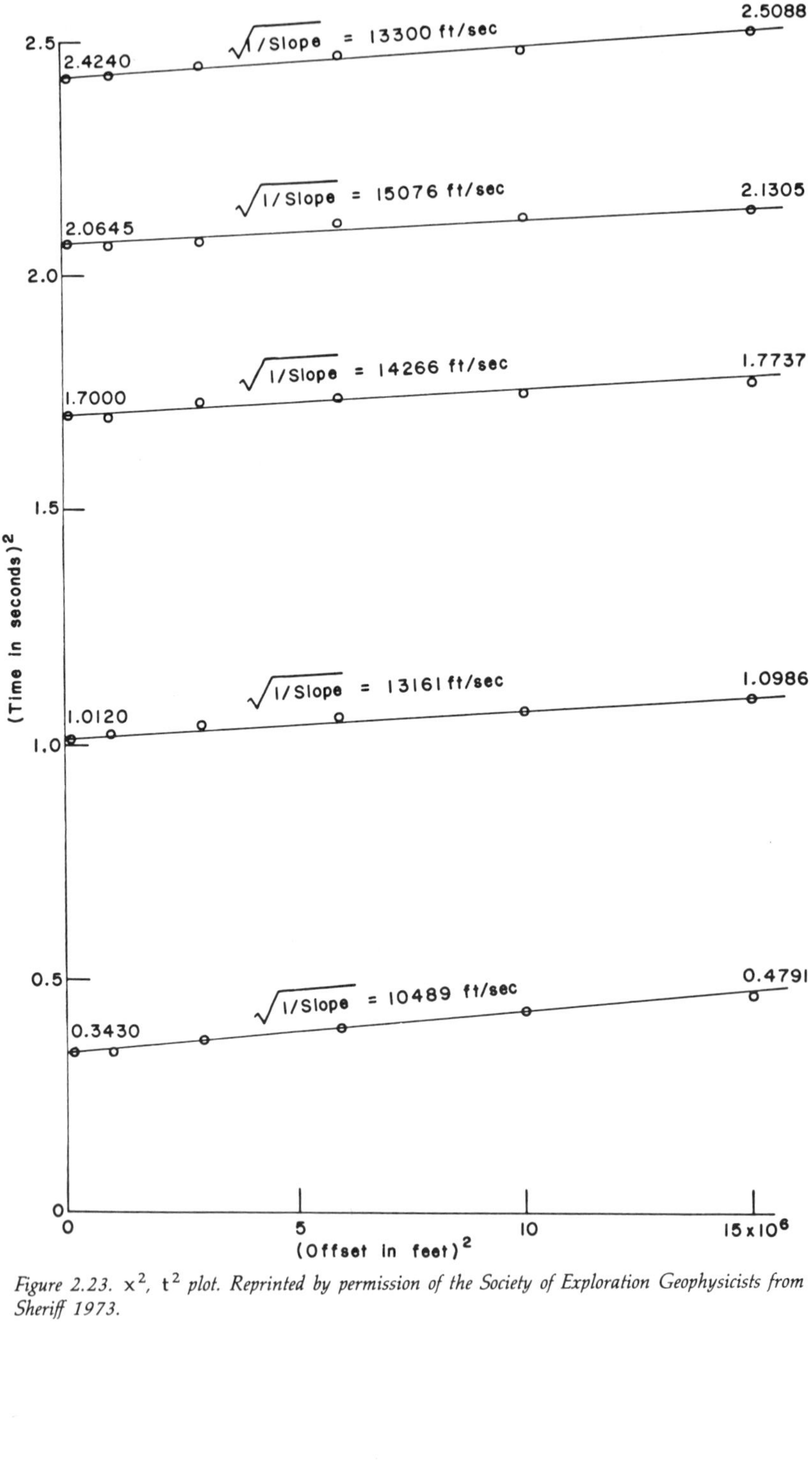

Figure 2.23. x^2, t^2 *plot. Reprinted by permission of the Society of Exploration Geophysicists from Sheriff 1973.*

which by the parametric equations is

$$\frac{dw}{du} = \frac{1}{v_{\text{srms}}^2}.$$

Therefore, we see that the slope of the slightly curved line is equal to the reciprocal of the square of the slant root-mean-square velocity.

Let us pick the required point as the w-intercept of the slightly curved line; that is, $u = 0, w = t_0^2$, where t_0 is the vertical two-way time (see fig. 2.24). The Snell parameter p for this point is zero, and v_{srms} reduces to the usual rms velocity

$$v_{\text{rms}} = \left[\frac{1}{t_0} \sum_{k=1}^{n} v_k^2 \, \Delta t_k(0)\right]^{1/2}.$$

The desired straight line, therefore, is the one that goes through the point $u = 0, w = t_0^2$ and has slope $1/v_{\text{rms}}^2$—that is, the straight line

$$w = \frac{u}{v_{\text{rms}}^2} + t_0^2.$$

Converting back to x and t, we see that this straight line becomes the hyperbola

$$t^2 - \frac{x^2}{v_{\text{rms}}^2} = t_0^2.$$

This hyperbola (see fig. 2.25) is the time-distance curve for a single-flat-interface model, with vertical two-way time t_0 and constant velocity v_{rms}; that is, the given approximation is one in which the n-layered medium is replaced by a single-layered medium whose velocity is given by the root-mean-square velocity of the n-layered medium.

As a second example, let us pick the required point as (u_1, v_1), where u_1 is not zero (see fig. 2.26). The Snell parameter p will not be zero, and there is a slant root-mean-square velocity corresponding to this value of p. Under the usual condition of velocity increasing with layer depth, v_{srms} will be greater than v_{rms}. The desired straight line is the one that goes through the point (u_1, w_1) and has slope $1/v_{\text{srms}}^2$; that is,

$$w - w_1 = \frac{(u - u_1)}{v_{\text{srms}}^2}$$

or

$$w = \frac{u}{v_{\text{srms}}^2} + w_1 - \frac{u_1}{v_{\text{srms}}^2}.$$

Converting back to x and t, we get the hyperbola (fig. 2.27)

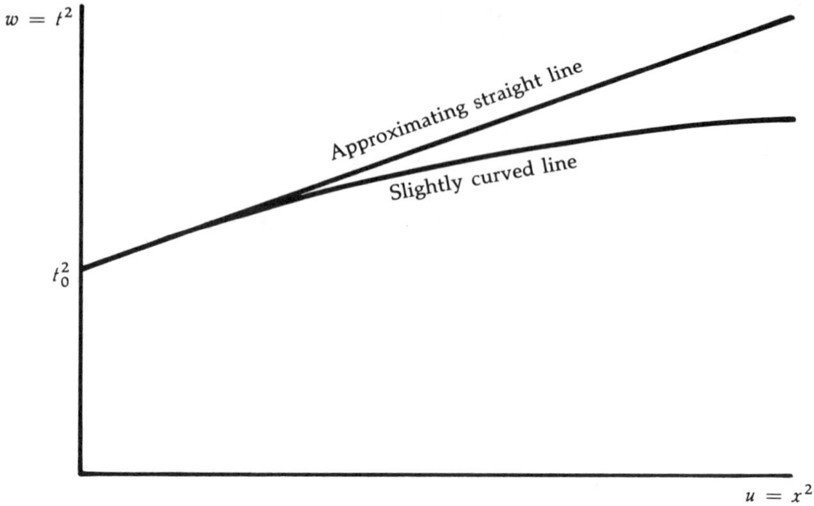

Figure 2.24. The approximating straight line is tangent to the slightly curved line at u = 0, w = t_0^2.

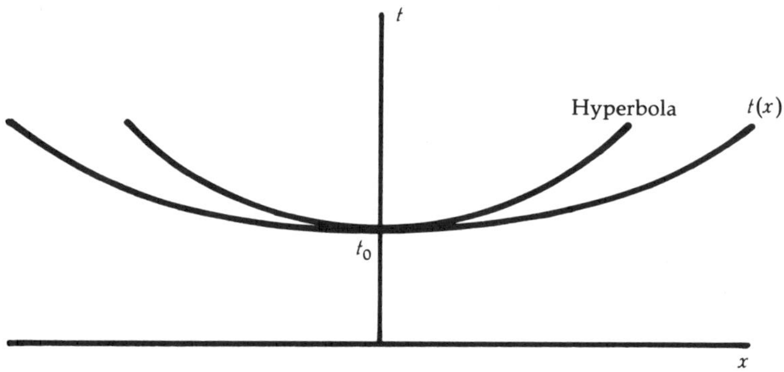

Figure 2.25. The approximating hyperbola is tangent to the time-distance curve t(x) *at* x = 0, t = t_0.

$$t^2 - \frac{x^2}{v_{\text{srms}}^2} = t_1^2 - \frac{x_1^2}{v_{\text{srms}}^2},$$

where x_1 and t_1 are defined by $u_1 = x_1^2$ and $w_1 = t_1^2$. This hyperbola is the time-distance curve for a single-flat-interface model, with vertical two-way time

$$T_0 = \left[t_1^2 - \frac{x_1^2}{v_{\text{srms}}^2} \right]^{1/2}$$

and constant velocity v_{srms}. In this single-layer approximation, the two-way vertical time is different from the two-way vertical time t_0 in the given n-layer model.

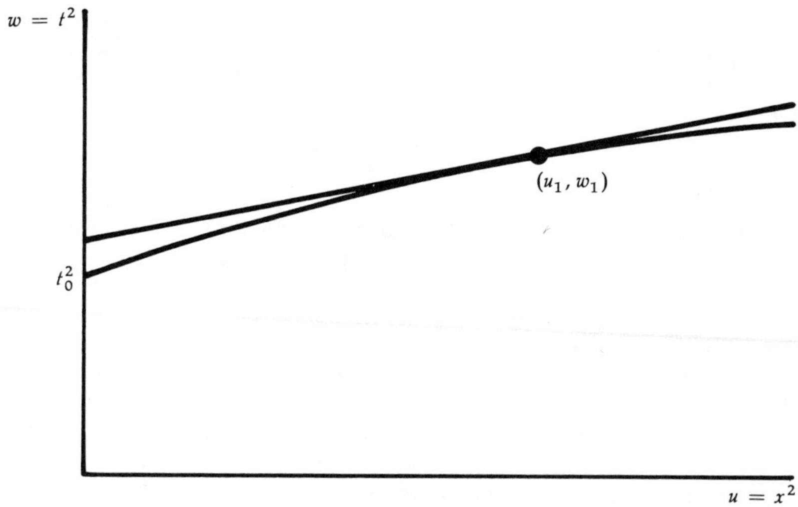

Figure 2.26. The approximating straight line is tangent to the slightly curved line at (u_1, w_1).

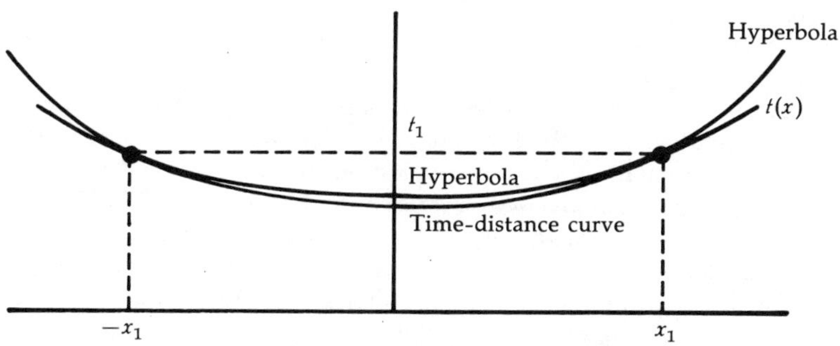

Figure 2.27. The approximating hyperbola is tangent to the time-distance curve $t(x)$ *at* $x = \pm x_1 = \pm\sqrt{u_1}$, $t = t_1 = \sqrt{w_1}$.

In seismic data processing, neither of the two foregoing approximations are used. Instead, a straight line

$$w = \frac{u}{v_s^2} + t_s^2$$

would be fitted according to a coherency criterion. Actually, the fitting is done in the (x, t) domain, so that, in fact, the corresponding hyperbola

$$t^2 - \frac{x^2}{v_s^2} = t_s^2$$

is fitted by a coherency criterion to each possible hyperbola-like reflected event appearing on the CMP gather of traces. The empirically determined quantities v_s and t_s are called the *stacking velocity* and *stacking time,* respectively. However,

v_s and t_s are usually interpreted as the rms velocity v_{rms} and two-way vertical time t_0, respectively. Such an interpretation can only be an approximation, as we have seen. In the next section, we will discuss how the stacking velocities are computed.

Velocity spectrum

Velocity is the most important variable in seismic prospecting, because we must know velocity in order to calculate the positions of the subsurface reflectors from the observed traveltimes. As discussed previously, seismic waves propagate with the velocity of sound in rock, which depends on many factors, such as chemical composition and local geology. The velocities of the rocks generally increase with depth and vary from values of 1100 ft/sec in air to values approaching 21,000 ft/sec in deep sedimentary basins. The objective of stacking is to increase signal-to-noise ratios to a level sufficient to ensure reliable identification of primary reflected events on the seismograms. Velocity as a function of seismic time, however, must be known quite accurately in order to apply proper NMO corrections prior to the processing method known as stacking. Only in this way can we obtain the correct hyperbola-shaped curves used in the stacking process. It is this hyperbolic characteristic of reflection time-distance curves that provides a means of establishing the necessary velocity-time relationship.

Velocity can be determined from seismic measurements of the variation of arrival time as the offset (i.e., the horizontal distance from source to receiver) changes. The measurement of velocity is based on the very simple single-flat-interface model—that is, a model of a seismic source and a horizontal reflecting interface, with rock of constant velocity v between the interface and the surface. For a receiver located at the source point, the reflected energy will travel down to the reflecting interface and back to the geophone, so that t_0, the arrival time, will be simply the vertical distance traveled divided by the velocity. This vertical distance is twice the depth, because the wave has to go both down and back. A receiver at horizontal distance x away from the source will detect reflected energy at a later arrival time t, because the energy has traveled farther. This greater distance will be vt (see fig. 2.28). The solution may be obtained by flipping the right triangle SAD so that it lies as triangle $S'AD$. Such a procedure in effect replaces the source point S with an image point S'. The Pythagorean theorem for the hypotenuse of a right triangle then gives the relationship

$$v^2t^2 = v^2t_0^2 + x^2,$$

which can be written as the equation of a hyperbola,

$$t^2 - \frac{x^2}{v^2} = t_0^2.$$

In the velocity search procedure to be described in this section, the vertical time t_0 is fixed at successive values t_s (called the *stacking times*), and the velocity v that is optimum for this stacking time is called the *stacking velocity* v_s.

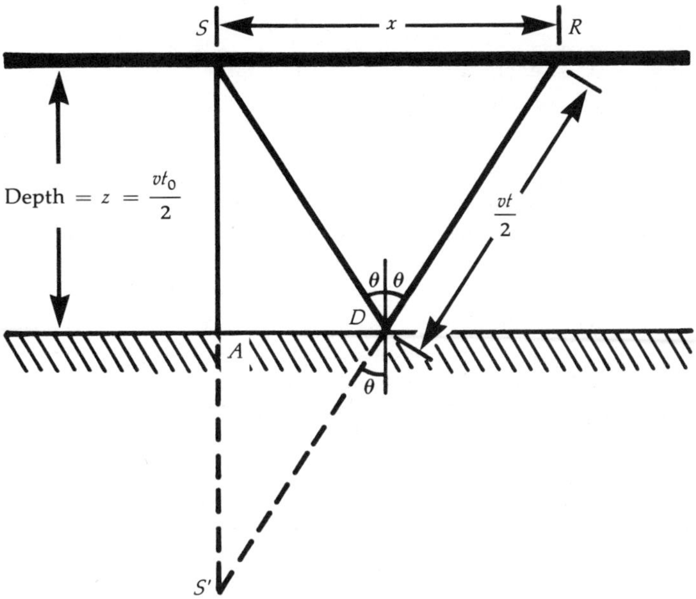

Figure 2.28. The single-flat-interface model.

Before we discuss the procedure of determining the value of v_s for each t_s, we will first introduce the idea of coherency. Trace-to-trace coherence or similarity can be given a quantitative measure in several ways. For two traces, the cross correlation can be used as a measure of coherence. For a large number of traces, we could make use of the fact that similar traces add (i.e., stack) in phase and hence reinforce each other, whereas dissimilar traces stack out of phase and hence cancel each other. Thus, when we stack several traces together, the resulting amplitude is generally large where the individual channels are similar (coherent) and small where they are unlike (incoherent). The ratio of the energy of the stack to the sum of the energies of the individual traces would therefore be a measure of the degree of coherence.

Let f_{it} be the amplitude of the individual trace i at time t. We suppose that there are M traces in the record section, so that the trace index i ranges from 1 to M. Furthermore we suppose that the time t follows some trajectory across the traces making up the record section. In velocity analysis this trajectory is the hyperbola specified by some particular value of the velocity v. Thus the actual value of the trajectory time t for any given trace depends upon the trace index i.

The amplitude of the stack for trajectory t is given by Σf_{it} where the summation is over $i = 1, 2, \ldots, M$. The square of this stack amplitude is the instantaneous energy of the stack for that particular trajectory. The instantaneous energy of each trace along the trajectory is f_{it}^2, so the sum of the energies of the input traces is Σf_{it}^2 where the summation is over $i = 1, 2, \ldots, M$. Let us now define the ratio E_t associated with trajectory t by means of the formula

$$E_t = \frac{\text{(instantaneous stack energy)}}{M \text{ (instantaneous input energy)}} = \frac{\left(\sum_{i=1}^{M} f_{it}\right)^2}{M \sum_{i=1}^{M} f_{it}^2}.$$

The reason for introducing the number M in the numerator is that it acts as a normalization factor. Thus the instantaneous energy ratio E_t is one when all the trace amplitudes are the same along the trajectory.

A coherent event extends over a time interval on each trace. In order to account for this effect, we usually use in place of E_t the semblance S_t (Neidell and Taner, 1971), which we will now define. Let us pick a time gate on each trace. This gate is centered on the trajectory t and extends the same width τ on each side. Thus the total width of the time gate is 2τ, which is the same for each trace. In summary, the time gate is an interval of length 2τ on each trace, and is symmetrically disposed about the trajectory t. This choice makes the gate boundaries parallel to the trajectory t. The *semblance* is a measure of the coherence of a signal which crosses the gate, and is defined as

$$S_t = \frac{\text{(stack energy within gate)}}{M \text{ (input energy within gate)}} = \frac{\sum_{q=t-\tau}^{t+\tau} \left(\sum_{i=1}^{M} f_{iq}\right)^2}{M \sum_{q=t-\tau}^{t+\tau} \sum_{i=1}^{M} f_{iq}^2}.$$

Not only does the semblance tend to be large when a coherent event is present, but the magnitude of the semblance is also sensitive to the amplitude of the event. Strong events exhibit large semblances and weak events exhibit moderate values of semblance, whereas incoherent data have very low semblances.

Semblance and other coherence measures are used to determine the values of parameters that will optimize a stack. The semblance is calculated for various time shifts between the component channels; the optimum time shift is one that maximizes the semblance. Semblance therefore can be used to determine static corrections. The most important application of semblance is in the computation of a velocity spectrum from which we determine stacking velocity as a function of stacking time. We will now explain this procedure.

The essential idea behind velocity analysis is that the data in the common midpoint (CMP) gather can be used to estimate velocity. The data-processing method used is called a *velocity search procedure.* At a fixed zero-offset time t_s, the common midpoint set of traces, with full offsets $x_1, x_2, \ldots, x_n$, is overlaid with a set of hyperbolas (see fig. 2.29). These hyperbolas are indexed by the velocity v—that is, a different velocity for each hyperbola. For each hyperbola, the semblance of the traces (within a small time gate centered around the hyperbola) is computed. The semblance is plotted for a set of values v for the given t_s. The velocity v that produces the maximum semblance represents the best-fitting hyperbola and thus produces the best stack at stacking time t_s. As a result, this optimum velocity is called the stacking velocity v_s at time t_s. This procedure is repeated for a collection of zero-offset (vertical) times t_s, generally at equally

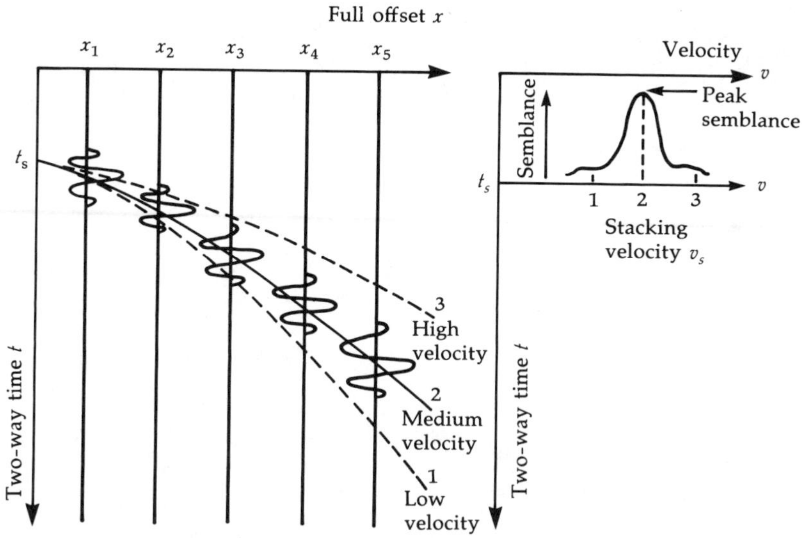

Figure 2.29. Hyperbolic sweep of CMP gather to determine stacking velocity v_s *at time* t_s*. This process is repeated at constant time intervals down the record.*

spaced time increments. The result is called the *velocity spectrum* at the common midpoint. The velocity spectrum consists of a three-dimensional display of semblance as a contour in (v, t_s) space. A curve showing the stacked semblance maxima as a function of zero-offset (stacking) time t_s identifies the coherent events and the stacking velocities v_s (see figs. 2.30, 2.31, and 2.32).

In summary, a velocity spectrum is obtained by scanning the CMP gather of traces along hyperbolic trajectories for signal coherence. These scans establish a velocity function (i.e., velocity versus vertical seismic time) that is used in calculating the NMO corrections prior to the stacking of the CMP traces. One further point that deserves mention is that the time-distance curves for flat interfaces are not hyperbolic, except in the case of the first interface. These curves are almost hyperbola-shaped, however, and the second-order hyperbolic approximation used in the velocity search works well even in areas of complex structures.

The Dix formula

In conventional seismic processing, normal moveout is computed by assuming the straight line propagation of the constant velocity model. The question addressed in this section is which mathematical expression for velocity we should use for the constant velocity appearing in this simple model. This is not a simple question to answer, but the answer given by Dix (1955) is used in 99.9 percent of petroleum exploration cases.

One of the most important problems in reflection seismology is how to handle and take into account the vertical variations of velocity. The simplest methods make use of various modifications of the constant velocity model, a model in which the actual earth layers existing between the earth's surface and an inter-

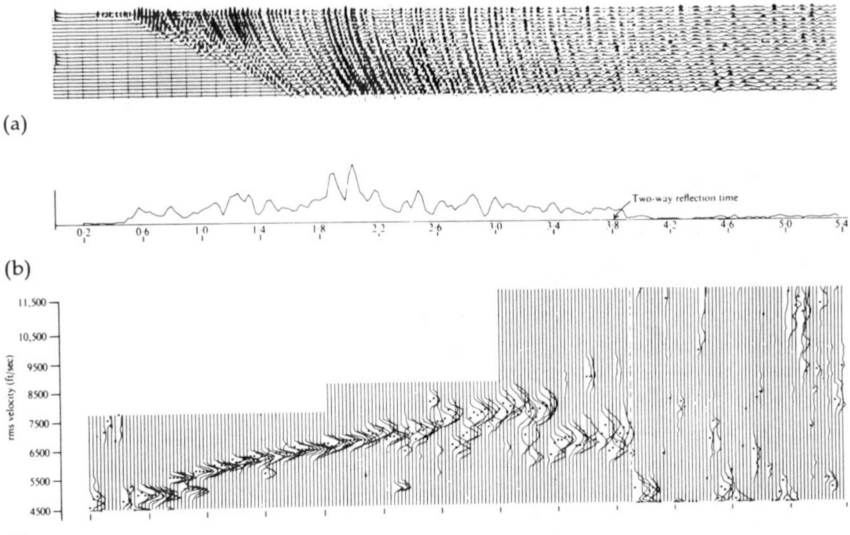

Figure 2.30. Velocity analysis: (a) common depth point (CMP) gather showing the data involved in the analysis; (b) maximum amplitude achievable in stacked traces (peaks indicate the stronger events); (c) amplitude of stacked trace as a function of stacking velocity at 100 msec intervals of t_s. *The dots indicate the stacking velocity that gives maximum amplitude. The stacking velocity increases with depth from about 5000 ft/sec at 0.6 sec to about 7700 ft/sec at 3.3 sec. The data between 2.7 and 3.9 sec that have velocities around 6600 ft/sec are undoubtedly multiples involving the strong reflections at 1.9–2.1 sec. There are probably few primary reflections below 3.3 sec. Courtesy Petty-Ray Geophysical. Reprinted by permission of Cambridge University Press from Telford et al. 1976.*

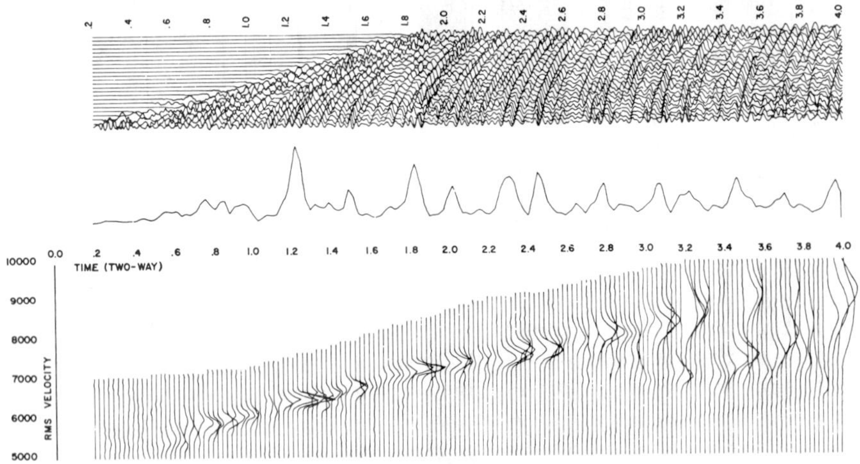

Figure 2.31. The upper diagram is a CMP gather, the center diagram shows the peak semblance, and the lower diagram is the velocity spectrum. Reprinted by permission of Geophysics from Taner and Koehler 1969.

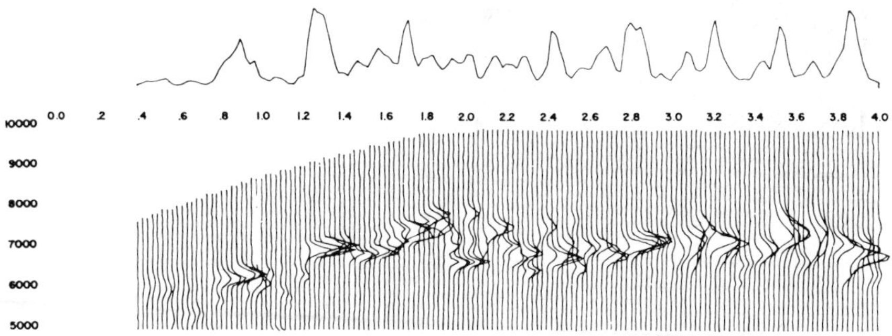

Figure 2.32. Velocity spectrum in an area of velocity inversion. Velocity analysis is presented as curves of semblance, which is simply the ratio of the energy resulting from a stacking velocity assumption compared to the amount of energy available to be stacked. Reprinted by permission of Geophysics from Taner and Koehler 1969.

face at some given depth are replaced by a single layer of some constant velocity V. By varying the depth of the interface, this assigned velocity, which is constant between the surface and interface, becomes a function $V(z)$ of the depth z of the interface (see fig. 2.33). If T_0 is the vertical traveltime to the interface, then, alternatively, V may be considered to be a function of T_0. Note that we are assuming throughout our presentation that the layers are horizontal (i.e., that they are not dipping). Thus, the constant velocity model is one in which the earth is assigned a different constant velocity $V(z)$ for each interface (where z is the depth of the interface).

Certainly, $V(z_n)$ must be some sort of average of the interval velocities v_1, v_2, . . ., v_n down to interface z_n. Thus, in figure 2.33, $V(z_1)$ would be some function of v_1, $V(z_2)$ some function of v_1 and v_2, and $V(z_3)$ some function of v_1, v_2, and v_3. The functional form we choose for $V(z)$ would depend on the purpose for which we want to use it. One purpose of a velocity function $V(z)$ is to compute normal moveout (NMO). Thus, we would want to pick a convenient functional form that will serve this purpose.

The time-distance curve for the single-flat-layer model is obtained by applying the Pythagorean theorem to the sides of a right triangle. The hypotenuse is the slanted straight line ray of length VT, one leg is the full offset x, and the other leg is the vertical distance VT_0. Here T is the two-way slant time and T_0 is the two-way vertical time. Thus, the Pythagorean equation is

$$x^2 + (VT_0)^2 = (VT)^2$$

or

$$T^2 = T_0^2 + \frac{x^2}{V^2}.$$

If we consider this equation as the equation of a curve of T versus x, we see that the curve is a hyperbola (see fig. 2.34). The T-intercept is equal to the vertical time T_0. If we consider this equation as the equation of a curve of T^2 versus x^2, however, we see that the curve is a straight line of slope $1/V^2$.

The NMO equation for the single-layer model is

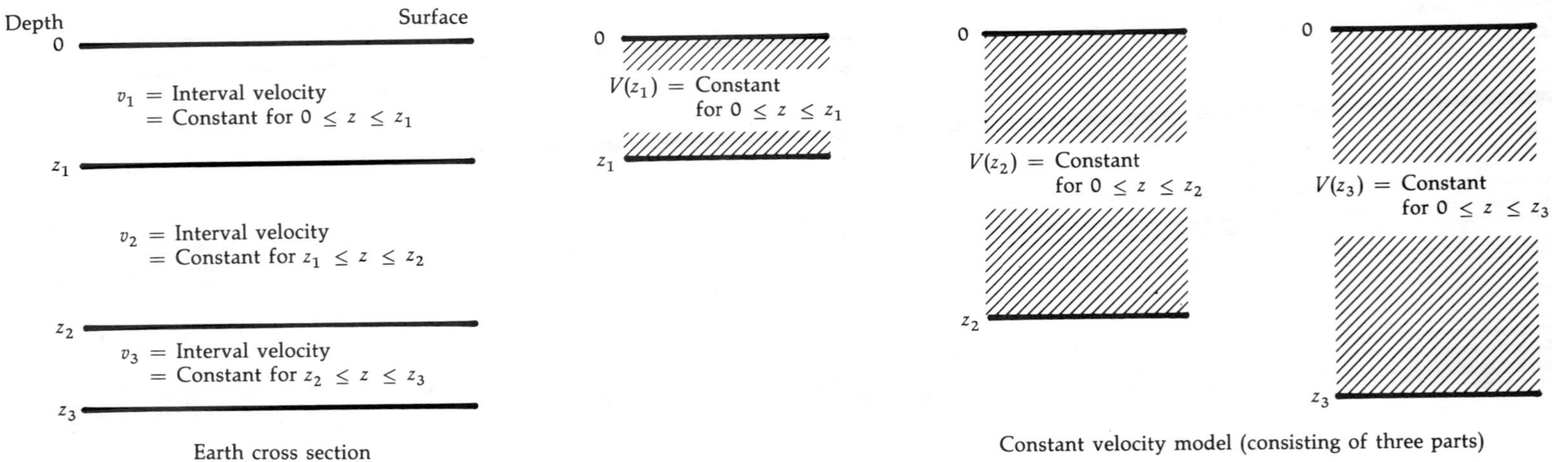

Figure 2.33. Replacement of the layered earth model with a constant average velocity model.

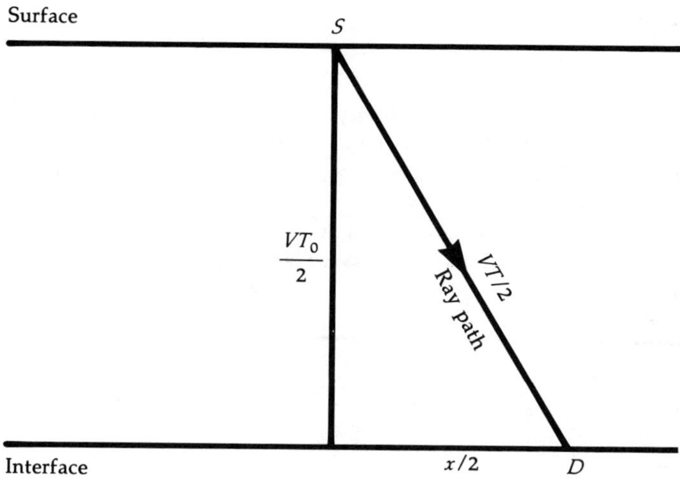

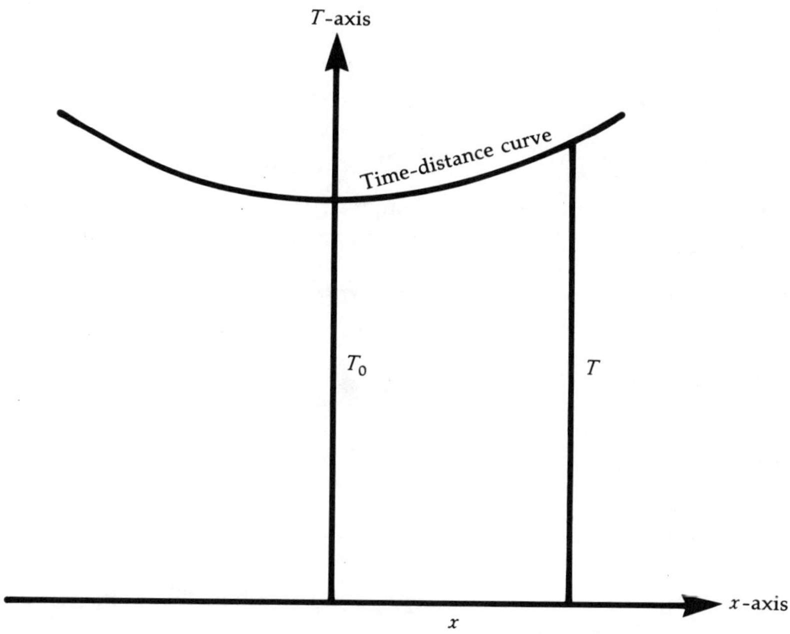

*Figure 2.34. (*top*) Constant velocity model. (*bottom*) Hyperbolic time-distance curve.*

$$\Delta T = \left[T_0^2 + \frac{x^2}{V^2} \right]^{1/2} - T_0,$$

where T_0 is the vertical time, V is the velocity, x is the offset distance, and T is the slant time. To gain insight, let us experiment with this equation by means of examples. The slant time T is an approximation to the slant time $t(x)$ of the layercake model. In the following two examples, we let T_0 be equal to the vertical time $t(0)$ of the layercake model and experiment with different definitions of the velocity V. In these examples, we let x denote full offset, and all times are two-way times.

Example 1. This example is the case of one layer with an interval velocity $v_1 = 8000$ ft/sec, a layer thickness $\Delta z_1 = 5000$ ft, and an offset distance $x = 6000$ ft (see fig. 2.35). We have

$$T_0 = t(0) = \frac{2(5000)}{8000} = 1.250 \text{ sec.}$$

We set $V = v_1 = 8000$ ft/sec:

$$T = \sqrt{T_0 + (x^2/V^2)} = \sqrt{(1.250)^2 + (6000^2/8000^2)} = 1.458 \text{ sec.}$$

The normal moveout ΔT is thus

$$\Delta T = T - T_0 = 1.458 - 1.250 = 0.208 \text{ sec.}$$

Example 2. This is the case of two layers with interval velocity $v_1 = 6000$ ft/sec for the first layer and $v_2 = 10{,}000$ ft/sec for the second layer (see fig. 2.36). Let us choose the thickness of the layers to be $\Delta z_1 = 1875$ and $\Delta z_2 = 3125$. The reason for this choice is that it makes the vertical time T_0 the same as for Example 1; that is,

$$T_0 = t(0) = \frac{2(1875)}{6000} + \frac{2(3125)}{10000} = 1.250 \text{ sec.}$$

The average velocity is

$$v_A = \frac{2(\Delta z_1 + \Delta z_2)}{T_0} = \frac{2(1875 + 3125)}{1.250} = 8000 \text{ ft/sec,}$$

which we now use for V in a single-layer approximation. The Pythagorean equation gives the slant time as

$$T = \sqrt{T_0^2 + (x^2/V^2)} = \sqrt{(1.250)^2 + (6000^2/8000^2)} = 1.458 \text{ sec,}$$

so the normal moveout is

$$\Delta T = T - T_0 = 1.458 - 1.250 = 0.208 \text{ sec.}$$

Instead of using the single-layer approximation, let us compute Δt for this second case by using the time-distance curve for the two-layer model as it is. The angles come out to be

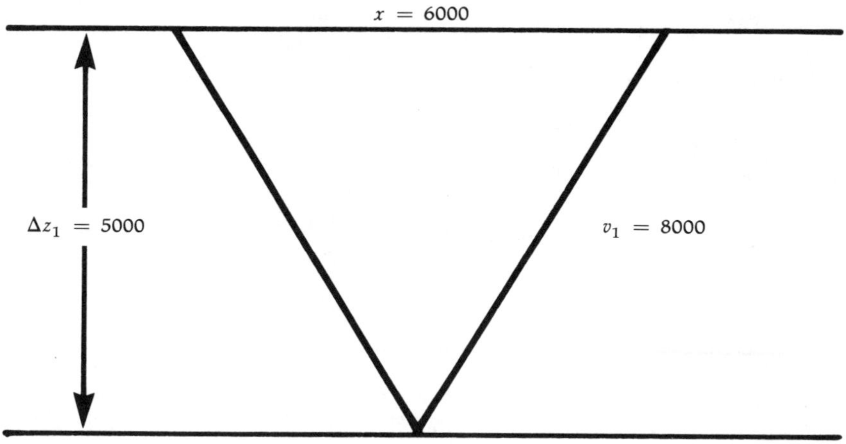

Figure 2.35. Illustration of Example 1.

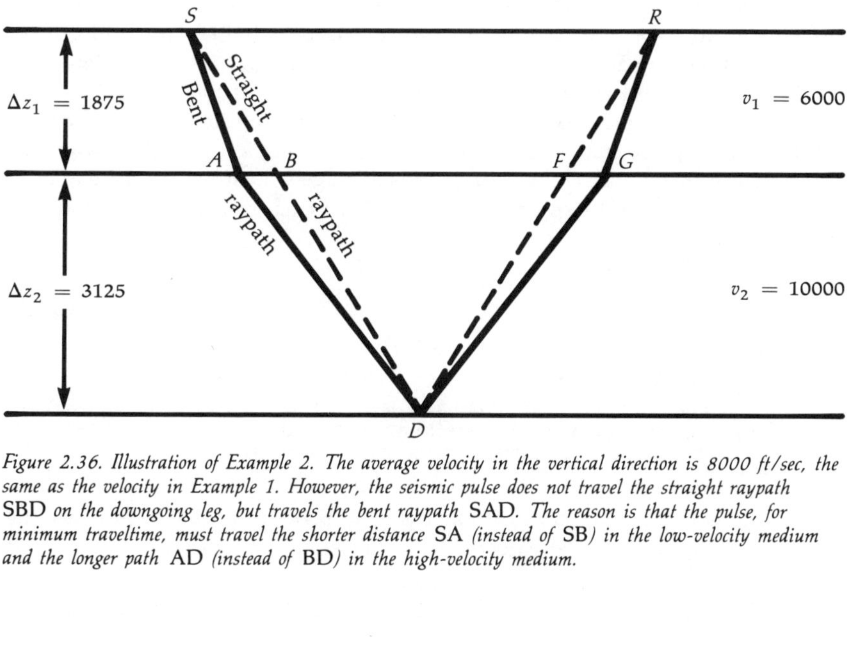

Figure 2.36. Illustration of Example 2. The average velocity in the vertical direction is 8000 ft/sec, the same as the velocity in Example 1. However, the seismic pulse does not travel the straight raypath SBD *on the downgoing leg, but travels the bent raypath* SAD. *The reason is that the pulse, for minimum traveltime, must travel the shorter distance* SA *(instead of* SB*) in the low-velocity medium and the longer path* AD *(instead of* BD*) in the high-velocity medium.*

$$\theta_1 = 20°46'', \qquad \theta_2 = 36°13'',$$

and the slant time is

$$\begin{aligned} t(x) &= \frac{2\Delta z_1}{v_1 \cos \theta_1} + \frac{2\Delta z_2}{v_2 \cos \theta_2} \\ &= \frac{2(1875)}{6000 \cos 20°46''} + \frac{2(3125)}{10000 \cos 36°13''} = 1.444 \text{ sec.} \end{aligned}$$

The resulting Δt is

$$\Delta t = 1.444 - 1.250 = 0.194 \text{ sec.}$$

This Δt is true normal moveout, as we have calculated it from the actual time-distance curve and not from approximation.

Discussion. Examples 1 and 2 have the same offset distance, average velocity, and vertical traveltime. As a result, ΔT as computed from the single-layer model is the same for both examples—namely, 0.208. In other words, the single-layer model does not take into account any effect caused by the layering in Example 2. If we do take into account the layering in Example 2, however, the slant time $t(x)$ is less than T. Thus, $\Delta t = 0.194$ is less than $\Delta T = 0.208$. The value 0.194 is the correct value, whereas the value 0.208 given by the single-layer approximation is wrong. To make the single-layer approximation give the correct Δt, or nearly the correct Δt, we cannot use the average velocity v_A in the equation but instead must use the rms velocity v_{rms}. For Example 1, the rms velocity is

$$v_{\text{rms}} = \sqrt{\frac{(8000)^2(1.250)}{1.250}} = 8000 \text{ ft/sec,}$$

which is the same as the interval velocity as well as the average velocity. For Example 2, the rms velocity is

$$v_{\text{rms}} = \sqrt{\frac{(6000)^2(0.625) + (10000)^2(0.625)}{1.250}} = 8246 \text{ ft/sec,}$$

which is greater than the average velocity $v_A = 8000$ ft/sec. Let us now use the rms velocity in the Pythagorean equation. The equation gives the slant time as

$$T = \sqrt{T_0 + (x^2/v_{\text{rms}}^2)} = \sqrt{(1.250)^2 + (6000^2/8246^2)} = 1.446 \text{ sec.}$$

This slant time is close to the true value of 1.444 sec computed earlier. Therefore, the resulting ΔT,

$$\Delta T = T - T_0 = 1.446 - 1.250 = 0.196 \text{ sec,}$$

is close to the true value $\Delta t = 0.194$ sec, given earlier.

In summary, if we use a multilayer raypath calculation, we get the correct Δt of 0.194 sec. If we use the average velocity in the single-layer approximation, the computed ΔT is 0.208 seconds; whereas, if we use the rms velocity, the computed ΔT is 0.196 sec. Thus, the rms velocity gives a better result than the average velocity when using the single-layer Pythagorean equation in any case in which there is layering. In fact, the rms velocity will differ from the average velocity more and more as the layering becomes more complex. In figure 2.37, we see the true time-distance curve and the approximating hyperbola with the same t-intercept and with the rms velocity.

As seen in the preceding section, the velocity spectrum as computed from the CMP data yields an empirical velocity function that gives velocity v_s versus time t_s. The quantities v_s and t_s are called *stacking velocity* and *stacking time,* respectively. The stacking velocity v_s gives the greatest semblance for the given stacking time t_s as compared to any other velocity v. Because v_s is empirically determined, we do not have any mathematical equation for it. However, we want to set v_s equal to a theoretical velocity V that does have a mathematical expression. The reason we need a mathematical formula is so that we can compute the interval velocities $v_1, v_2, \ldots, v_n$ from the observed stacking velocities at their respective stacking times.

Before 1955, the empirically determined velocity v_s was set equal to the weighted average velocity v_A, defined as

$$v_A = \frac{\text{total distance}}{\text{total time}} = \frac{2(\Delta z_1 + \Delta z_2 + \cdots + \Delta z_n)}{t(0)}.$$

This choice was poor, as demonstrated by the foregoing examples. In 1955, Professor C. H. Dix said: "Set the empirical velocity v_s equal to rms velocity v_{rms}" (see fig. 2.37). As our examples show, this gives good results, especially for small offset distances x, and this choice has been used by the seismic industry ever since then.

Once the stacking velocities have been determined (for example, from the velocity spectrum) at two successive stacking times t_s, the interval velocity can be found from the Dix formula.

Let us derive the Dix formula, which shows how the rms velocities for two successive interfaces can be used to find the intervening interval velocity. Writing $v_{\text{rms}}(n)$ for the rms velocity for the nth interface and $v_{\text{rms}}(n-1)$ for the rms velocity for the interface above it, we have

$$\sum_{k=1}^{n} v_k^2\, \Delta t_k(0) = \sum_{k=1}^{n-1} v_k^2\, \Delta t_k(0) + v_n^2\, \Delta t_n(0) = v_{\text{rms}}^2(n) \sum_{k=1}^{n} \Delta t_k(0),$$

$$\sum_{k=1}^{n-1} v_k^2\, \Delta t_k(0) = v_{\text{rms}}^2(n-1) \sum_{k=1}^{n-1} \Delta t_k(0).$$

(Note that $\Delta t_k(0)$ denotes the two-way *vertical* time in layer k.) Subtracting, and then dividing both sides by $\Delta t_n(0)$, we obtain the Dix formula for the interval velocity v_n. The celebrated formula is

$$v_n^2 = \frac{1}{\Delta t_n(0)} \left[v_{\text{rms}}^2(n) \sum_{k=1}^{n} \Delta t_k(0) - v_{\text{rms}}^2(n-1) \sum_{k=1}^{n-1} \Delta t_k(0) \right].$$

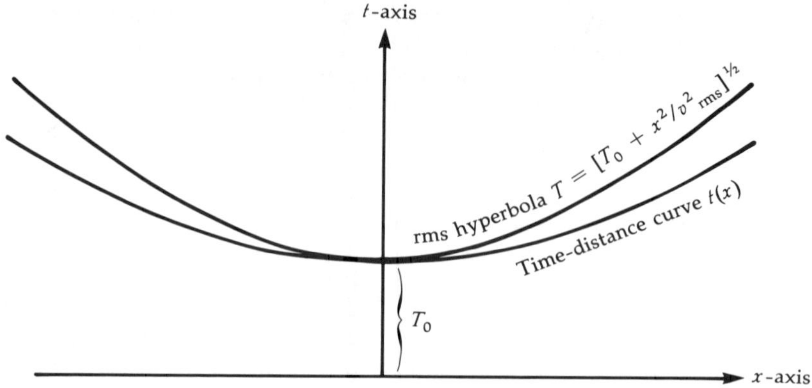

Figure 2.37. The rms hyperbola has the same vertical time as the true time-distance curve, but otherwise always has a greater traveltime.

The basic assumption of velocity analysis is now invoked. Given two successive stacking times $t_{s,n-1}$ and $t_{s,n}$ and the corresponding empirically determined stacking velocities $v_{s,n-1}$ and $v_{s,n}$, the basic assumption is to set the stacking velocities equal to the rms velocities and the stacking times equal to the vertical traveltimes appearing in the Dix formula. Thus, the Dix formula becomes

$$v_n^2 = \frac{1}{t_{s,n} - t_{s,n-1}}[v_{s,n}^2 t_{s,n} - v_{s,n-1}^2 t_{s,n-1}].$$

From the velocity spectrum, we can measure the empirical values of all the quantities on the right. Using the Dix formula, we therefore determine the interval velocity v_n.

Figure 2.38 shows a velocity analysis, and figure 2.39 shows an interpretation of this analysis. The observed stacking times and stacking velocities are as follows:

t_s	v_s
0.660	8,200
0.760	8,400
0.920	10,200
1.050	11,100
1.280	11,200
1.860	12,500
2.090	13,300

Let us now compute the interval velocities. The first is obtained by the Dix formula as

$$v = \left[\frac{8400^2(0.760) - 8200^2(0.660)}{0.760 - 0.660}\right]^{1/2} = 9600,$$

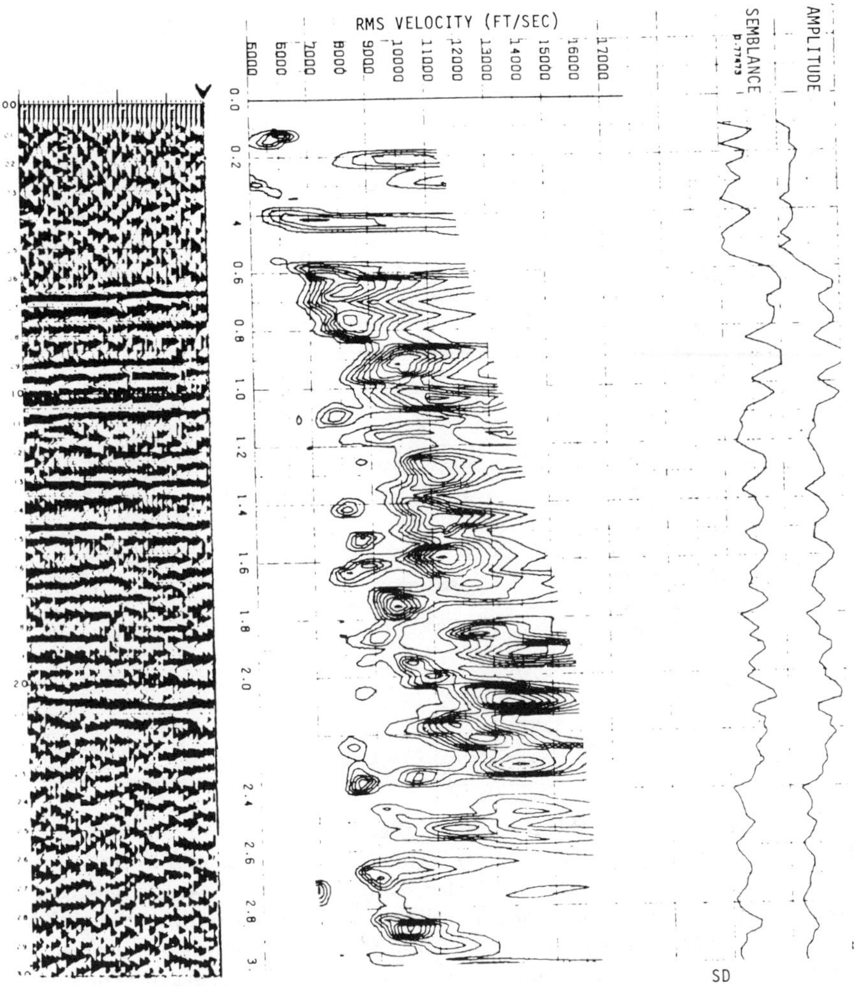

Figure 2.38. Velocity spectrum. Reprinted by permission of IHRDC from Sheriff 1978.

and the others are obtained similarly. We thus obtain the sequence of interval velocities: 9,600, 16,200, 16,100, 11,600, 15,000, 17,200. This is the classic way of finding interval velocities from seismic data.

Offset velocity

Let us now approach the velocity problem from the opposite direction. From Example 2 of the preceding section, let us take the actual Δt of 0.194 seconds and then calculate what velocity in the equation will give this Δt. The result of this calculation is the velocity v_x, given by

$$v_x = \sqrt{\frac{x^2}{t^2(x) - t^2(0)}} = \sqrt{\frac{6000^2}{1.444^2 - 1.250^2}} = 8310 \text{ ft/sec.}$$

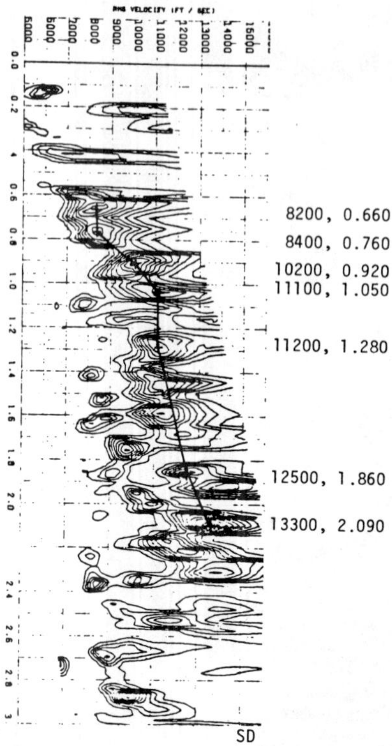

Figure 2.39. Interpretation of velocity spectrum. Reprinted by permission of IHRDC from Sheriff 1978.

A velocity calculated by this method is called the *offset velocity* v_x. By definition, the offset velocity (for a given x) gives the exact Δt when used in the single-layer model for that particular offset x. Thus, the offset velocity yields the correct result when making a Δt correction on a seismic record for that specific offset. The average velocity yields poor results unless the interval velocities are nearly the same. The rms velocity yields nearly the correct result. The offset velocity is a velocity chosen to fit the data at one fixed value of x, so that as we move away from that particular x value, the results become poorer. If we use a shorter offset in Example 2, for example, such as 4000 ft instead of 6000 ft, the offset velocity is different. For $x = 4000$ ft, the true slant time is

$$t(x) = \left[\frac{1875}{6000 \cos 15^\circ} + \frac{3125}{10000 \cos 25^\circ 23''}\right] = 1.340 \text{ sec},$$

so that the offset velocity is

$$v_x = \sqrt{\frac{x^2}{t^2(x) - t^2(0)}} = \sqrt{\frac{4000^2}{1.340^2 - 1.250^2}} = 8280 \text{ ft/sec}.$$

Thus, the offset velocity is a function of the offset distance as well as the layering. We can summarize the velocities in Example 2 as

Average velocity $= v_A = 8000$ ft/sec
rms velocity $= v_{rms} = 8246$ ft/sec
Offset velocity $= v_x = 8280$ ft/sec for $x = 4000$ ft
Offset velocity $= v_x = 8310$ ft/sec for $x = 6000$ ft

The average velocity should be used for depth conversion, because it is the true vertical velocity in the ground. The average velocity should never be used for NMO calculations, however, because it gives a ΔT that is too large. If the interval velocities are available from a well survey, then the rms velocity can be calculated and can be used for a ΔT correction. The rms velocity will always give a better result for ΔT than the average velocity calculated from the same survey.

Let (x_1, t_1) be a point on the time-distance curve. The minimum of the time-distance curve is the point $(0, t_0)$, and the slope there is zero. (This minimum point corresponds to $p = 0$, as we know that the slope dt/dx of the time-distance curve is equal to p.) The approximating hyperbola

$$T = [t_0 + x^2/V^2]^{1/2}, \qquad \text{where } t_0 = t(0),$$

also goes through the point $(0, t_0)$ and likewise has a zero slope at that point. The offset velocity v_x is the value of V that makes this hyperbola also go through the point (x_1, t_1) on the time-distance curve. Thus, v_x is found by solving the equation

$$t_1 = [t_0 + x_1^2/v_x^2]^{1/2},$$

and the value of v_x so found is called the value of the offset velocity for offset x_1 (see figs. 2.40 and 2.41). Offset velocity increases with increased spread length or with increased dip of the reflecting horizon. The offset velocity is larger than the rms velocity, and the rms velocity is larger than the average velocity.

The offset velocity v_x as just defined might be called a naive stacking velocity as compared to the sophisticated statistical determination of empirical stacking velocities by means of the velocity spectrum, as described previously. The empirical stacking velocity as derived from the seismic data should always be used for ΔT corrections, but the stacking velocity should never be used for depth conversion.

Another approach to finding a mathematical expression for the empirical stacking velocity does not require that the approximating hyperbola go through the minimum point $(0, t_0)$ of the time-distance curve. The hyperbola will still have a minimum and a zero slope at $x = 0$, but its intercept T_0 will not necessarily equal t_0. Thus, the approximating hyperbola is

$$T = [T_0^2 + x^2/V^2]^{1/2}.$$

Now we have two arbitrary constants T_0 and V available in order to fit the hyperbola to the time-distance curve.

Next, let us describe the use of the slant rms velocity as an approximation to the stacking velocity. At some point (x_1, t_1) on the time-distance curve, as evidenced by the seismic data, we measure the slope $(dt/dx)_1$. As we know, this

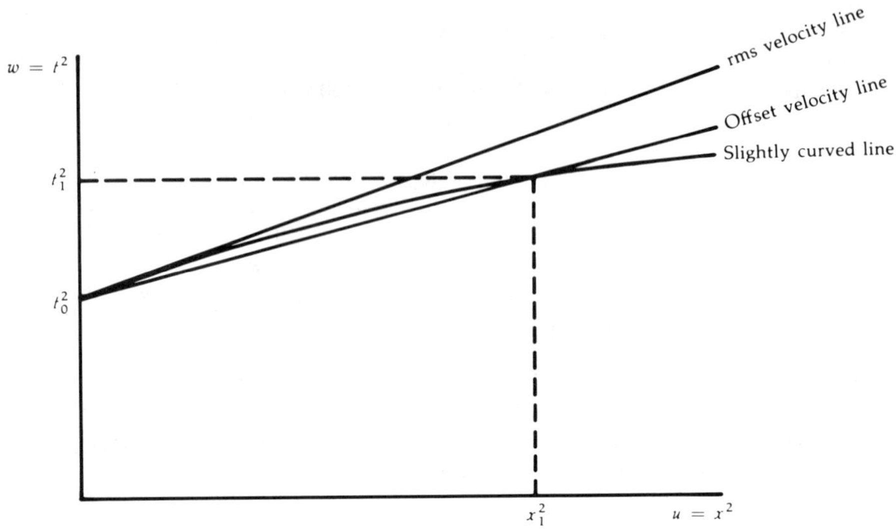

Figure 2.40. The slightly curved line is the x^2, t^2 *transformation of the true time-distance curve. The two straight lines are the approximations for offset velocity and rms velocity, respectively.*

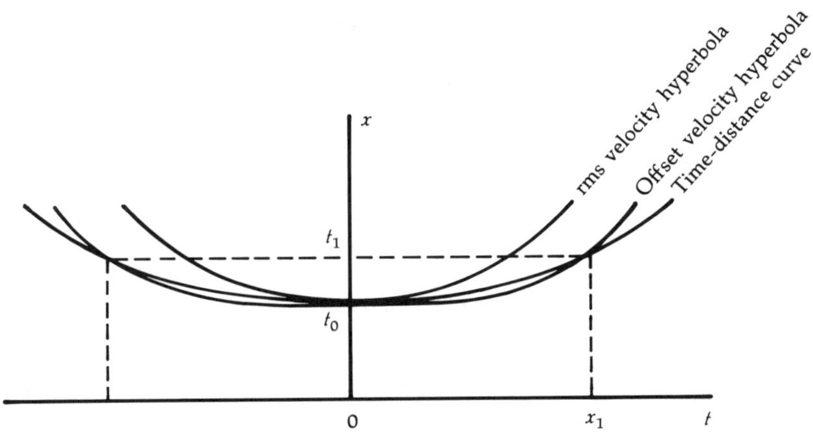

Figure 2.41. The time-distance curve and the approximating hyperbolas for offset velocity and rms velocity.

slope is equal to Snell's parameter p. Let $p_1 = (dt/dx)_1$. From our study of the layercake model, we recall that the slant rms velocity for parameter p_1 is

$$v_{\text{srms}} = \sqrt{\frac{x_1}{p_1 t_1}}.$$

We now let $V = v_{\text{srms}}$ and we solve for T_0 by requiring that the hyperbola go through the point (x_1, t_1). Thus, T_0 is given by

$$T_0^2 = t_1^2 - x_1^2/v_{\text{srms}}^2$$

This method, in which we choose the slant rms velocity to approximate the empirical stacking velocity, provides a good approximation of the hyperbola to the time-distance curve near (x_1, t_1).

As we have seen, the mathematical expression chosen to approximate the empirical stacking velocity provides a hyperbola that in some sense best fits the time-distance curve. The time-distance curve is the curve of a reflected event on the seismic data, so that the final result is a velocity determined directly from the data. The best-fit criterion varies according to the method used in the mathematical approximation to the stacking velocity. Al-Chalabi (1973) gives the following least-squares criterion. Let the approximating hyperbola be

$$T^2 = a + bx^2,$$

where $a = T_0{}^2$ and $b = 1/V^2$ are the two arbitrary constants to be determined. From the seismic data, we obtain m points (x_i, t_i) on the time-distance curve. According to the principle of least-squares, we want to minimize

$$\sum_{i=1}^{m} (a + bx_i^2 - t_i^2)^2.$$

If we set the partial derivatives with respect to a and b to equal zero, we obtain the normal equations

$$am + b \sum_{i=1}^{m} x_i^2 = \sum_{i=1}^{m} t_i^2,$$

$$a \sum_{i=1}^{m} x_i^2 + b \sum_{i=1}^{m} x_i^4 = \sum_{i=1}^{m} t_i^2 x_i^2.$$

We now solve this set of two simultaneous equations for a and b. The best-fitting hyperbola then has vertical time $T_0 = \sqrt{a}$ and velocity $V = 1/\sqrt{b}$.

In summary, velocity analyses are made by determining how much stacked energy (or the quantity called *semblance*) results from a sequence of trial stacking velocities. This determination is made for a number of stacking velocities over a series of narrow windows of data. The variation of arrival time as we change the offset distance from source to receiver carries the information that allows us to determine the stacking velocity. This variation is embodied in the form of the time-distance hyperbola for the single-flat-interface model. An estimated stacking velocity v_s is a velocity determined from the seismic data by requiring a good fit of the hyperbola to the seismic event (the manifestation of the time-distance curve). The stacking velocities so computed are empirical quantities and thus do not have a mathematical expression in terms of the interval velocities. In most cases, the stacking velocity is set equal to the rms velocity, which does have a nice mathematical expression. Thus, if we know the rms velocity $v_{\rm rms}$, the interval-velocity can be calculated by the Dix formula:

$$v_k^2 = [v_{\rm rms}^2(k)t_k(0) - v_{\rm rms}^2(k-1)t_{k-1}(0)]/[t_k(0) - t_{k-1}(0)],$$

where $v_{\rm rms}(k)$ and $v_{\rm rms}(k-1)$ indicate rms velocities for interfaces k and $k-1$, respectively, and $t_k(0)$ and $t_{k-1}(0)$ are the corresponding vertical traveltimes. (Note: To repeat, we use the notation $t_k(0)$ for the vertical traveltime because the slant time, using true raypath determined by Snell's law, is denoted by $t_k(x)$ when x is offset. Hence, $t_k(0)$ would be the zero-offset or vertical time.)

Stacking

Deconvolution and common midpoint stacking are the two most important applications of data processing in improving data quality. In this section, we will describe the CMP stacking method. Because of its intimate connection with velocity estimation, we will quickly review the salient features. The variation of normal moveout with velocity and record time has already been discussed. Velocity analysis techniques utilize the variation of normal moveout with record time to find velocity. The major objective of velocity analysis is to find the amount of normal moveout that should be removed to maximize the stacking of events that are considered to be primaries. However, this does not necessarily optimize the primary-to-multiple energy ratio. In the velocity analysis procedure in common use, we assume a stacking velocity at a given stacking time, apply the normal moveouts appropriate for the offsets of the CMP traces as a function of arrival time, and then measure the coherence (degree of match) among all the CMP traces to be stacked. Several measures of coherence may be used, but the most common measure is the semblance. Another stacking velocity is then assumed and the calculations repeated, and so on, until the coherence has been determined as a function of both stacking velocity and stacking time. (Sometimes normal moveout is the variable rather than stacking velocity.) A velocity analysis display, called a velocity spectrum, shows the coherency measure (semblance) plotted against stacking time and stacking velocity. Contour lines can be used to depict semblance, or a three-dimensional (3-D) display can be used (see fig. 2.42). The peak amplitude of semblance versus stacking time is also usually displayed as a separate curve. Peaks on the peak amplitude curve correspond to events. The locations of the peaks on the semblance surface yield the velocities (or normal moveouts) that optimize the stack (hence the name *stacking velocity*). Multiples as well as primaries will give rise to peaks, and the results must be interpreted to determine the best values with which to stack the data. In many areas where the velocity increases more-or-less monotonically with depth, the peaks associated with the highest reasonable stacking velocities represent primary reflections, and the peaks associated with lower velocities represent multiples of various sorts. In some areas, however, the relationships

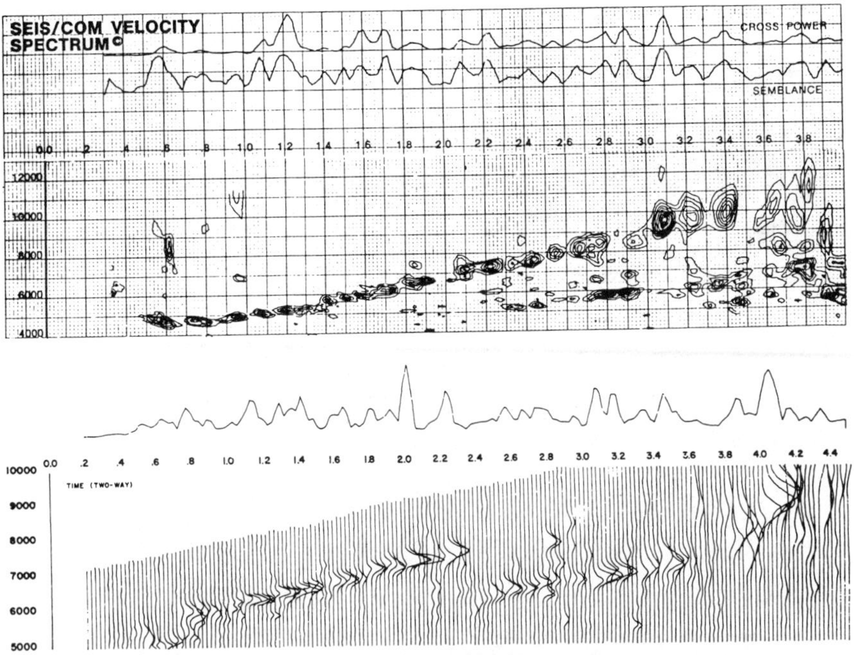

*Figure 2.42. Contoured velocity spectrum (*top*) and 3-D velocity spectrum (*bottom*). The peak amplitude curve is shown above each velocity spectrum. Reprinted by permission of Geophysics from Taner and Koehler 1969.*

are not so obvious; even where the velocity relationships are generally regular, difficulties are encountered (see fig. 2.43).

We will now discuss the important technique in the sequence of data processing operations known as *stacking.* The CMP method of stacking was originated by Mayne (1962). Because of the redundancy inherent in the multiple coverage method of seismic prospecting, unwanted interference and noise can be attenuated by stacking. As we have seen, the dynamic corrections put all primary reflections in phase on the traces in each CMP gather. Because the raypaths of multiple reflections are different, the dynamic corrections do not put the multiples in phase (see figs. 2.44, 2.45, and 2.46). Hence, if we add together (i.e., stack) all the corrected traces in a CMP gather, we severely attenuate the multiples as well as other incoherent noise (see fig. 2.47). Thus, we obtain one output trace for each midpoint; this composite trace is called the stacked trace for that midpoint. This type of stack is usually called the *common depth point (CDP) stack,* although the more accurate term *common midpoint (CMP) stack* is gaining acceptance (see fig. 2.48). The reason for using the term *midpoint* instead of *depth point* is that for slanting subsurface horizons, the depth points (i.e., points of reflection) do not lie directly under the source-receiver midpoints. The CMP stack is based on the fact that the reflection times of multiple reflections generally increase faster with increasing shot-receiver distances than do those of primary reflections with the same vertical traveltime. The static corrections necessary before stacking are time-independent time shifts. They are supposed to compensate irregularities in the elevation of the earth's surface and in the character of

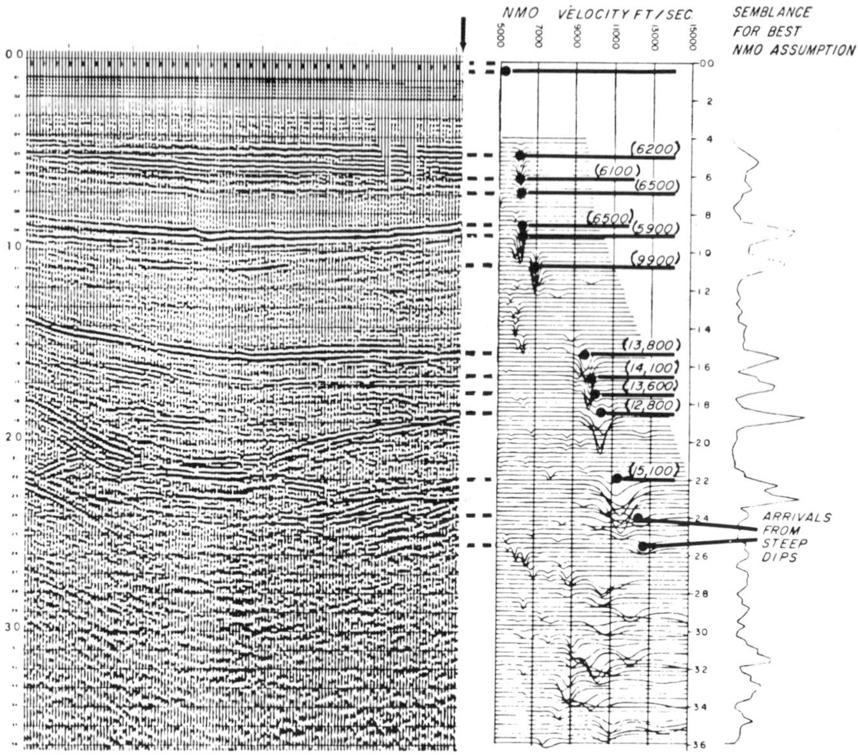

(a) (b)

Figure 2.43. Velocity analysis: (a) seismic record section; (b) velocity analysis of data at the right side of the seismic section. The graph shows semblance as a function of the velocity that normal moveout implies. Interval velocities calculated from Dix's formula are shown in parentheses. Velocity picks have been related to specific seismic events. Reprinted by permission of Seiscom Delta Report and the Society of Exploration Geophysicists from Sheriff 1973.

the so-called weathering layer, at least down to the depth of the groundwater.

The velocity function used in making the NMO corrections (i.e., the dynamic corrections) required before the CMP traces are stacked is the empirical stacking velocity function obtained from the velocity spectrum. Optimum velocity estimation is therefore essential for good signal alignment and for subsequent stack efficiency. A sketch of the stack response as a function of stack velocity is presented in figure 2.49.

We will now describe the process of stacking in more detail. The traces with a common midpoint make up the CMP gather. In the case of horizontal interfaces between the subsurface layers, each such trace would receive reflections from depth points directly under the common midpoint. Thus, for flat layers, CMP traces also have common points on reflecting interfaces called common reflection points or common depth points (CDP). For dipping interfaces, however, no such common depth points exist, and the spread between the depth points for the various traces in the CMP gather becomes greater as the structures become steeper. Despite this, complex geometries in the subsurface frequently

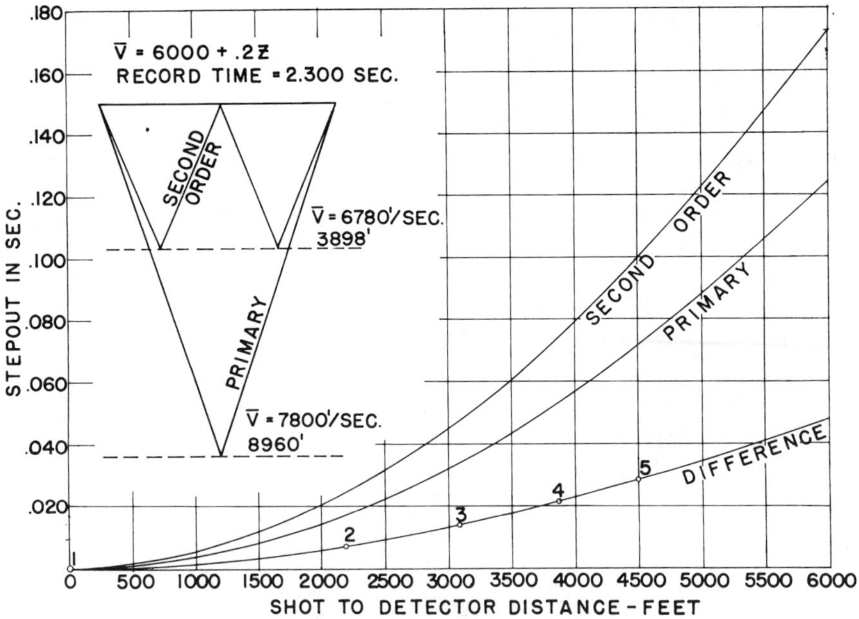

Figure 2.44. Normal moveout (labeled stepout in the figure) as a function of offset (labeled shot to detector distance) for a primary reflection (labeled primary) and a multiple reflection (labeled second order), each with the same vertical traveltime. The indicated velocity function is typical of the Gulf Coast of Texas. The lower curve (labeled difference) gives the difference in normal moveout between the multiple and primary for each offset. Reprinted by permission of Papers for the Field Geophysicist from Mayne 1962.

cause small dispersion of the depth points (reflecting points) because of statistical cancellation due to the many interfaces of different dips through which the waves travel on their round trip through the earth. Hence, the consideration of the CMP traces as recording reflections from common depth points, as required in stacking, is still successful. The process of stacking is conceptually simple. Stacking consists of simply summing the traces contained in a time-corrected CMP gather to produce a single composite trace called the stacked trace (see figs. 2.50 and 2.51). The geographic location of the stacked trace is equated with the position on the earth's surface given by the common midpoint (CMP). The time corrections provide that the summation is carried out with respect to the arrival times of the primary reflections. Stacking is useful because the primary reflections are added in phase and thus are constructively reinforced on the stacked trace, whereas ambient noise and other seismic signals that are not in phase cancel. Such compositing increases reflection signal-to-noise ratios by factors approaching $\sqrt{N}$, where N is the number (called the *fold*) of CMP traces summed. Stacks of 12-, 24-, or 48-fold are routinely produced.

Stacking can be described in various ways. The conventional way is in terms of normal moveout (NMO). Each trace on a CMP gather is corrected for normal moveout; that is, an event on a trace that appears at time t is time-corrected by subtracting the NMO from t so that the corrected event will now appear at vertical time t_0. As we have seen, the NMO for a given trace decreases as t_0

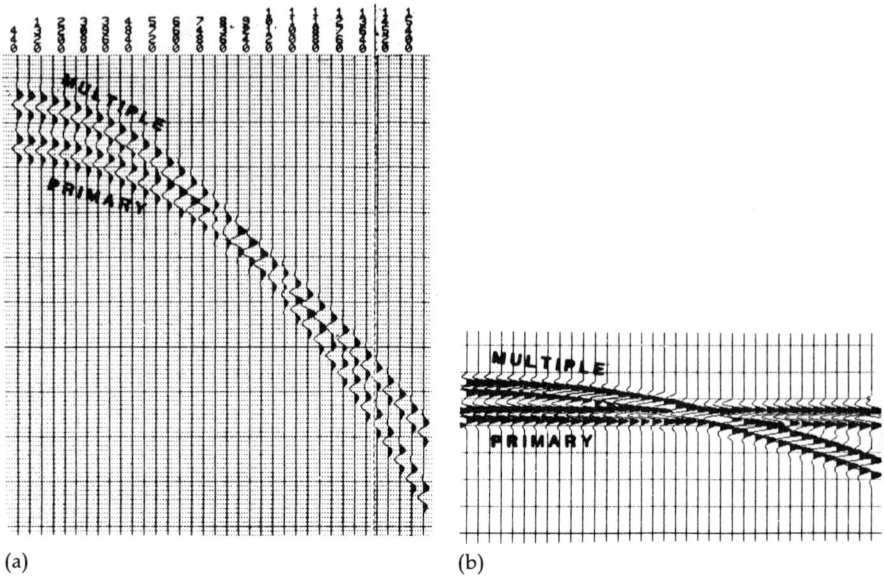

*Figure 2.45. CMP gather: (*left*) before NMO correction; (*right*) after NMO correction. When these traces are stacked, the primary will be reinforced and the multiple will be attenuated because of canceling. Reprinted by permission of IHRDC from Sengbush 1983.*

increases. Thus, the NMO correction changes with record time, so that the correction is said to be dynamic (as opposed to static). After correcting the set of traces for normal moveout, the event due to a horizontal reflector will line up on all the traces of the CMP gather at the same time t_0. Thus, the traces can be added together, and such reflections are reinforced. The process is called stacking, and the resulting output trace is called the stacked trace. One stacked trace results from each CMP gather; the collection of all the stacked traces makes up the stacked section.

Stacking can also be described in engineering terms. Suppose that the velocity were constant for the entire geologic region, so that the same hyperbola applies over the entire CMP gather. Let a unit spike be placed at every point on this hyperbola, and let us perform the two-dimensional correlation of this spike hyperbola with the CMP gather. We place the hyperbola on a transparency and slide the transparency to a position on the CMP gather. We then add all the values on the CMP gather that fall on the hyperbola, and the resulting sum is the required correlation value. In stacking, we want only a single output trace for each CMP gather, so we want only the correlation values that correspond to hyperbolas with their axes along the vertical line through the common midpoint M. Thus, stacking may be described as the output on the midpoint vertical axis of the correlation of the spike hyperbola with the CMP gather. If velocity v varies as a function of depth, the correlations process would be a space-varying one. In seismic prospecting, of course, velocity does vary as a function of depth, so stacking is a space-varying process.

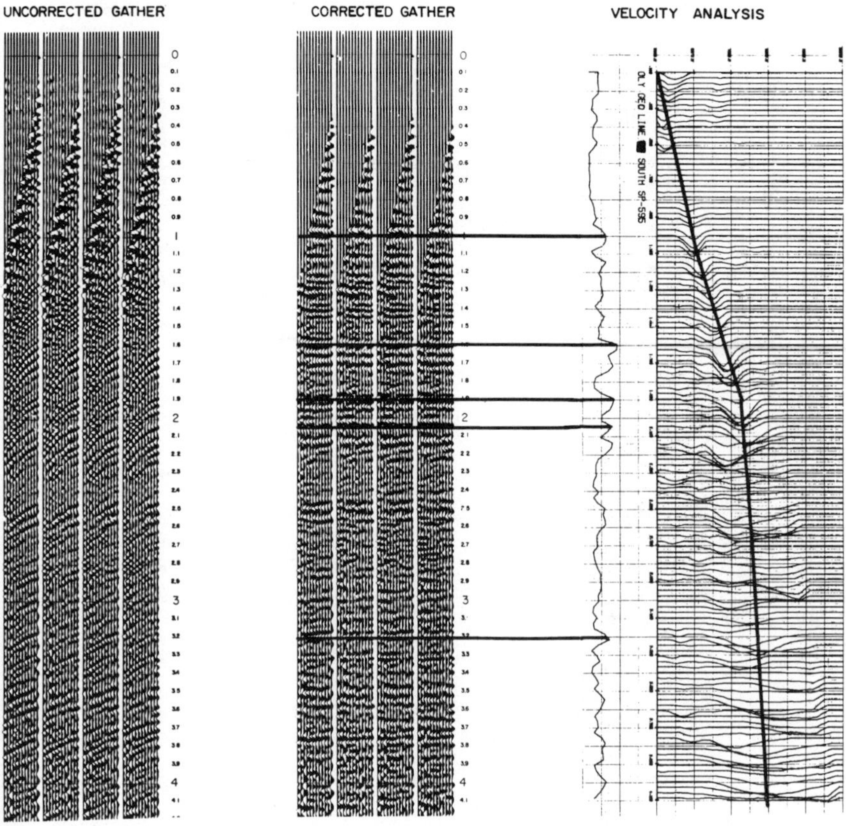

Figure 2.46. Stacking velocity applied to correct the CMP gathers. Reprinted by permission of Grant Geophysical from Cosgrove et al. 1969.

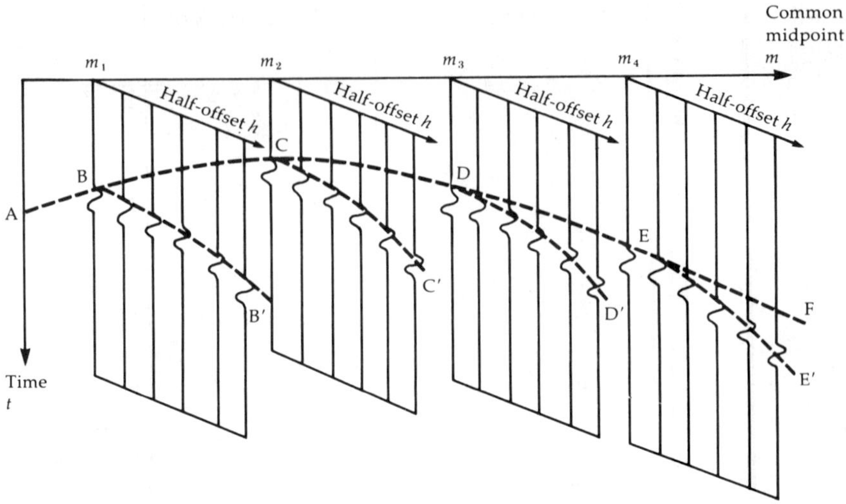

Figure 2.47. A multioffset section. In the diagram, the three axes are orthogonal. This display contains all the traces along a particular surface traverse line. In the process of stacking, all the traces (after static and dynamic corrections) along each offset axis are added together. The result is one stacked trace for each common midpoint. The plot of the stacked traces versus common midpoint is the stacked seismic section. Reprinted by permission of IHRDC from Robinson 1983.

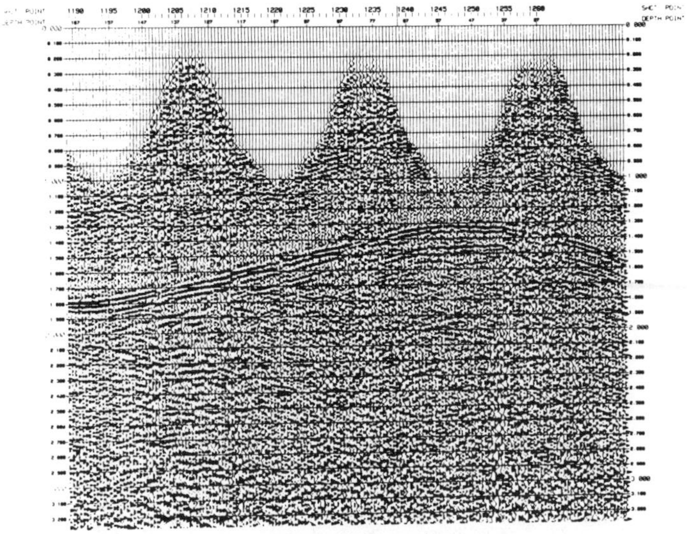

Singlefold section

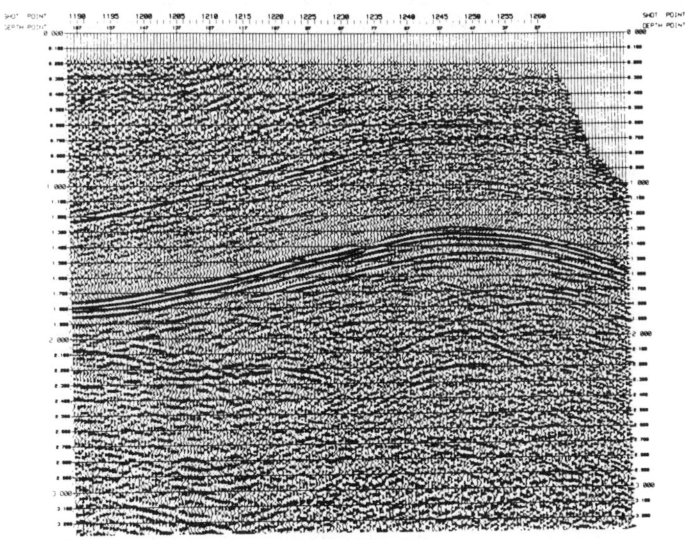

Stack section

*Figure 2.48. Singlefold section after deconvolution (*top*) and stack section (*bottom*). Reprinted by permission of Birkhauser-Verlag from Al-Sadi 1980.*

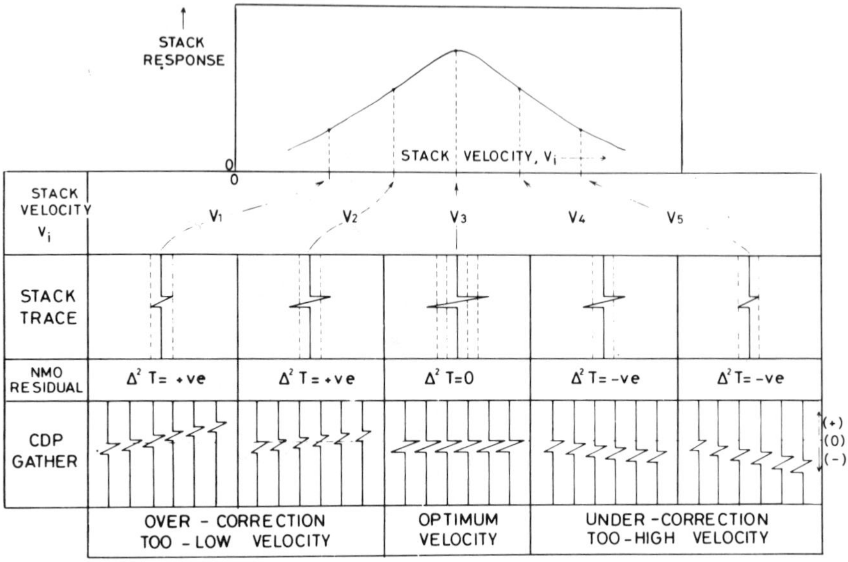

Figure 2.49. Schematic representation of stack response as a function of stack velocity. Reprinted by permission of Birkhauser-Verlag from Al-Sadi 1980.

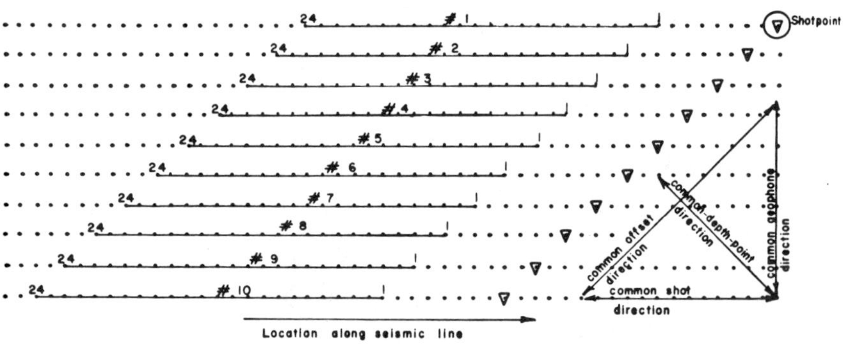

Figure 2.50. Stacking chart, a diagram showing the interrelationships among the traces from CDP shooting. This is used to determine the proper traces for stacking and to determine parameters for shifting traces. Components along various alignments represent common depth point, common shot, common geophone, or common offset. Each horizontal line shows the location of geophone groups and shot point for a single record. In the right corner are shown the directions for finding all traces with various elements in common. Reprinted by permission of the Society of Exploration Geophysicists from Sheriff 1973.

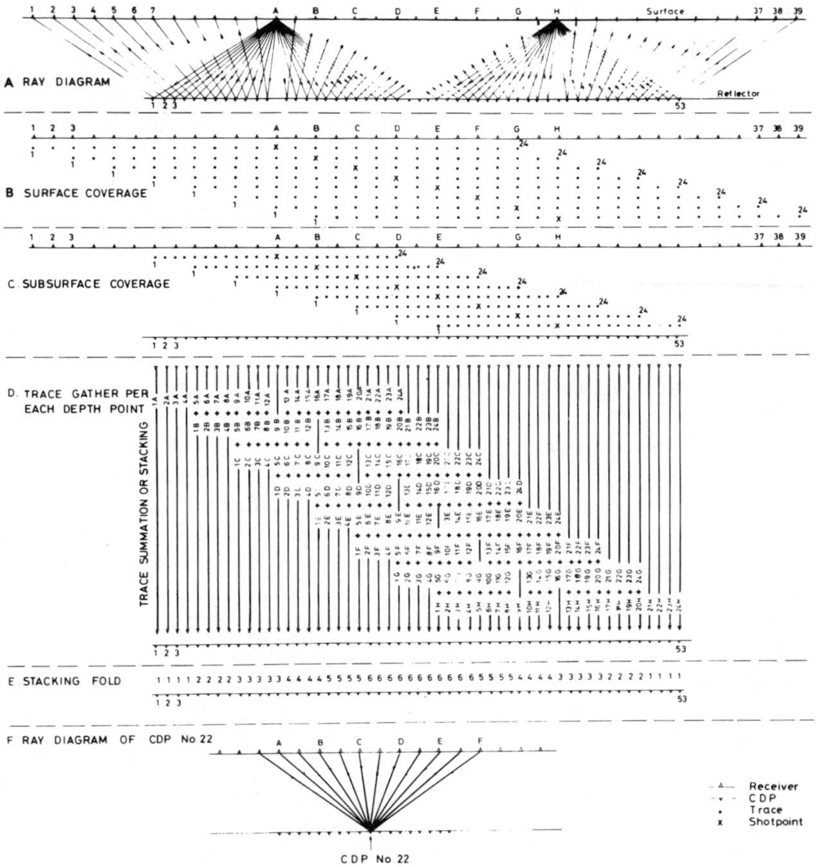

Figure 2.51. Configuration of a 24-trace, center-shooting spread of a constant offset and constant moveup rate. In this example, the shooting system is designed so as to ensure continuous 6-fold coverage. (A) Raypath diagram drawn only for the first and last shotpoints. (B) Actual surface locations of the shot points and their respective traces. (C) Locations of each reflection point that belongs to each shot point. (D) The trace gather of each common depth point; the arrows indicate the traces that are summed up to produce the stacked trace for each CDP. (E) The coverage-fold map. (F) Raypath diagram of reflections belonging to depth point No. 22, which is taken at random. Reprinted by permission of Birkhauser-Verlag from Al-Sadi 1980.

Exercises

1. In the investigation of the earth with geophysical methods, two kinds of information are generally desired. One relates to the structure of subsurface formations, the other to their composition. In seismic exploration for oil and gas, the emphasis until recently was on mapping the geometry of lithologic boundaries to study structural conditions. Recently, however, the seismic reflection method is also being used to identify lithology. What seismically determined parameter is most useful for making such an identification?
2. What are the two basic methods in current use for obtaining interval velocities from reflection records?
3. What are isovelocity surfaces?
4. What is the difference between a long-path multiple reflection and a short-path multiple reflection? Why does a long-path multiple show more normal moveout than a primary reflection with the same arrival time?
5. Velocity estimation becomes slightly more complicated when the reflecting interface in the single-layer model is dipping. Suppose that the reflecting interface is planar and that the seismic traverse is in the dip direction. Consider the seismic traces in a CMP gather, and let x, as usual, represent the offset distance. A common reflection point vertically below the common surface point no longer occurs; rather, the reflection points are displaced in an updip direction, the amount of displacement increasing with distance between the source and seismometer stations. The raypath for $x = 0$ is perpendicular to the reflecting interface, just as for the horizontal reflecting interface; however, the travel distance and traveltime are reduced. For this single-layer model, the travel distance d of the perpendicular raypath is simply $z \cos \phi$, where z is depth to the reflecting interface below the common ground position and ϕ is the dip angle. Let v be the constant velocity. Find an expression for the traveltime $t(x)$.
6. We may estimate the velocity by a best fit of a hyperbola to the reflection time-distance curve in the case of a dipping interface. What differences occur? In the case of a single dipping interface, will the NMO correction align the events?
7. In the foregoing exercise it is assumed that the depth model is two-dimensional; that is, the seismic traverse is at right angles to the strike of the dipping reflecting interface. In the event that the seismic traverse is at an angle θ with the dip direction of the reflecting interface, the apparent velocity again differs from the true velocity. Describe the main features.
8. The methods given in this book are based on the assumption of reflecting surfaces that are essentially horizontal or inclined planes. Some structures, however, are too irregular for the assumption of plane boundaries to be tenable, and a different approach is needed for determining interval velocities. Faults, diapirs, and large-angle convergences across unconformities are typical of geologic situations where the plane-boundary approximation might be expected to break down. What alternative approaches are there for such cases?
9. Coherency functions may or may not be sensitive to amplitude changes between time gates, this sensitivity depending on normalization procedures. Show that the following normalized coherency function employing zero-lag

values of autocorrelation and cross-correlation functions is not sensitive to rms signal amplitude variations between channels:

$$C = \frac{2}{M(M-1)} \sum_{i=1}^{M} \sum_{j>i} \frac{R_{ij}(0)}{\sqrt{R_{ii}(0)R_{jj}(0)}},$$

where M is the number of CMP traces, $R_{ii}(0)$ is the unnormalized zero-lag value of the autocorrelation function of the ith trace, and $R_{ij}(0)$ is the unnormalized zero-lag value of the cross-correlation function between the ith and jth traces. This coherency measure varies between -1 and 1, where 1 corresponds to perfect signal coherence. Another useful quantity for measuring multichannel coherence is semblance, which is defined in the text as the normalized output/input energy ratio. The semblance coefficient S is sensitive to channel amplitude differences and varies between 0 and 1, with 1 denoting identical signals. Show that it can be expressed in terms of zero-lag values of the unnormalized correlation functions

$$S = \frac{\sum_{i=1}^{M} \sum_{j=1}^{M} R_{ij}(0)}{M \sum_{i=1}^{M} R_{ii}(0)}.$$

Construct a simple numerical example to demonstrate that the coherency measures C and S are closely related. Explain why it might be said that the optimum statistic for extracting velocity information from CMP trace gathers may not yet have been found.

10. Explain why a large value of coherence merely means that an event is brought out by an appropriate assumption, not that the event is a primary reflection.
11. Let $X(k, n)$ be a CMP gather for receiver positions $k = -K, -K+1, \ldots, K$ and for time $n = 0, 1, \ldots, N$. The common midpoint is $k = 0$. Let $d/2$ be the spacing between receiver positions, so that the offset for trace k is $x = kd$. Write a FORTRAN program to compute the velocity spectrum $S(j, n)$, where the velocity v is $j\,\Delta v$. Also let the program compute the velocity function and the stacked trace.
12. Show that the normal moveout ΔT is approximately equal to $x^2/2V^2T_0$.
13. For the flat layercake model, find the derivatives $t'(0)$ and $t''(0)$ at $x = 0$ on the time-distance curve. Find the derivatives at $x = 0$ up to the sixth derivative, and hence show that the Taylor series of seven terms (where $h = x/2$ is half-offset and $\tau = t/2$ is one-way time) is

$$\tau = A + (1/B)h^2 - (C/8B^4)h^4 + [(C^2/8B^7) - (D/16B^6)]h^6,$$

where $A = \Sigma v_k^{-1}\,\Delta z_k$, $B = \Sigma v_k\,\Delta z_k$, $C = \Sigma v_k^3\,z_k$, $D = \Sigma v_k^5\,\Delta z_k$.
14. In seismic practice, full offset $x = 2h$ and two-way traveltime $t = 2\tau$ are used. Write the Taylor series in Exercise 13 in terms of these quantities.
15. The average velocity to the nth interface is defined as

$$v_A = \frac{1}{t(0)} \sum_{k=1}^{n} v_k \, \Delta t_k(0).$$

Let $z = \Sigma \Delta z_k$ be the depth to the nth interface. Show that

$$v_{\text{rms}}/v_A = (1 + G)^{1/2},$$

where the heterogenity factor G is defined as

$$G = z^{-2} \sum_{k=1}^{n-1} \Delta z_k \sum_{j=k+1}^{n} [(v_k - v_j)^2/v_k v_j] \Delta z_j.$$

The heterogeneity factor is a nonnegative quantity and is equal to zero only in the case of a homogeneous model (i.e., all layers with the same velocity). The value of G is independent of the order of the layering. The foregoing equation shows that the rms velocity equals the average velocity when the earth is homogeneous and progressively exceeds it as the earth becomes more heterogeneous.

16. Let the model consist of five layers, with thicknesses 400, 1000, 2000, 2500, 3000 ft, respectively, and velocities 5000, 6000, 8000, 6500, 10000 ft/sec, respectively.
    (a) Find the average velocities and rms velocities.
    (b) For the full offsets 0, 1800, 3000, 4200, 6000, compute the two-way traveltimes by the parametric time-distance curve and by the rms hyperbolic approximation.
17. Let the point 0 be the origin $x = 0$, $z = 0$. A point scatterer is located at $x = 0$ and at depth z. Let the horizontal coordinate of the source S be s and the horizontal coordinate of the receiver R be r. Let the medium have constant velocity v. Let t be the total traveltime from the source S to point scatterer P to receiver R.
    (a) Show that

    $$vt = [z^2 + s^2]^{1/2} + [z^2 + r^2]^{1/2},$$

    which represents a generalized time-distance curve. This curve gives the traveltime t as a function of depth z, source coordinate s, and receiver coordinate r (see fig. 2.52).
    (b) Let $s = m - h$ and $r = m + h$, where m is the coordinate of the midpoint M and h is the half-offset. Show that the generalized time-distance curve is also given by

    $$vt = [z^2 + (m - h)^2]^{1/2} + [z^2 + (m + h)^2]^{1/2}.$$

    (c) Define t_0 as the traveltime for $s = r = 0$, so that $t_0 = 2z/v$. Plot contours (of equal t) of the generalized time-distance curve (see fig. 2.53). Show that the point scatterer P lies on an ellipse with foci S and R and center M (see fig. 2.54). Let A and B be the points where this ellipse cuts the horizontal axis. Draw the circle with center M that also cuts the horizontal axis at the same points. Let Q be the point where this circle cuts the

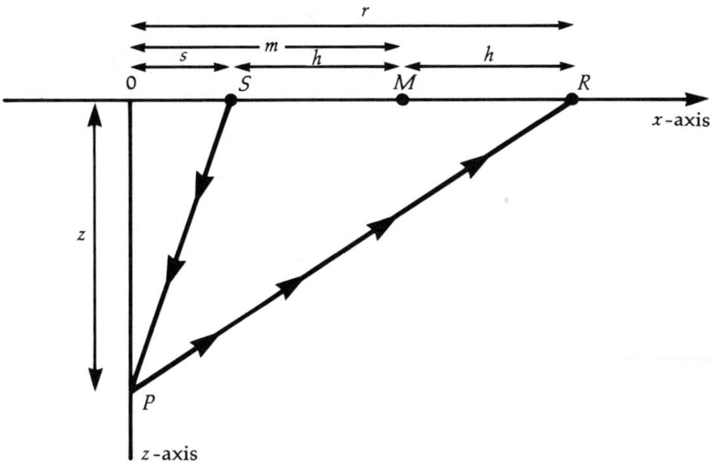

Figure 2.52. Geometry of point scatterer P.

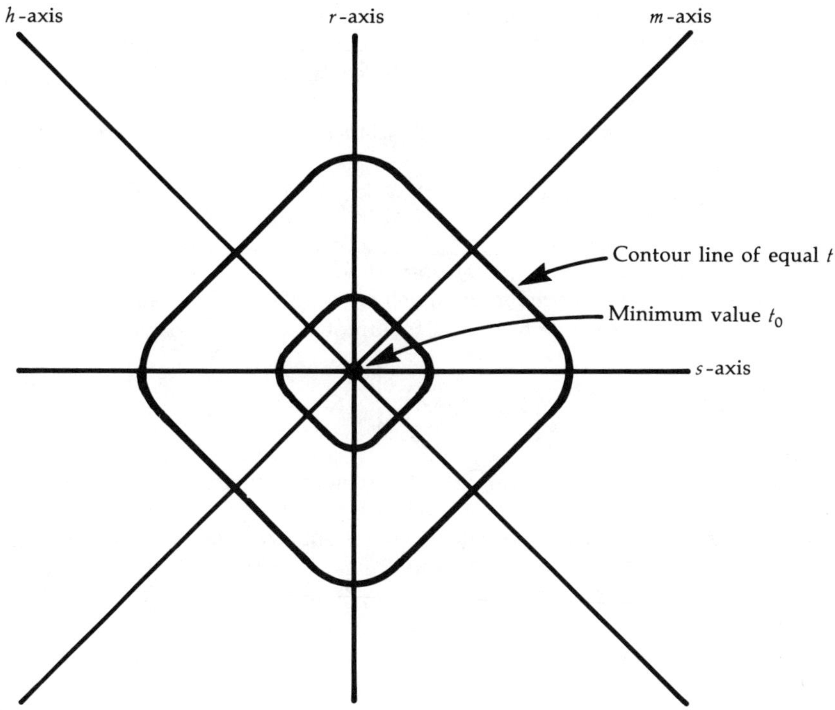

Figure 2.53. Bowl-shaped surface of generalized time-distance curve.

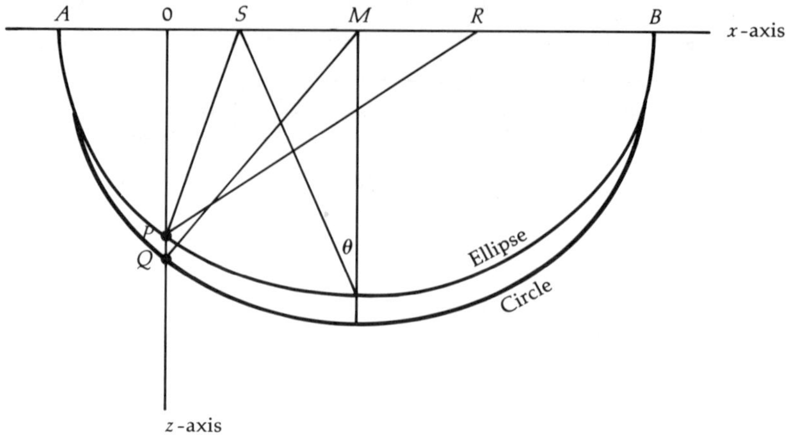

Figure 2.54. Ellipse and circle associated with point scatterer P.

z-axis. Show that the traveltime from S to P to R is the same as the traveltime from M to Q to M.

(d) Given that line OP has length z, show that line OQ has length $z/\cos\theta$, where $\cos\theta$ is defined as the ratio of the minor axis to the major axis of the ellipse.

(e) Show that the generalized time-distance curve can be expressed as

$$vt = 2\left[\frac{z^2}{\cos\theta} + m^2\right]^{1/2}$$

(f) Show that $\sin\theta = h/(vt/2)$, so that $\cos\theta = [1 - h^2/(vt/2)^2]^{1/2}$.

18. If v_m is the greatest of the interval velocities $v_1, v_2, \ldots, v_n$, then show that the time-distance curve has as its asymptote the line whose equation is

$$v_m t - x = 2v_m \sum_{k=1}^{n} \frac{1}{v_k}\left[1 - \left(\frac{v_k}{v_m}\right)^2\right]^{1/2} \Delta z_k.$$

19. Show that the velocity over a depth increment calculated from the top and bottom rms velocities is the rms velocity of the depth increment.
20. Show that the velocity over a depth increment calculated from the top and bottom average velocities is the average velocity of the depth increment.
21. Show that if the stacking velocity is taken as the slant rms velocity, then the stacking velocity is greater than the (conventional) rms velocity.
22. Recall that the time-distance curve is given by the parametric equations

$$x = ptv^2_{\text{srms}},$$

$$t = \Sigma(\cos\theta_k)^{-1}\,\Delta t_k(0),$$

and that the slope of the time-distance curve is $dt/dx = p$. Thus, show that at a given point (x_1, t_1) on the curve, the slant rms velocity is the geometric mean of the slope x_1/t_1 to the curve and the slope $(dx/dt)_1$ at that point.

Also show that the slant rms velocity is the geometric mean of the slope x_1/t_1 and the slope $(dx/dt)_1$ of the approximating hyperbola using the slant rms velocity.

23. Show that Snell's parameter p satisfies

$$0 \leq p \leq \min_{k=1}^{n} v_k^{-1}.$$

24. Given a flat-layered earth model with layer thicknesses Δz_k and layer velocities v_k, let the two-way traveltime to interface n for offset x be

$$t = \sum_{k=1}^{n} \Delta t_k,$$

where Δt_k is the two-way time in layer k. Let the Snell's parameter for this travel path be p. Recall that the slant root-mean-square (srms) velocity for this path is defined as

$$v_{\text{srms}} = \left[\frac{1}{t} \sum_{k=1}^{n} v_k^2 \, \Delta t_k \right]^{1/2}.$$

Show that v_{srms} reduces to v_{rms} for $p = 0$ (vertical travelpath). Also show that

$$x = ptv_{\text{srms}}^2.$$

25. Show that the slope of the time-distance curve $t(x)$ for the case described in Exercise 24 is

$$\frac{dt}{dx} = \frac{x}{tv_{\text{srms}}^2}.$$

26. Plot $t^2(x)$ versus x^2. Sketch this curve. Show that the slope of this curve is $1/v_{\text{srms}}^2$.
27. Show that the usual hyperbolic fit to $t(x)$ corresponds to a straight line in the (x^2, t^2) plane. Show that this straight line has slope v_{rms}^{-2} and is tangent to the curve of Exercise 26 at the point $(0, t^2(0))$. Hence show that the usual hyperbolic fit is a hyperbola that has apex at $[0, t(0)]$ and that lies entirely above curve $t(x)$.
28. Fit a hyperbola to the time-distance curve $t(x)$ with the following specifications: it passes through the point (x_1, t_1), and it has the same slope as $t(x)$ at the point (x_1, t_1). What is the equation for this hyperbola? Show that it lies entirely above the curve $t(x)$. In particular, show that it intercepts the t-axis at a point above the vertical two-way time $t(0)$.
29. The earth model consists of two flat layers such that $\Delta z_1 = 1$ km, $v_1 = 2$ km/sec, $\Delta z_2 = 1.5$ km, $v_2 = 3$ km/sec. Consider three rays with $p = 0$, $p = 1/12$, $p = 1/6$, respectively. For each raypath, find x, t, and v_{srms} and plot $t(x)$. Plot the approximating hyperbola for the vertical path (Exercise

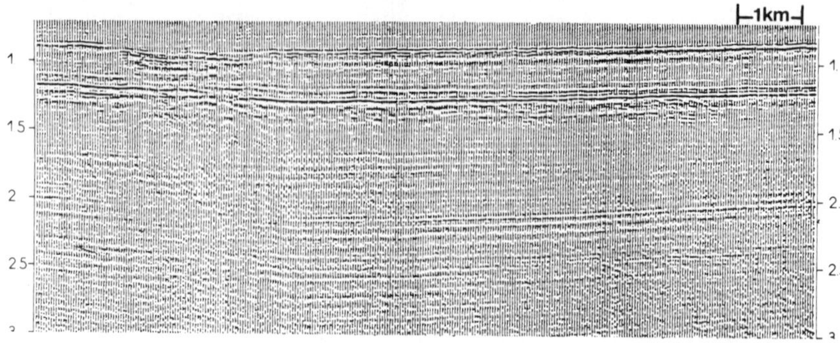

(a)

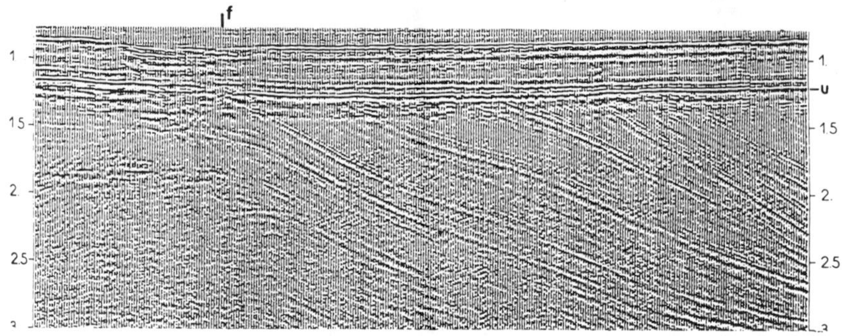

(b)

Figure 2.55. Two versions of the same seismic section, differing only in choice of stacking velocity. Reprinted by permission of IHRDC from Sheriff 1980.

27) and the approximating hyperbolas for the raypaths $p = 1/12$ and $p = 1/6$ (Exercise 28). Also plot the corresponding curves in the (x^2, t^2) domain.

30. Figure 2.55 shows two versions of the same seismic section, differing only in the choice of stacking velocity.
    (a) What is a possible explanation of the difference between the sections?
    (b) Which section was processed with the higher stacking velocity at depth? Why?

3. Seismic wavelets and reflectivity

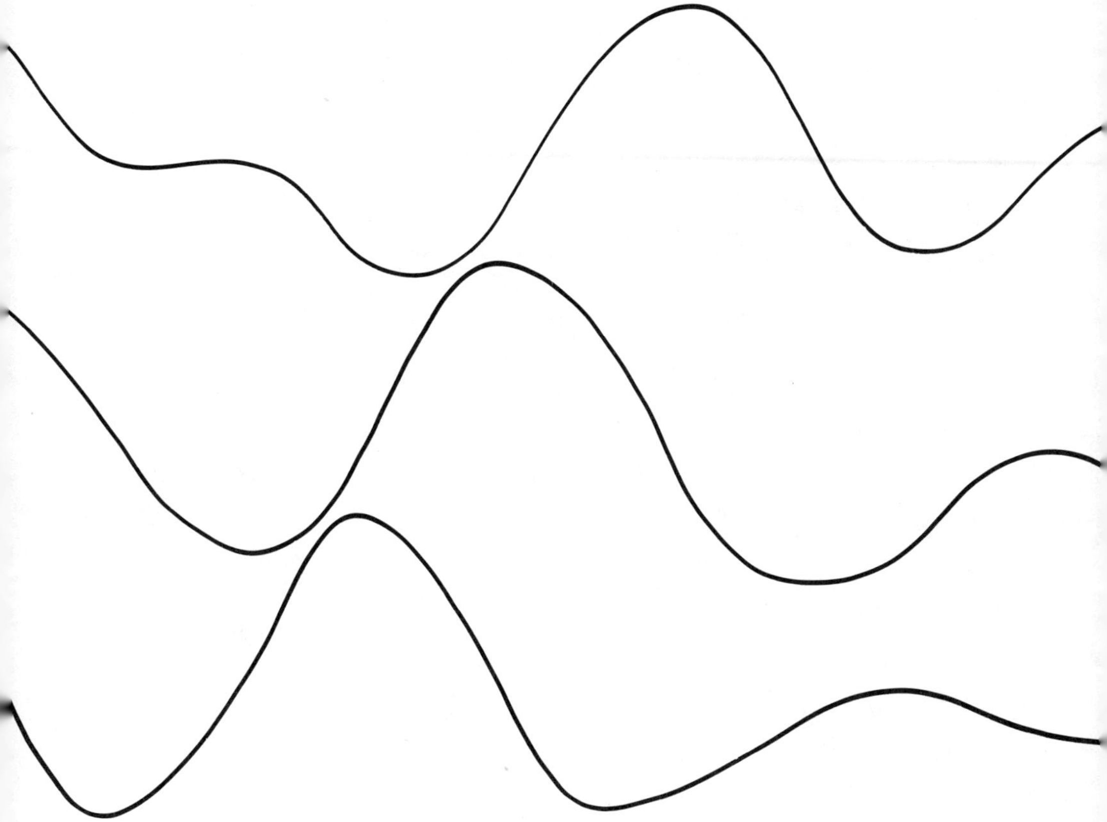

Time series and wavelets

Seismic records are in the form of traces, one trace for each recording channel (see fig. 3.1). Such traces are functions of continuous time. As a rule, continuous-time functions such as these appear in engineering and the physical sciences. Continuous-time functions may be recorded (or measured, or observed, or sampled) at equal intervals of time, thereby generating *time series.* The time series provides only a limited description of the function; that is, the time series represents sampled data of the continuous-time function, the sampling instants being discrete, equally spaced time points. By taking the sampling instants sufficiently close together, the amount of information that is lost by replacing a well-behaved continuous-time function by a time series can be made small. The time spacing must be chosen relative to the frequency characteristics of the continuous-time function being sampled. A time spacing that is too gross would mean substantial information lost in the sampling process. At the other extreme, a time spacing that is too fine would mean substantial redundancy in information produced by the sampling process. Thus, in determining the time interval, we must always balance redundant information versus lost information, with consideration given to their relative costs.

We shall let k designate the index of discrete time. The time variable t is

$$t = k\,\Delta t,$$

where Δt is the sampling increment.

A time series, therefore, is simply a time sequence of discrete data values, each data value $x(k)$ being identified by a fixed time index k. According to our convention, these indices are equally spaced, and each one is labeled by a whole number k. (Note that we now use the symbol x for the amplitude of a time series [i.e., seismic trace], whereas in previous chapters we used x for offset distance.)

Digitizing, or analog-to-digital conversion, is the transformation of analog data to digital form. Often, this step occurs at the point where the data are being recorded. At other times, the data are recorded in analog form and then converted later to digital form at a processing center.

Unless otherwise stated, we shall always assume that continuous-time functions are reduced to the form of *time series.* Such time series may appear to our senses in any physical form; for example, a time series may appear as a train of pulses on magnetic tape, as a sequence of numbers written on paper, or as a sequence of numbers stored in the memory of a computer. Thus, when we

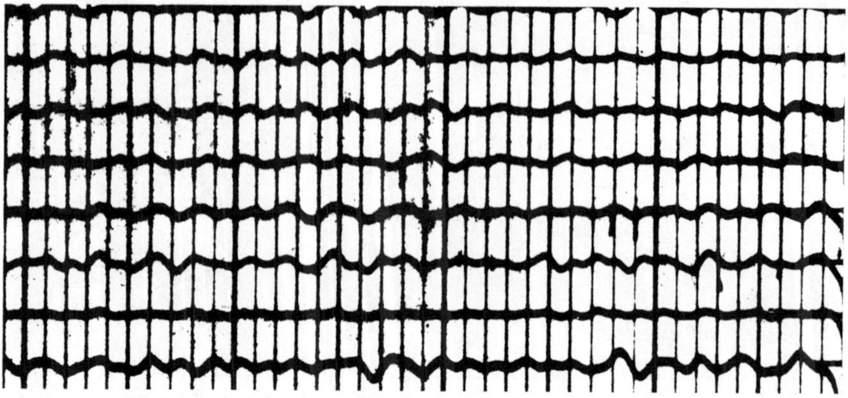

Figure 3.1. Seismic traces.

speak of a seismic trace, we shall mean the time series representing the continuous-time trace.

Time series is a generic term. Unless otherwise specified, however, the term *time series* is usually used to represent a time phenomenon continuing over all time, from the remote past to the distant future. Thus, we may say that a time series has infinite time duration, or infinite length.

Good examples of such continuing time series are provided by natural geophysical phenomena, such as microseismic noise that has been under way from the remote past history of the earth and will continue to the distant future. An exploration seismic record, however, is properly a transient phenomenon, although, as we shall see, for some purposes it is expedient to think of such a record as a portion of a continuing series.

In any actual situation, we can only obtain the values of a time series over a finite interval of time. Because a finite number of data values represents a sample of an infinite-length time series, we shall call a finite portion of a time series a *sample time series.* Here we must distinguish between (1) the sampling process that extracts a time series (over all discrete time) from a continuous-time function (over all continuous time) and (2) the sampling process that extracts a sample time series (over a finite interval of discrete time) from a time series (over all discrete time). A sample time series, then, is a finite portion of a time series—that is, the portion from some fixed time to a later time. For a given time series, for example, we may have available only the portion from time $k = 1$ to time $k = 25$, which consists of the 25 values

$$x(1), x(2), \ldots, x(25).$$

We now introduce the concept of a wavelet. The physical phenomenon we have in mind is a transient signal with a definite time origin. A wavelet is characterized by two properties:

(1) The *one-sided property:* A wavelet has a definite origin (or arrival) time; that is, all values of the wavelet before its origin time are zero.

(2) The *stability property:* A wavelet has finite energy; that is, it is a transient or dying-out phenomenon as time progresses.

We must always be careful to distinguish between the concepts of *wavelet* on the one hand and *sample time series* on the other. The point of difference is that a wavelet is a self-contained or entire entity, and this entity is a transient phenomenon with a definite arrival time. In contradistinction, a sample time series is only a portion of an entire entity—namely, the time series—and this entity is a continuing phenomenon (see figs. 3.2 and 3.3).

We can plot a wavelet as its values from the time it arrives until the time it damps out. Its origin time, or arrival time, is called time index 0, which serves as the reference date for the wavelet. At time index 0, we plot the amplitude of the wavelet at its origin. At time index 1, we plot the amplitude of the wavelet one time unit from its origin. At time index 2, we plot the amplitude of the wavelet two time units from its origin, and so on (see fig. 3.4).

Suppose, for example, that the wavelet has amplitude 5 at time index 0, amplitude -3 at time index 1, amplitude 1 at time index 2, and zero amplitude for all succeeding times. We may summarize this wavelet by the row vector $(5, -3, 1)$, where it is understood that 5 is the initial amplitude (i.e., at time index 0), so that all preceding amplitudes would be zero; -3 is the amplitude at time index 1; and 1 is the amplitude at time index 2, or the final amplitude, so that all following amplitudes are zero (see fig. 3.5).

Wavelets may be represented algebraically. We may write the wavelet a, for example, as

$$a = (a_0, a_1, a_2).$$

This notation means that wavelet a has amplitude a_0 at time index 0, amplitude a_1 at time index 1, and amplitude a_2 at time index 2. The equation

$$(a_0, a_1, a_2) = (5, -3, 1)$$

means that $a_0 = 5$, $a_1 = -3$, $a_2 = 1$.

It is possible to consider wavelets with complex amplitudes—that is, amplitudes of the form $u + iv$, where u and v are real numbers and $i = \sqrt{-1}$. An example is the wavelet $(i, 0.5)$, where i is the amplitude at time index 0 and 0.5 is the amplitude at time index 1. Another example is the wavelet $(2 + i, 1 - i, i)$, where $2 + i$ is the amplitude at time index 0, $1 - i$ is the amplitude at time index 1, and i is the amplitude at time index 2. Geophysicists consider complex wavelets when they study the Hilbert transform and signal envelopes.

The amplitudes of a wavelet are also called the coefficients of the wavelet. For the wavelet (a_0, a_1, a_2), for example, we may call a_0, a_1, and a_2 either the amplitudes or the coefficients of the wavelet. Because the wavelet (a_0, a_1, a_2) has three coefficients, we say that it has *length* (or *time duration*) equal to 3 or that it is a 3-length wavelet.

Thus far we have considered only finite wavelets—that is, wavelets with a finite number of coefficients, or wavelets with finite length. These wavelets

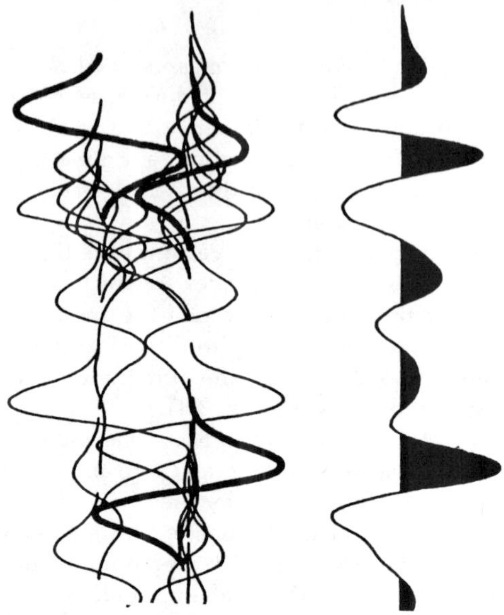

*Figure 3.2. Wavelets (*left*), which when superimposed produce time series (*right*). The time axis is vertically downward. Reprinted by permission of IHRDC from Sheriff 1980.*

become zero after a certain time index. Wavelet $(5, -3, 1)$, for example, becomes zero at time index 3.

It is possible to have infinite-length wavelets, but for stability we must require that they have *finite energy.* Thus, an infinite-length wavelet may be represented by

$$(a_0, a_1, a_2, \ldots),$$

where a_0 is the coefficient for time index 0, a_1 is the coefficient for time index 1, a_2 is the coefficient for time index 2, and the three dots indicate that the coefficients continue on for all positive times to form an infinite sequence. For stability, it is required that the wavelet's energy, given by

$$a_0^2 + a_1^2 + a_2^2 + \cdots,$$

be finite. If the coefficients of the wavelet are complex, then the energy is given by

$$|a_0|^2 + |a_1|^2 + |a_2|^2 + \cdots.$$

An example of an infinite-length wavelet is

$$a = \left(1, \frac{1}{2}, \frac{1}{4}, \frac{1}{8}, \frac{1}{16}, \frac{1}{32}, \frac{1}{64}, \ldots\right),$$

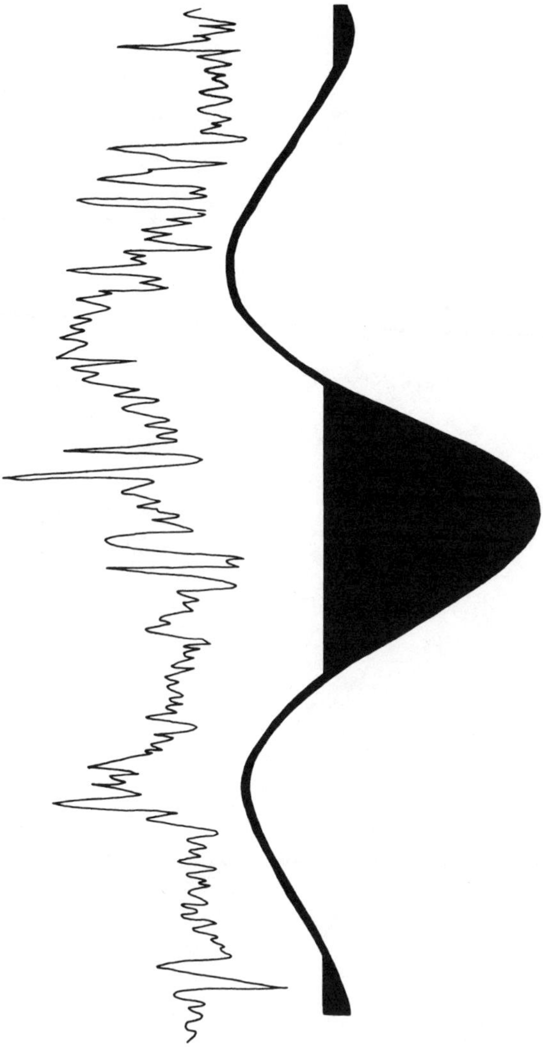

*Figure 3.3. The relative dimensions of the earth layering, as shown by the well log time series (*top*) and the effective interpreter's wavelet (*bottom*). Reprinted by permission of IHRDC from Anstey 1982.*

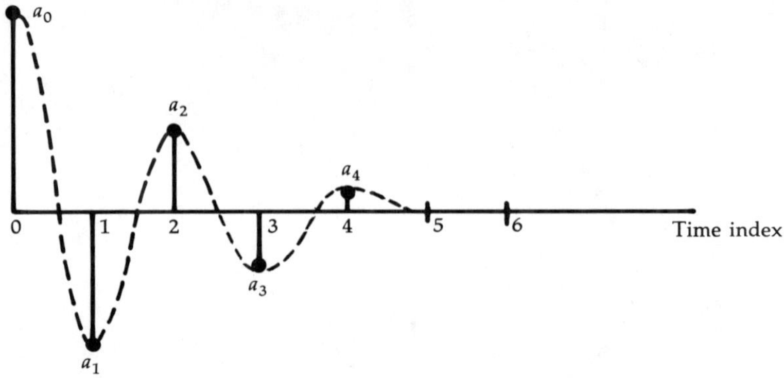

Figure 3.4. Plot of the wavelet $(a_0, a_1, a_2, a_3, a_4)$. *The dashed line connecting the discrete points is drawn to improve clarity.*

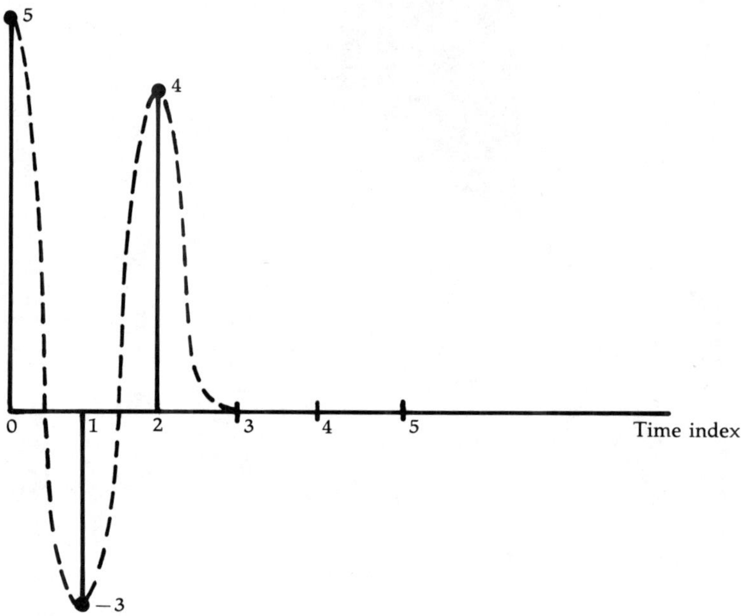

Figure 3.5. Plot of the wavelet (5, −3, 4).

where the first coefficient, 1, is the coefficient for time index 0, the next coefficient, ½, is the coefficient for time index 1, and so on. An alternative way of representing this wavelet would be

$$a = (a_0, a_1, a_2, \ldots), \qquad \text{where } a_n = \left(\frac{1}{2}\right)^n \quad \text{for} \quad n = 0, 1, 2, \ldots .$$

Because of the stability property, we see the magnitude of the coefficients asymptotically approach zero as time increases (see fig. 3.6).

We have defined a wavelet as a one-sided, stable time function; that is, a wavelet is a transient waveform with a definite arrival time. This definition is in line with our physical understanding of what a seismic wavelet is. Because seismic energy is transported at a certain finite velocity, the energy at some distant point cannot arrive infinitely fast, and so there must be a definite arrival time. As a result, a physical wavelet cannot have a precursor that occurs before its arrival time. We can sum up this property by saying that a physical wavelet obeys the causal laws of physics. For this reason, the one-sided property of a wavelet is often described as the causal property. Now we will discuss the stability property. In our seismic experiment, we consider the dynamite explosion, or other man-made source of energy, as the only source of energy, except for such things as wind noise. We do not consider any natural sources, such as earthquakes, nor do we consider ocean wave activity or other sources of microseisms. As a result, our model considers the earth as a completely passive medium with no internal energy sources. We insert a transient burst of energy (the dynamite explosion) into this medium, the medium responds, and the energy decays and dies out over time. For this reason, the physical seismic wavelet is a transient or dying-out phenomenon.

The term *wavelet* as used in this sense (i.e., a causal, transient waveform) has its origin and usage in exploration seismology. In the English language, the word *wavelet* means "little wave." An example of this usage would be a description of wavelets (i.e., little waves) on a pond. It is only in this sense that the term *wavelet* is used in physics, for the sole place it appears in physics texts is in regard to Huygens' principle. Huygens' principle states that, as a wave propagates, each point in the medium may be considered as the origin of secondary wavelets (i.e., little waves), which when superimposed give the traveling wave motion. The physics term that most closely corresponds to the geophysics term *wavelet* is *wave packet.* Electrical engineers use the term *causal transient* or simply *transient waveform,* since the causality property is usually understood when working with physical systems. Geophysics literature, however, is deeply imbued with the word *wavelet,* and we see no reason not to continue this tradition.

Having said this, we must now make an exception and say that geophysicists also use the term *wavelet* in another sense, that of the so-called *interpreter wavelet.* An interpreter wavelet is not a physical entity but merely an artifact introduced in computer processing. An interpreter wavelet is defined as a transient waveform that can be two-sided. Because such waveforms are not causal, they cannot be transmitted by the earth. The reason they are introduced in the data-processing stage is that they make the final seismic display more intelligible to the seismic interpreter. An interpreter wavelet usually is a zero-phase waveform—that is, a symmetrical waveform that has its maximum at its arrival

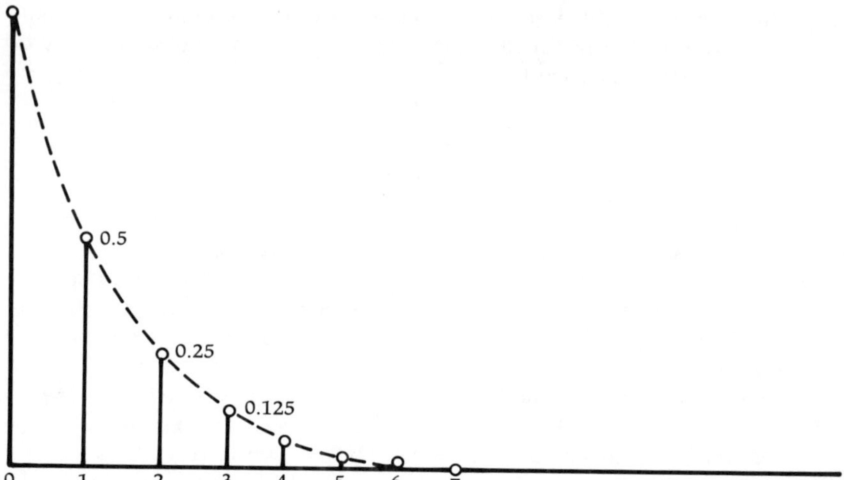

Figure 3.6. Plot of the geometrically damped wavelet (1, 0.5, 0.25, 0.125, . . .). This wavelet is infinitely long, but soon its values are so close to zero that they cannot be distinguished from zero. This wavelet is the discrete counterpart of the exponentially damped continuous wavelet.

time and that damps quite rapidly in both the positive and the negative time direction. Thus, half of the interpreter wavelet represents a precursor that arrives before the arrival time of the wavelet. The data-processing scheme that converts physical seismic wavelets on the seismic trace to interpreter wavelets is called *wavelet processing.* Such a conversion cannot be carried out in real time, of course, and is done in the data-processing operations after the seismic experiment has been completed.

The z*-transform*

The z-transform provides a method for calculating the effect of various operations on digital data. If the sample values of a wavelet at successive (integer-valued) times are

$$b_0, b_1, b_2, \ldots, b_n,$$

then the z-transform of the wavelet is the polynomial

$$B(z) = b_0 + b_1 z + b_2 z^2 + \cdots + b_n z^n.$$

The z-transform of the time series $x(k)$ is

$$X(k) = \sum_k x(k) z^k,$$

where the index k runs over all values of the time series.

Consider the digital unit-impulse function (i.e., Kronecker delta function) δ_k defined as

$$\ldots, 0, 0, \underset{\uparrow}{1}, 0, 0, 0, \ldots,$$

where the vertical arrow indicates the time origin $k = 0$. Then its z-transform is

$$\sum_{k=-\infty}^{\infty} \delta_k z^k = 1.$$

There is a clear analogy between the Kronecker δ_k and the Dirac $\delta(t)$ as they are utilized in the theory of digital and analog systems, respectively. Suppose that the impulse is delayed by, for example, three sampling increments (fig. 3.7):

$$\ldots, 0, 0, \underset{\uparrow}{0}, 0, 0, 1, 0, 0, \ldots .$$

Then the z-transform is

$$\sum_{k=-\infty}^{\infty} \delta_{k-3} z^k = z^3.$$

The unit-step function

$$\ldots, 0, 0, \underset{\uparrow}{1}, 1, 1, \ldots$$

has z-transform

$$\sum_{k=0}^{\infty} z^k = \frac{1}{1-z} \quad \text{for } |z| < 1.$$

The same sequence, advanced by two time units, is (fig. 3.8)

$$\ldots, 0, 0, 1, 1, \underset{\uparrow}{1}, 1, \ldots,$$

which has z-transform

$$z^{-2} + z^{-1} + 1 + z + z^2 + z^3 + \cdots = \frac{z^{-2}}{1-z} \quad \text{for } |z| < 1.$$

We see that the delay of a signal by n time units corresponds to multiplying its z-transform by z^n. Also, the advance of a signal by n time units corresponds to multiplying its z-transform by z^{-n}. Thus, the operation represented by z may be regarded as a unit-delay operator. The operation z^{-1} is the unit-advance operator.

Physical description of seismic waves

A solid may be considered to be made up of many small particles. Elastic waves transfer energy from one point to another by the movements of these small

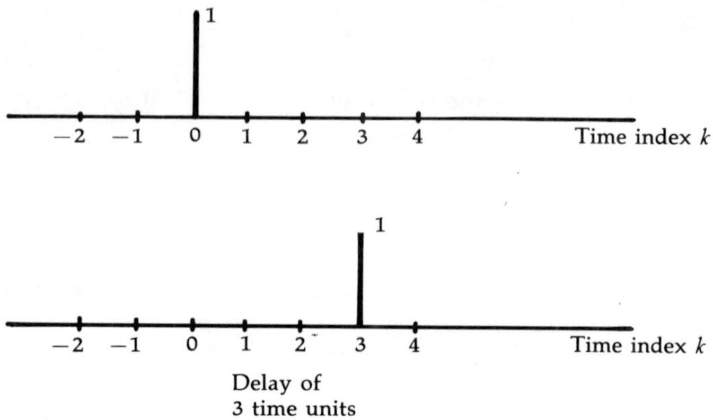

Figure 3.7. (top) *The unit spike* δ_k. (bottom) *The delayed unit spike* δ_{k-3}.

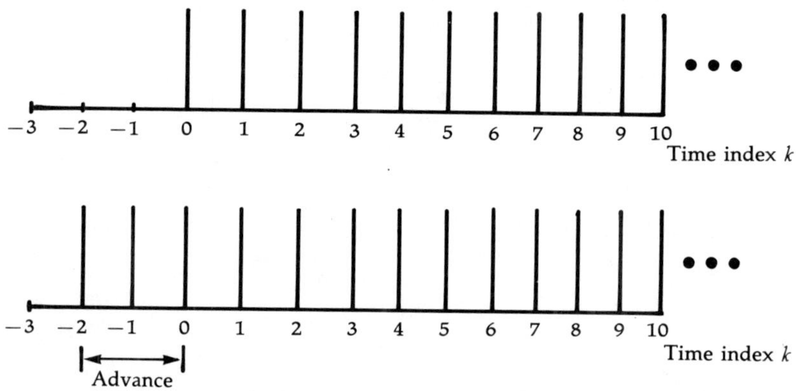

Figure 3.8. (top) *The unit-step function.* (bottom) *The unit-step function advanced by two units. The unit amplitude continues indefinitely to the right for the unit-step function.*

particles. The particles themselves are not transported, of course, but only vibrate with small displacements about their respective equilibrium positions. Their movements can propagate energy in the form of a wave, however, and the transfer of energy by an elastic wave in a solid such as rock or steel involves millions of particles. The dimensions of the elastic wave, such as its amplitude and wavelength, are large relative to the dimensions of the vibrations of the particles. Elastic waves can also propagate in a fluid, in which case they are called acoustic waves. Our speech is transported by acoustic waves in the air; in this case, the particles are air molecules. The elastic waves that propagate in the solid earth, as well as the acoustic waves that propagate in the seas and other bodies of water, are known as *seismic waves.*

A wave is nature's way of transporting energy from one place to another. Electromagnetic waves, which include the visible light as well as the infrared heat waves, transfer energy from the sun to the earth and thus make life possible. Electromagnetic waves can propagate in a vacuum, and so do not involve the motion of small particles. Their theory must be explained by quantum mechanics, whereas the theory of seismic waves can be fully handled by the classical mechanics of material particles.

Most seismic interpretation is done in terms of models involving P-wave propagation through a sequence of rock formations. It is usual to make the assumption that each formation is homogeneous and isotropic. A seismic wave, as well as other types of waves, represents an interchange between kinetic and potential energy. The kinetic energy comes from the physical motion of the particles, and the potential energy comes from their relative positions with respect to the interparticle elastic (or restoring) forces. The rest position of each particle can be taken as the origin of the coordinate system describing that particle. The particle vibrates around this rest position. At any instant, we can in principle measure the particle's velocity and also, if we wish, the particle's acceleration. We can also measure the degree of compression of the particles in the form of a pressure or stress. In the case of a homogeneous, isotropic solid, physical theory tells us that we need only two quantities for a complete specification of the wave motion resulting from the particle motion. It is convenient to take for these two quantities the *particle velocity* and the *pressure.*

The subsurface geologic structure of the earth can be labeled by a system of coordinates with an origin, perhaps, at a point on the surface of the earth. The wave motion is then described in terms of this coordinate system. In two dimensions, we let x denote the horizontal distance and z denote the depth into the earth. At each point (x, z), a wave is specified by its particle velocity and pressure. These quantities vary with time, and in so doing give rise to the propagating wave motion. One way to look at the wave motion is to take a slice through the earth and take a "snapshot" at one instant of time (see fig. 3.9). We see waveforms, and the spatial distances between peaks are indicative of the wavelengths λ. Another way to look at waves is to stay at one point within the earth and record the wave motion over time (see fig. 3.10). Again we see waveforms, and the time distances between peaks are indicative of the periods T. The reciprocal of the wavelength is called the wavenumber κ; the reciprocal of the period is called the frequency f. Usually we work with sines and cosines, so it is convenient to use instead the radian wavenumber and radian frequency, which are, respectively,

$$k = 2\pi\kappa = \frac{2\pi}{\lambda}, \qquad \omega = 2\pi f = \frac{2\pi}{T}.$$

Usually, we do not use the word *radian* but simply call k the *wavenumber* and ω the *frequency.* When we do not have a pure sine wave, but instead have some complicated wave motion (which is usually the case), then we must carry out a Fourier analysis to determine the spectral distribution (or spectrum) of the wavenumbers and frequencies present.

In exploration seismology, we are usually concerned with P-wave propagation in layers of homogeneous, isotropic rocks. Such a rock may be described by two physical parameters: its *density* ρ and the *velocity* v of propagation of P-waves in the rock. The velocity v is called the wave velocity or the material velocity. These parameters can actually be measured in an oil well by borehole instruments in a well survey.

We have already emphasized the distinction between particle velocity and

Figure 3.9. Wave profile, or "snapshot," of the wave motion at one instant of time. The horizontal axis x *measures distance from the shot. This figure shows that for a fixed instant of time, the high amplitude has traveled the farthest.*

Figure 3.10. Wavelet, or motion at a fixed point in space, as a function of time t. *Here we have plotted wave amplitude versus* vt, *where* v *is a constant velocity. Thus, the horizontal axis has the dimensions of distance, and the wavelet shown appears as the reverse of the wave profile shown in the figure 3.9. This figure shows that for a fixed spatial point, the high amplitude appears first.*

wave velocity. Wave velocity (more precisely, group velocity) represents the speed at which seismic energy is transported. Typical wave velocities in rock formations are on the order of thousands of meters per second. Particle velocities represent the velocities of the very small rock particles about their stationary equilibrium positions. A typical particle velocity is expressed in terms of millionths of a meter per second. To get some idea of particle velocity, we can consider the baseplate of the mechanical vibrator used in the Vibroseis method. The displacement of the baseplate is no more than 1 or 2 cm, so the particle velocity imparted to the earth at a 10 Hz frequency is on the order of 1 m/sec. This velocity is entirely different, both in concept and in magnitude, from the wave velocity of the waves leaving the baseplate. The wave velocity would be on the order of 1000 m/sec, or more. The geophones placed on the surface of the ground pick up and record the particle velocity, which, when amplified, appears as the seismic trace. These received signals have much smaller particle velocities than the source signals. The direct wave from a dynamite source might have a particle velocity of 0.001 m/sec, whereas the wave corresponding to a deep reflection might have a particle velocity of 10^{-8} m/sec.

We have characterized the physical properties of a homogeneous, isotropic rock formation by its density and velocity. Another important parameter is the *acoustic impedance* Z, defined as the product of density times velocity:

$$Z = \rho v.$$

To convey a physical feeling for the acoustic impedance, Anstey (1977) calls it *acoustic hardness.* This is not exactly the same as what the geologist refers to as hardness, but it is still a useful characterization. Suppose, for example, that we take a soft shale. Such a rock would have a low material velocity v, so its acoustic impedance, or hardness, would be small. A hard limestone, however, would have a high velocity, and so its acoustic impedance, or hardness, would be large.

We have now introduced the concepts of particle velocity and pressure, which describe the wave motion, and acoustic impedance, which describes the rock formation. How are they related? First, an analogy might be helpful. The best known one is the electric analogy. In electrical work, we have current (amps), voltage (volts), and resistance (ohms). Ohm's law states:

voltage $= \pm$(current)(resistance).

Thus, for a given voltage, we would have a large current in a material of low resistance (a conductor, such as copper) and a small current in a material of high resistance (a dielectric, such as wood).

We now make the analogy:

particle velocity $\leftrightarrow$ current

pressure $\leftrightarrow$ voltage

acoustic impedance $\leftrightarrow$ resistance

Corresponding to Ohm's law, we thus have the equation

pressure $= \pm$(particle velocity)(acoustic impedance).

Thus, for a given pressure, we would have a large particle velocity in a material of low acoustic impedance (soft shale) and a small particle velocity in a material of high acoustic impedance (hard limestone).

Just as an electric wave may be described in terms of either its voltage or its current, so a seismic P-wave may be described in terms of either its particle velocity or its pressure. A seismic trace is a graph of amplitude versus time. In marine work, the hydrophone measures pressure, so the amplitude of a marine seismic trace is in terms of pressure. In land work, the geophone measures particle velocity, so the amplitude of a land seismic trace represents particle velocity.

In closing this section, we want to emphasize that it is the abstraction "the wave" that is propagated through the material, not the material itself. Any particular small particle simply vibrates about its own rest position while the wave is passing, and returns to its rest position and remains there after the wave has passed. Since motion from the rest position is a necessary characteristic of wave motion, the initial energy that created the wave spreads out with the wave. The purpose of a wave occurring in nature is to carry energy from one place to another.

Polarity

In geophysical work, it is important to have standards that everyone follows. Because so many seismic sections are traded and sold, and because one geographic area may require the use of geophysical data recorded and processed by several companies, it is important that everyone follow the same conventions. One such convention is *polarity,* which refers to which direction is used as the positive direction for a seismic amplitude. When two seismic lines intersect, we want the reflection observed on the near traces (i.e., the small offset traces) to be identical. For large offsets, the same event may be due to reflections from different depth points because of the dips of the reflecting interfaces. Thus, we should not expect far traces to have identical reflections. If one company displaces positive amplitudes in one direction and another company in the opposite direction, then a great deal of confusion can result. The convention we shall use in this book is the Anstey (1980) convention (fig. 3.11):

(1) A positive seismic signal is defined as a positive acoustic pressure on a hydrophone in the water or an upward motion on a geophone on the land.
(2) A positive seismic signal produces a positive number on tape, a positive deflection (upswing) on a monitor record, and a peak (black area) on a seismic section.

Let us now consider an explosion in a shot hole buried a hundred feet below the surface of the earth or, alternatively, an explosion below the surface of the water (see fig. 3.12). We want to describe the upgoing signal from this explosion. Because the initial motion of this signal is upward, it will produce a positive particle velocity as recorded by a geophone above the shot, or a positive pressure as recorded by a hydrophone above the shot. To describe the downgoing signal, suppose that a geophone in the case of a solid earth, or a hydrophone in the case of water, is placed below the shot. Because the initial motion of this signal is downward, it will produce a downward motion on the geophone and thus a negative particle velocity, but it will still produce a positive pressure on the hydrophone. In summary, according to our convention, a shot has this characteristic: the upgoing pulse has positive particle velocity and positive pressure, and the downgoing pulse has negative particle velocity but still a positive pressure.

Let us now describe the flow of energy, called the *energy flux.* This flow can be either upward or downward. We will assign a positive sign to energy flux for an upward flow and a negative sign for a downward flow. The energy flux is defined as the product of particle velocity and pressure; that is,

Energy flux = (particle velocity)(pressure).

Let us consider our explosion in a shot hole. The resulting energy flow upward is

Energy flux = (positive particle velocity)(positive pressure).

We see that the energy flux is a positive quantity, in agreement with our convention for an upgoing flow. The energy flow downward from the shot is

Energy flux = (negative particle velocity)(positive pressure).

This is a negative quantity in agreement with our convention for a downgoing flow. Energy is a positive quantity, of course, but the flow of energy has a direction, and hence a sign is attached to it.

Positive Reflection	Negative Reflection
Soft rock to hard rock (for example):	Hard rock to soft rock (for example):
• shale to carbonate	• carbonate to shale
• shale to tight sand	• tight sand to shale
• gas sand to shale	• shale to gas sand

Figure 3.11. The above convention for display polarity is the one adopted by Anstey, though it is at odds with the implications of the SEG recommendation on polarity. The industry is increasingly accepting the polarity shown above, particularly because it alone is compatible with pseudo-log display. All sections, whether commissioned from a processing house or received in trade, should be annotated to show the display polarity. The Anstey convention is that the reflection is positive if the contact is from soft to hard rock, and negative if it is from hard to soft. Note that a reflection does not come from a formation, but from a contact. Reprinted by permission of IHRDC from Anstey 1980.

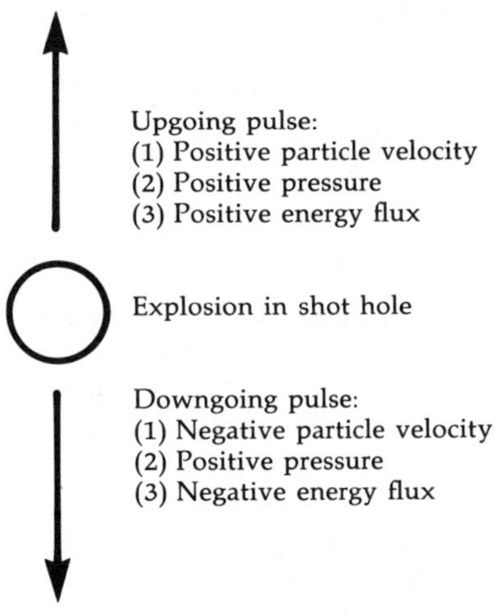

Figure 3.12. The Anstey (1980) convention.

Reflection and transmission

The basic concepts of the reflection seismic method are relatively simple and straightforward. In land exploration, dynamite is still one of the most commonly used materials for a source of energy. A charge of dynamite is exploded at a point just under the surface of the earth, and the resulting seismic pulse travels outward as a wave in all directions. The dynamite explosion itself produces an intense pressure pulse of very short duration, measured in fractions of a millisecond. The rock immediately surrounding the explosive charge is subjected to intense stresses (pressures), exceeding the elastic limit for a certain distance out from the charge. In most cases, the material flows plastically, and the physical condition of the rock material close to the shot point suffers a permanent change. Because of these factors, the initial short duration pulse of the dynamite explosion is transformed into a seismic pulse of considerably longer duration. This seismic pulse is called the *source signature* or *source wavelet* (see figs. 3.13 and 3.14). The form of the source wavelet has been carefully studied by theoretical methods, and a reasonable approximation to it often can be estimated by use of recording apparatus close to the explosion. Thus, in such cases, we can consider the source signature wavelet as known data. Often, however, we cannot or do not directly measure the source wavelet.

For our present purposes, let us consider the process of the reflection of the downward-traveling seismic pulse. Once the seismic pulse leaves the immediate area of the shot, we can assume that the seismic pulse, now of a definite form, propagates through the earth and reflects from rock interfaces in accordance with the linear laws of elastic wave propagation.

We now want to consider essentially a one-dimensional problem, in which we treat only variations along the depth axis z, which points straight down into the earth. We thus disregard the generally spherical form of the wavefront and treat the problem as that of a plane wave traveling vertically downward in a direction perpendicular to the flat-lying (horizontal) rock strata. Suppose at some instant a wave is traveling in a rock layer of density ρ_1 and seismic wave velocity v_1, and then the wave encounters a different rock layer of density ρ_2 and velocity v_2. The result is that a portion of the energy in the wave will be reflected at the interface and the remainder will be transmitted. This splitting of the incident wave into a reflected wave and a transmitted wave at the interface is caused by the abrupt change in rock density and velocity. For the case of normal incidence (which we are treating here), the reflected and transmitted wavelets have identically the same shape and breadth as the incident wavelet, but they differ in amplitude. The ratio of the amplitude of the reflected wave to that of the incident wave is termed the *reflection coefficient.* Similarly, the ratio of the amplitude of the transmitted wave to that of the incident wave is called the *transmission coefficient.* However, the reflection coefficient defined for a wave in which the amplitude is measured in terms of particle velocity is different from the reflection coefficient in terms of pressure. The same statement also holds for the transmission coefficient. Let us now find this difference.

Suppose that a wave propagating in shale comes to an interface with a limestone layer. For the rocks not to rupture at the interface, the particle velocity on both sides of the interface must be the same; that is, at the interface, the particle velocity on the shale side must be the same at all times as the particle

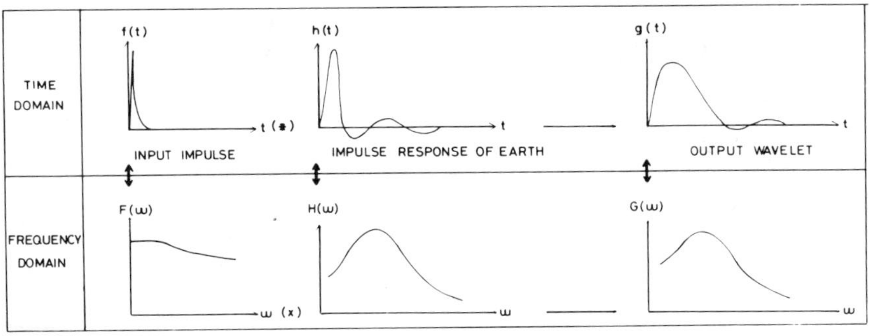

Figure 3.13. Earth filtering effect on an impulsive source function represented in both time and frequency domains. Reprinted by permission of Birkhauser-Verlag from Al-Sadi 1980.

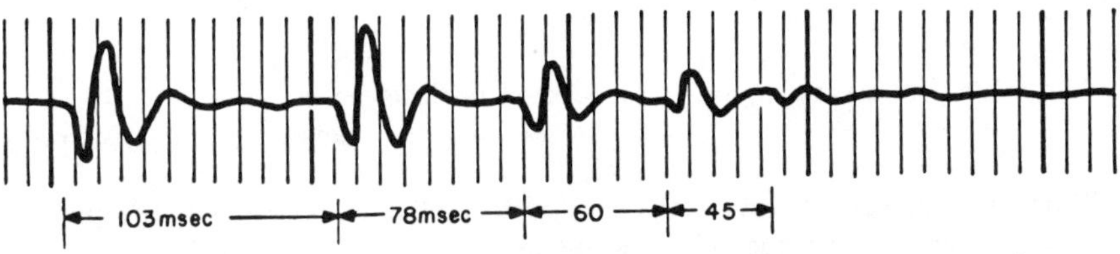

Figure 3.14. Bubble pulses from a small underwater explosion. The time between successive implosions decreases as energy is dissipated. Most high-energy marine seismic sources generate bubbles; the bubble oscillation frequency is given by the Rayleigh-Willis relation. Various arrangements are used to attenuate bubble effects or to correct for them. Bubble oscillations can be prevented if the bubble is generated close enough to the surface so that the gases vent before the bubble collapses. Reprinted by permission of the Society of Exploration Geophysicists from Sheriff 1973.

velocity on the limestone side. We express this physical fact in mathematical terms by the statement that particle velocity must be continuous at an interface. Similarly, pressure must be continuous at an interface.

An explosion gives a large downgoing pulse, and we assign to it a negative amplitude in the case of particle velocity and a positive amplitude in the case of pressure. Suppose the explosion is set off in soft shale (Z_1 is small). The source pulse travels down and hits hard limestone (Z_2 is big) (see fig. 3.15). The incident pulse has a fixed amount of (negative) energy flux, given by the product of its (negative) particle velocity times its (positive) pressure. Since the soft shale has a low impedance Z_1, the particle velocity is relatively large in magnitude, and its pressure is correspondingly small. At the interface, some of the incident energy is transmitted into the limestone and the rest is reflected back into the shale. The total particle velocity must be the same on both sides of the interface. The hard limestone, because of its higher impedance, cannot take up all the particle velocity present in the soft shale. Hence, only some of the particle velocity will be transmitted into the limestone. To keep the bookkeeping straight (i.e., continuity of particle velocity), a reflected wave must be thrown upward, back into the shale. The particle velocity of this reflected wave must be a positive quantity, so as to cancel out some of the incident negative particle velocity; that is, we have the bookkeeping entry (where p.v. is particle velocity)

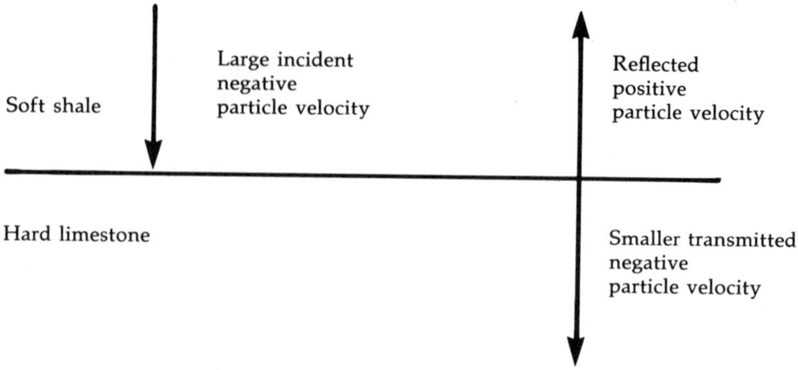

Figure 3.15. Continuity of particle velocity at a soft-to-hard interface. The arrows indicate the raypath direction (i.e., direction of travel).

(Large incident negative p.v.) + (reflected positive p.v.)
= (smaller transmitted negative p.v.).

Thus, the particle velocity reflection coefficient is negative, and the particle velocity transmission coefficient is positive.

Next let us look at pressure (see fig. 3.16). Since the soft shale has a low impedance, the downgoing incident pressure is relatively small. By continuity, the pressure must be the same on both sides of the interface. The limestone is hard and can support more pressure than the soft shale. Thus, the amplitude of the transmitted pressure wave is greater than the amplitude of the incident pressure wave. To keep the bookkeeping straight (i.e., continuity of pressure), a reflected wave must be thrown upward, back into the shale. The pressure of this reflected wave must be a positive quantity, so as to augment the incident pressure; that is, we have the bookkeeping entry

(Small incident positive pressure) + (reflected positive pressure)
= (larger transmitted positive pressure).

Thus, the pressure reflection coefficient is positive, and the pressure transmission coefficient is also positive.

Let us now look at energy flux (i.e., energy flow). Energy is a positive quantity, of course, but a flow has a direction assigned to it. Since the incident wavelet is downgoing, we assign to it the negative energy flux given by

(Incident energy flux) =
(large incident negative p.v.)(small incident positive pressure).

The transmitted wave is still a downgoing wave, of course, so it has the negative energy flux given by

(Transmitted energy flux) =
(smaller transmitted negative p.v.)(larger transmitted positive pressure).

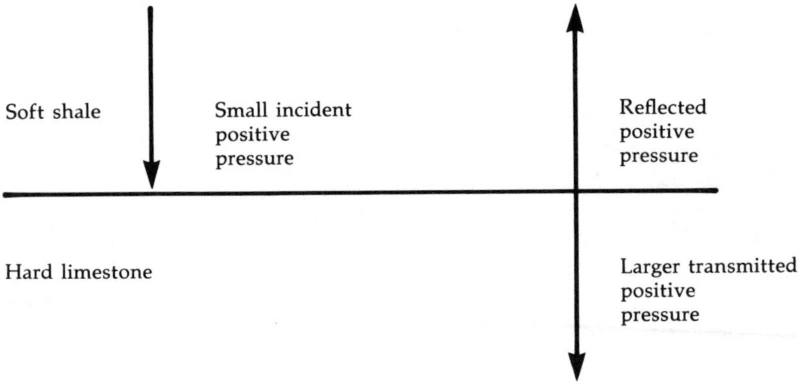

Figure 3.16. Continuity of pressure at a soft-to-hard interface. The arrows indicate the raypath direction.

The reflected wave is an upgoing wave, however. Thus, it must have a positive sign for its energy flux. Its energy flux is

(Reflected energy flux) =
(reflected positive particle velocity)(reflected positive pressure),

which we see is indeed a positive quantity.

Energy, not energy flux, is conserved. Energy is positive, so the downgoing energy would be the negative of the downgoing energy flux. Thus, the law of conservation of energy (fig. 3.17),

(Incident energy) = (reflected energy) + (transmitted energy),

becomes, in terms of energy flux, the equation

−(Incident energy flux) =
(reflected energy flux) − (transmitted energy flux).

This equation is, however, the equation (fig. 3.18)

(Incident energy flux) + (reflected energy flux) = (transmitted energy flux).

Thus, the total downgoing energy flux in each rock layer (i.e., the soft shale and the hard limestone) must be the same. The interface cannot change the total flow of net downgoing energy, although a sequence of layers can change the time distribution of this flow.

Let us now prepare some equations (see fig. 3.19). Let an incident downgoing wave in a formation with impedance Z_1 strike a formation with impedance Z_2. If we are measuring the wave amplitudes in terms of particle velocity (p.v.), the reflection coefficient is

$$c_{\text{p.v.}} = \frac{Z_1 - Z_2}{Z_1 + Z_2}.$$

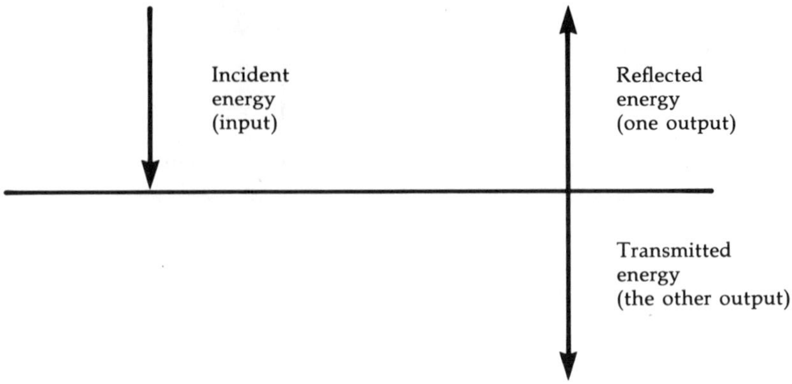

Figure 3.17. Law of the conservation of energy. The input energy is equal to the sum of the output energies. The arrows indicate the raypath direction.

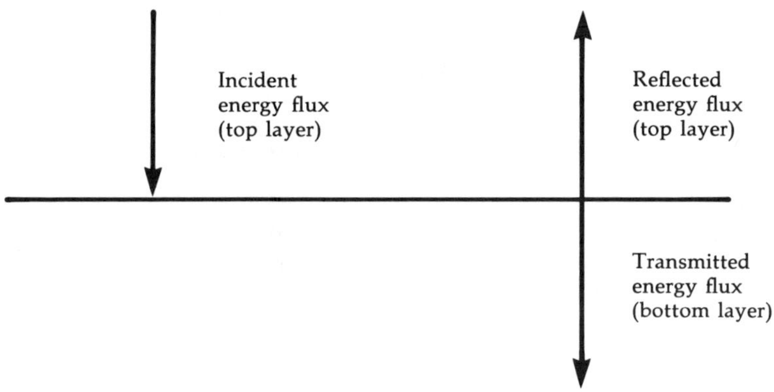

Figure 3.18. The net downgoing energy flow is the same in each layer. The sum of the fluxes in the top layer equals the flux in the bottom layer. The arrows indicate the raypath direction.

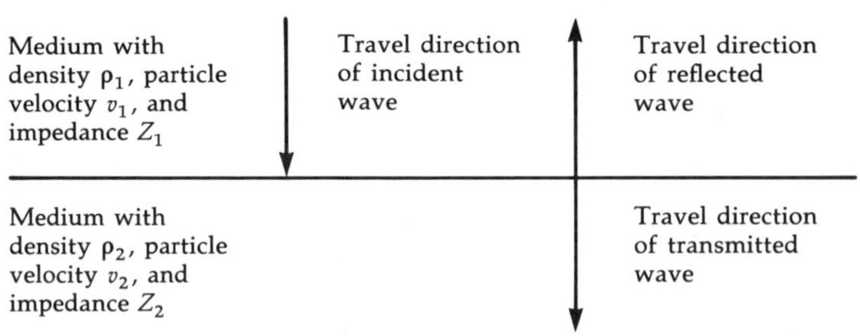

Figure 3.19. Setup for the definition of the reflection coefficients $c_{\text{p.v.}}$ *and* c_{pressure} *and the transmission coefficients* $t_{\text{p.v.}}$ *and* t_{pressure}.

The corresponding pressure reflection coefficient is its negative; that is,

$$c_{\text{pressure}} = -c_{\text{p.v.}} = \frac{Z_2 - Z_1}{Z_1 + Z_2}.$$

The transmission coefficient t is always equal to $1 + c$. Thus, the particle velocity transmission coefficient is

$$t_{\text{p.v.}} = 1 + c_{\text{p.v.}} = \frac{2Z_1}{Z_1 + Z_2},$$

and the pressure transmission coefficient is

$$t_{\text{pressure}} = 1 + c_{\text{pressure}} = \frac{2Z_2}{Z_1 + Z_2}.$$

What are the reflection coefficient C and transmission coefficient T for an upgoing wave in a medium with impedance Z_2 when the wave strikes an interface with a medium with impedance Z_1 (see fig. 3.20)? Now the incident wave is upgoing, and the change of impedance is from Z_2 to Z_1. Thus, the roles of Z_1 and Z_2 are interchanged, so the upgoing coefficients (designated by capital letters as opposed to small letters for the downgoing coefficients) are

$$C_{\text{p.v.}} = -c_{\text{p.v.}} = \frac{Z_2 - Z_1}{Z_1 + Z_2},$$

$$C_{\text{pressure}} = -c_{\text{pressure}} = \frac{Z_1 - Z_2}{Z_1 + Z_2},$$

$$T_{\text{p.v.}} = 1 + C_{\text{p.v.}} = \frac{2Z_2}{Z_1 + Z_2},$$

$$T_{\text{pressure}} = 1 + C_{\text{pressure}} = \frac{2Z_1}{Z_1 + Z_2}.$$

Example 1. Let an incident downgoing spike have amplitude -1 for particle velocity and amplitude $+1$ for pressure (see fig. 3.21). The reflected spike has amplitude $(-1)c_{\text{p.v.}}$ for particle velocity and amplitude $(+1)c_{\text{pressure}}$ for pressure. Thus, the energy flux of the incident wave is $(-1)(1) = -1$, and the energy flux of the reflected wave is $-c_{\text{p.v.}}\, c_{\text{pressure}}$. The transmitted spike has amplitude $(-1)t_{\text{p.v.}}$ for particle velocity and amplitude $(+1)t_{\text{pressure}}$ for pressure, so its energy flux is $-t_{\text{p.v.}}\, t_{\text{pressure}}$. The equation for equality of flux in each layer is

$$-1 - c_{\text{p.v.}} c_{\text{pressure}} = -t_{\text{p.v.}} t_{\text{pressure}}.$$

This is the fundamental equation relating the coefficients. It is more usual to write this equation as

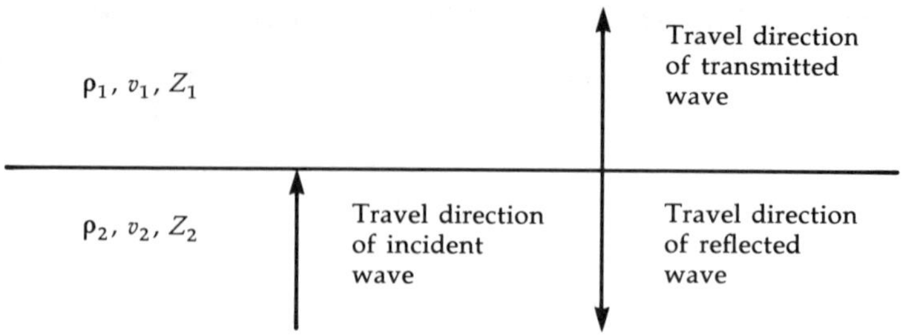

Figure 3.20. Setup for the definition of the reflection coefficients $C_{\text{p.v.}}$ *and* C_{pressure} *and the transmission coefficients* $T_{\text{p.v.}}$ *and* T_{pressure}.

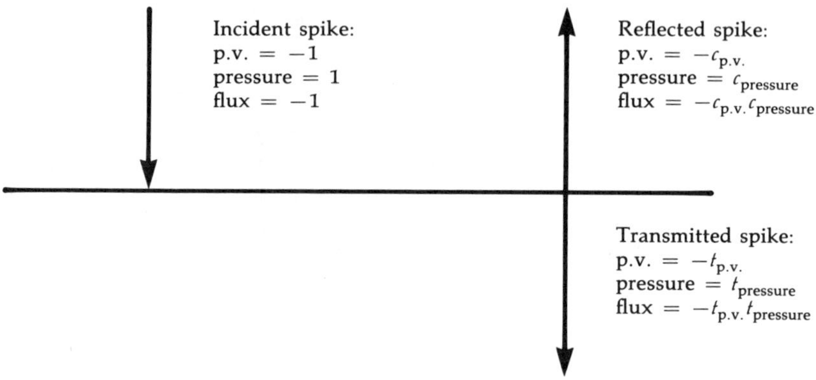

Figure 3.21. Relationships among the coefficients.

$$1 + c_{\text{p.v.}} c_{\text{pressure}} = t_{\text{p.v.}} t_{\text{pressure}} .$$

Because

$$c_{\text{p.v.}} = -c_{\text{pressure}} ,$$

$$t_{\text{p.v.}} = (Z_1/Z_2) t_{\text{pressure}} ,$$

this equation becomes either

$$1 - c^2_{\text{pressure}} = (Z_1/Z_2) t^2_{\text{pressure}}$$

or

$$1 - c^2_{\text{p.v.}} = (Z_2/Z_1) t^2_{\text{p.v.}} .$$

Note that regardless of whether $Z_1 > Z_2$ or $Z_2 > Z_1$, the energy flux of each downgoing wave (i.e., the incident wave and the transmitted wave) is always negative, so that for each wave the particle velocity amplitude and the pressure amplitude have the opposite polarity (180° out of phase):

$$\text{Incident p.v.} = -1, \quad \text{Incident pressure} = +1,$$

$$\text{Transmitted p.v.} = -t_{\text{p.v.}}, \quad \text{Transmitted pressure} = +t_{\text{pressure}}.$$

The energy flux for the reflected wave, however, is positive. This means that, for a reflected wave, the particle velocity amplitude always has the same polarity as (i.e., is in phase with) the pressure amplitude

$$\text{Reflected p.v.} = -c_{\text{p.v.}}, \quad \text{Reflected pressure} = c_{\text{pressure}}.$$

A reflected pulse, whether measured by a geophone or a hydrophone, will always have the same polarity. If $Z_2 > Z_1$ (i.e., top layer soft, bottom layer hard), the polarity of the reflected pulse will be positive. If $Z_1 > Z_2$, the polarity of the reflected pulse will be negative.

Example 2. An explosion in medium 2 (lower medium) initiates an upgoing spike and a downgoing spike. The upgoing spike is reflected at the interface with medium 1, and the downgoing reflected spike (the ghost) forms a composite waveform with the primary downgoing spike. What will this composite wavelet look like? Suppose that the top layer is very soft (for example, air with $Z_1 = 0$). The upgoing reflection coefficients are

$$C_{\text{p.v.}} = \frac{Z_2 - 0}{0 + Z_2} = 1, \qquad C_{\text{pressure}} = \frac{0 - Z_2}{0 + Z_2} = -1.$$

On both sides of the shot point (i.e., above and below), the pressure starts out as a high positive value and the particle velocity starts out at a large magnitude. There is a polarity difference, however, between the particle velocity above the shot and the particle velocity below the shot, because the explosion forces material away from the shot point. We therefore must consider particle velocity as a directional quantity. According to the convention we have introduced, we take the particle velocity of the downgoing explosive wave as negative and the upgoing one as positive. Thus, the energy flux of the upgoing wave from the explosion is the negative of the energy flux of the downgoing explosive wave, as the energies of the two waves are flowing in opposite directions. In summary, we have

Explosion → upgoing wave (positive particle velocity, positive pressure),
Explosion → downgoing wave (negative particle velocity, positive pressure).

The reflected upgoing wave is the ghost. The particle velocity reflection coefficient is $+1$, and the pressure reflection coefficient is -1. Thus, the ghost (a downgoing wave) has positive particle velocity and negative pressure. The composite downgoing waveform has the following form (fig. 3.22):

In terms of p.v.: negative direct pulse followed by a positive ghost pulse.
In terms of pressure: positive direct pulse followed by a negative ghost pulse.

Thus, whether we measure pressure or particle velocity, the ghost reflection is of opposite polarity to the direct pulse. The delay time between the direct and

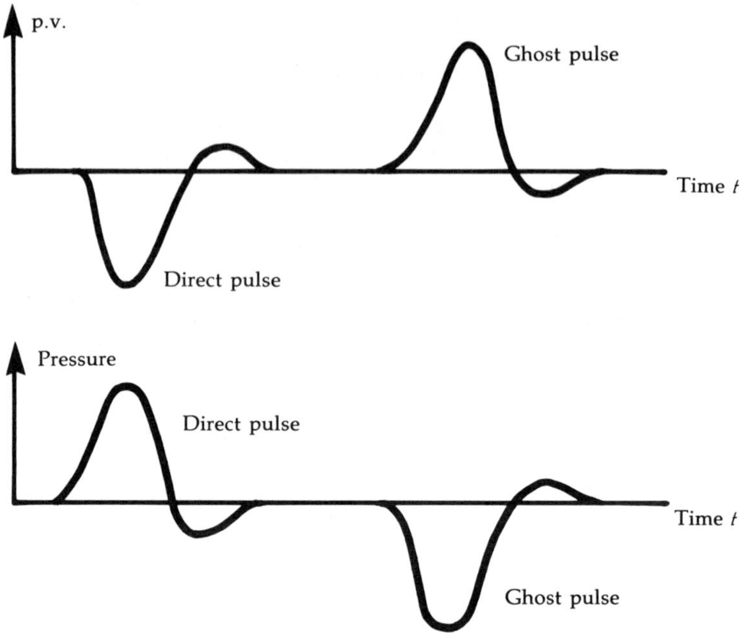

*Figure 3.22. The downgoing direct-wave/ghost-wave complex with amplitude in particle velocity (*top*) and pressure (*below*).*

the ghost is equal to the two-way traveltime from the shot to the upper interface. If this delay is represented by n discrete time units, and if the explosive pulse has unit amplitude, then the composite wavelet has the form

$\pm\,(1, 0, 0, \ldots, 0, -1, 0, 0, \ldots),$
↑ $n-1$ zeros ↑
primary ghost

where the leading sign ($\pm$) is minus for p.v., and plus for pressure.

Example 3. For a rigid sea bottom ($Z_2 = \infty$ for the bottom), the reflection coefficients are

$$c_{\text{p.v.}} = -1, \qquad c_{\text{pressure}} = 1.$$

Because particle velocity cannot be transmitted into the hard bottom when a downgoing wave strikes it, the particle velocity at the interface must be zero. Thus, a p.v. sensitive receiver dragging on the bottom from a cable would lose its effectiveness for a rigid sea bottom.

Example 4. Consider the water surface as a free surface ($Z_1 = 0$ for air). The reflection coefficients are

$$C_{\text{p.v.}} = 1, \qquad C_{\text{pressure}} = -1.$$

The particle velocity at the surface is twice the particle velocity of the upgoing incident wave, whereas the pressure at the surface is zero. Thus, the hydrophone must be kept below the surface of the water, and its effectiveness diminishes as the streamer floats to the surface.

The seismic trace

Let us assume that the source of seismic energy is an explosion of dynamite. Ideally, the source signal would be a sharp spike, but because of nonlinear effects about the shot point, it soon becomes modified into an oscillatory waveform of some breadth. This waveform is called the source wavelet, or source signature, and in certain cases it can be monitored and recorded on magnetic tape. The source waveform is transmitted into the earth, where it suffers various effects, one of which is absorption. The most important of these effects, however, are the modifications caused by multiple reflections due to the earth layering. The earth layering itself is characterized by the primary reflections due to the sequence of reflection coefficients of the interfaces. The sequence of reflection coefficients is called the *reflectivity function,* or the *reflection coefficient (RC) log* (fig. 3.23). Finally, the reflected energy is recorded by means of detectors and recording instruments (i.e., the receivers). To a first approximation, all of these successive changes wrought upon the source wavelet can be modeled as a sequence of convolutions. A convolution is a mathematical operation denoted by an asterisk; we will describe it in detail in Chapter 4. For the time being, however, let us just think about it as the operation performed by the various agencies. Thus, the successive effects that produce the final seismic trace can be envisaged as the equation

$$\text{(source wavelet)}*\text{(multiples and absorption)} *\text{(reflectivity function)}*\text{(receiving effects)} = \text{(seismic trace)}.$$

The reflectivity function (i.e., the sequence of reflection coefficients) is a characteristic of the earth; it represents the geologic information we desire. The other components of this equation—the source wavelet, multiple reflections, absorption, and receiving effects—represent characteristics of the seismic experiment with which we probe the earth. The seismic trace is the convolution of all these components. All the components on the left, except the reflectivity function, are grouped together and called the seismic wavelet. Hence, we make use of the fact that the operation of convolution is commutative (i.e., we can interchange the order of the components). Thus, the seismic wavelet is defined as

$$\text{(seismic wavelet)} = \text{(source wavelet)}*\text{(multiples and absorption)}*\text{(receiving effects)}.$$

The foregoing two equations reduce to the *convolutional model* of the seismic trace given by the equation (see fig. 3.24)

$$\text{(seismic trace)} = \text{(seismic wavelet)}*\text{(reflectivity function)}.$$

Let us discuss this model in terms of the mathematical concepts introduced in the preceding sections. First, as we know, all the wave motion with which we deal has been digitized, so we will regard all the signals as being in digital (i.e., discrete time) form. The complete and entire seismic trace as recorded by

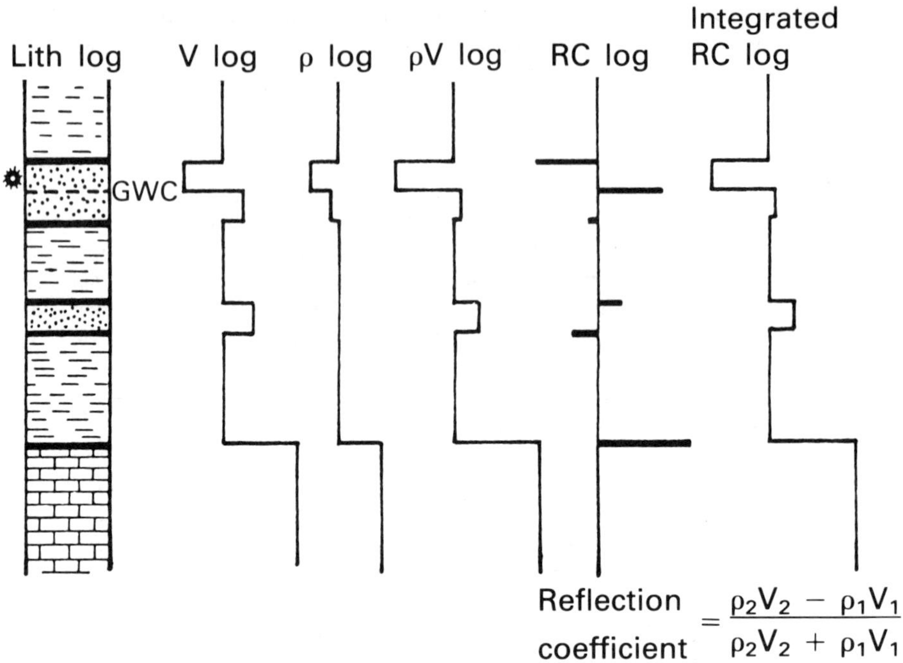

Figure 3.23. The derivation of the reflection coefficient log (i.e., the reflectivity function). We postulate the lithology shown at the left: a thick shale overlaying a massive limestone, with two sand units within the shale. The lower sand is completely water-saturated, while the upper contains both water and free gas. The velocity is shown under V-log in the figure: high velocity in the incompressible carbonate, a velocity a little higher than shale velocity in the wet sand, and a major depression of velocity in the gas sand. The porosity of the lower sand is such that, for water saturation, the density is the same as that of the shale. The porosity of the upper sand yields a slightly lower density where the sand is wet and a much reduced density in the gas zone. The density is shown as the ρ *log. The acoustic impedance log, or* ρV *log, is found as the product of density* ρ *and velocity* V. *The* ρV *log looks much the same as the* V *log, except in the few cases where the velocity and density change in opposite directions. Here this occurs in the upper wet sand and is minor; in other rocks—particularly salt, gypsum, anhydrite, and coal—the distinction can be significant. However, at the first level of concern we often think of the* ρV *log as being of substantially the same shape as the* V *log. Now we invoke again the fact that reflections are generated at the contact of two rocks of different acoustic impedance (which Anstey calls "hardness"). The formula for computing the effective strength of the reflections is the one for* reflection coefficient, *as shown—the difference of hardness across the contact, divided by the sum. Obviously, it is the numerator—the difference of hardness—that is important; a big difference of* ρV *makes a strong reflection. If we work down the RC log making a running sum of all the values in a moving window, we obtain the integrated RC log at the right of the figure; we have restored the appearance of the* ρV *log, and now our readings indicate formations rather than contacts. Reprinted by permission of IHRDC from Anstey 1982.*

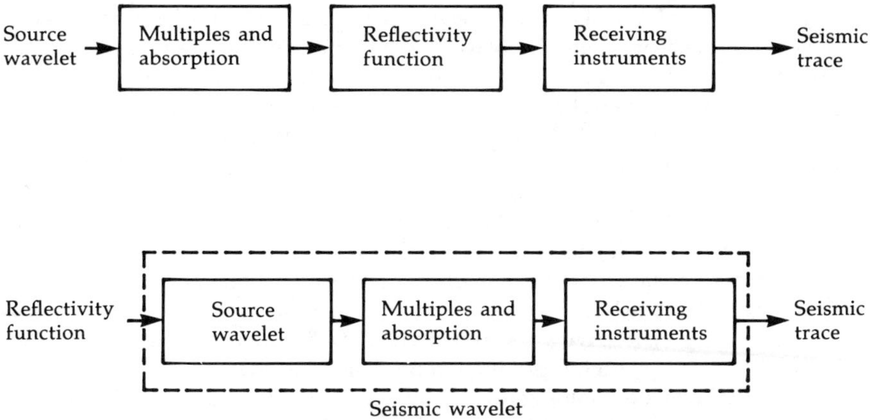

Figure 3.24. The top diagram shows the generation of the seismic trace. Because convolution is commutative, we can rearrange the order to give the bottom diagram. The components within the dashed box make up the seismic wavelet. Such a convolutional model is an approximation to reality, of course, but has proved to be extremely useful in the analysis of seismic records.

a group of geophones represents a transient phenomenon with a definite time origin; therefore, in a precise sense, the seismic trace is a wavelet. The seismic trace is corrected, however, for amplitude decay due to spherical spreading and other factors. As a result, the trace does not appear in our displays as a transient or dying-out phenomenon. Instead, it appears as a continuing phenomenon over the seismic time scale of interest (from 0 to 6 sec). Thus, it is traditional to regard the seismic trace not as a wavelet but as a time series, or more precisely as a time sample, from 0 to 6 sec, of a continuing time series.

The seismic wavelet (i.e., the component of the trace due to source, multiples, absorption, and receiving effects) is indeed a one-sided transient with respect to the seismic time scale. Usually, the seismic wavelet is only a few hundred milliseconds in length, which is only a fraction of the 6 sec time scale. In our discussion here we are excluding the Vibroseis wavelets, which are much longer, but of course the seismic time scale is accordingly increased.

The seismic reflectivity (i.e., the component of the trace representing the deep earth structure) is a continuing mathematical function with depth. We thus regard reflectivity as a continuing phenomenon. In a precise sense, the reflectivity function is a function of depth. Depth may be converted by a velocity function into seismic traveltime, however, and hence the reflectivity component of the seismic trace may be regarded as a continuing time series sampled over the seismic time scale from 0 to 6 sec.

For our purposes, we deal only with a single-dimensional model, so we regard every seismic trace, with its wavelet and reflectivity components, as a single-channel time function.

The multiple reflection component includes such things as ghosts, reverberations, short-period multiples, and long-period multiples. The data-processing operations of deconvolution and stacking try to remove this component as much as possible. In addition, signature deconvolution tries to remove the source wavelet and receiving effects as much as possible.

The makeup of the seismic wavelet can be analyzed in several important ways. Variations occur in the wavelet from trace to trace, and such lateral wavelet variations are harmful, since the interpreter wants to detect and map lateral variations in the subsurface. On land, the source waveform and the effect of the near-surface layer can change markedly from trace to trace. Ghosting with explosive sources is also a problem. The recording system is the only component that does not change. In deep water (beyond the continental shelf) the seismic wavelet is quite constant, assuming that the cable depth and source performance are maintained constant. Only the deep multiple reflection effects are variable. On the continental shelf, the effect of reverberations (i.e., short-period multiples) within the water layer is so destructive that in many areas exploration was ineffective prior to the widespread introduction of deconvolution and stacking as digital production operations in the early 1960s.

The difficulty in determining the components that make up the seismic wavelet is somewhat due to to lateral variability. The response of the recording system can be measured directly and stored in computer memory for future use. The marine ghosting effect can be estimated if the source and receiver depth are known. In the ideal case, we need to determine the marine source pulse only once, or only once in a while, for a given configuration of the source of energy. Estimation of the response of the laterally variable components is the most significant problem for successful application of the convolutional model. We have introduced the convolutional model in this section; much more will be said about it in Chapter 4.

Thin beds

Vertical resolution refers to the ability to distinguish the reflections from the top and the bottom of a thin sedimentary layer. Clearly, it is easier to determine the top and bottom reflections of a bed 100 ft thick than one 10 ft thick (assuming, of course, that all other things are equal). The seismic wavelet is an important factor in vertical resolution. The vertical resolution depends on the sharpness of the wavelet. We can see a thin layer better with a sharp wavelet than with a broad wavelet. The sharpness of the wavelet depends primarily on its bandwidth—that is, on the effective spread of frequencies in its amplitude spectrum. A wavelet with a wide band of frequencies (large bandwidth) will be sharper than one with a narrow band of frequencies. In fact, a very narrow band of frequencies would correspond to a wavelet with many high side lobes—that is, a broad, "ringing" wavelet (see fig. 3.25). In seismic work, we try to generate wavelets with wide bandwidth by using a well-designed source, good spread geometry, the best recording instruments, and by carefully executed deconvolution. However, there are bandwidth limitations imposed by the filtering effects of the sedimentary column in the earth over which we have no effect. In addition to deconvolution, we can, as a final step, carry out the data-processing method known as *wavelet processing.* In such an operation, we filter the seismic traces in such a manner that we change the shape of the basic seismic wavelet in a predetermined way. This filtering operation is also known as a *wave-shape operator,* or simply a *shaping filter* (see Figs. 3.26 and 3.27). The seismic wavelet is a one-sided (or causal) wavelet, with its initial nonzero value occurring at the arrival time of the reflected event. Since the earth necessarily acts as a causal

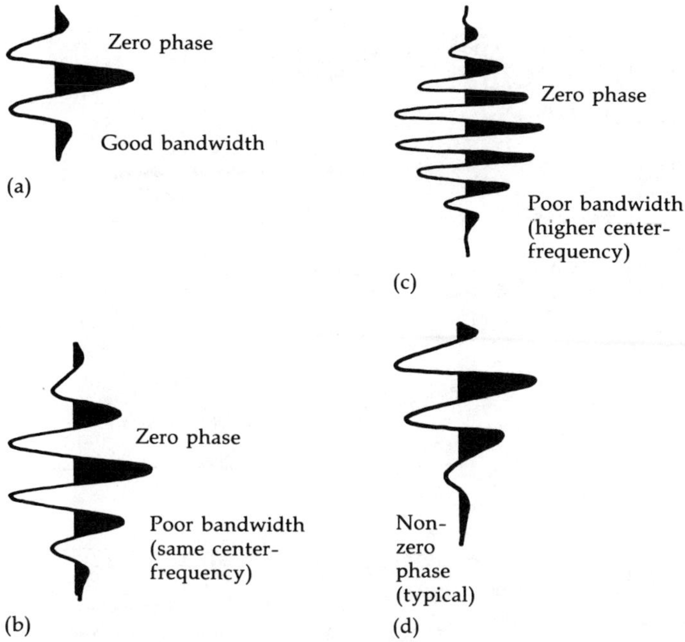

Figure 3.25. Some typical seismic wavelets. The vertical resolution depends on the sharpness of the seismic pulse; the sharper the pulse, the thinner the layer we can see. The sharpness of the pulse depends primarily on its bandwidth; the wider the band of frequencies the pulse represents, the better (pulses a and b in the figure). Good bandwidth is obtained (to within limits imposed by the filtering in the earth itself) by using a well-designed source, by appropriate spread geometry and recording filters, and by deconvolution. The sharpness of the pulse also depends on its phase; the best condition is the zero-phase wavelet (pulses a and d). In this example, the pulses are symmetric about a peak (black) for a positive reflection and about a trough (white) for a negative reflection; to bring the wavelet to the zero-phase condition requires a phase-zeroing filter in the wavelet processing step. The secret of resolution is bandwidth, and not just high frequencies. Of course, we can always raise the frequencies apparent on a section merely by filtering out the low frequencies, but this does not improve the resolution; it just makes the pulse leggy, losing sharpness because of its profusion of cycles (pulses a and c). Reprinted by permission of IHRDC from Anstey 1980.

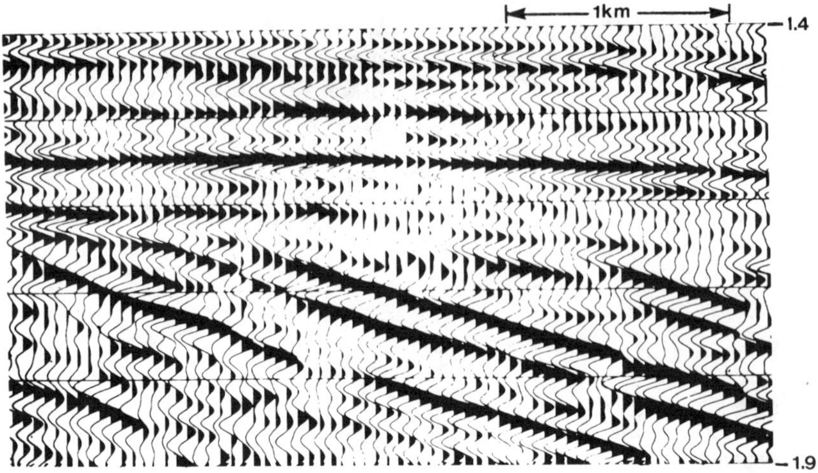

(a)

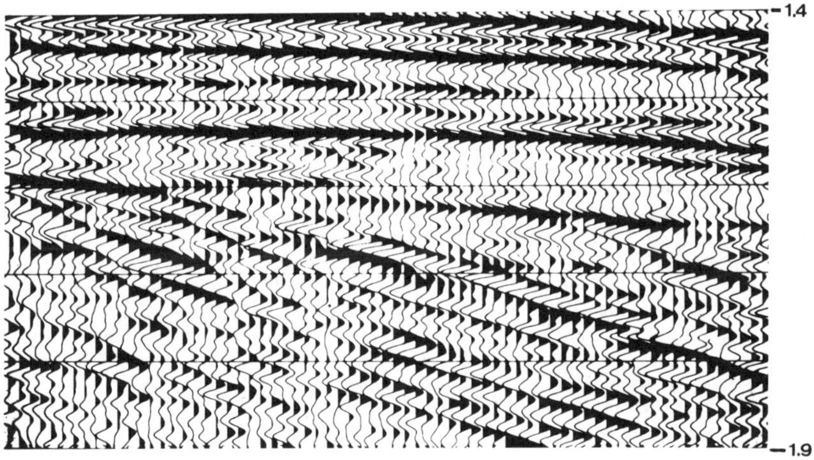

(b)

Figure 3.26. Effects of wavelet processing: (a) without wavelet processing and (b) with deterministic waveshape correction based on near-field measurements at each source point (marine air gun source). Reprinted by permission of Seiscom Delta.

(a)

(b)

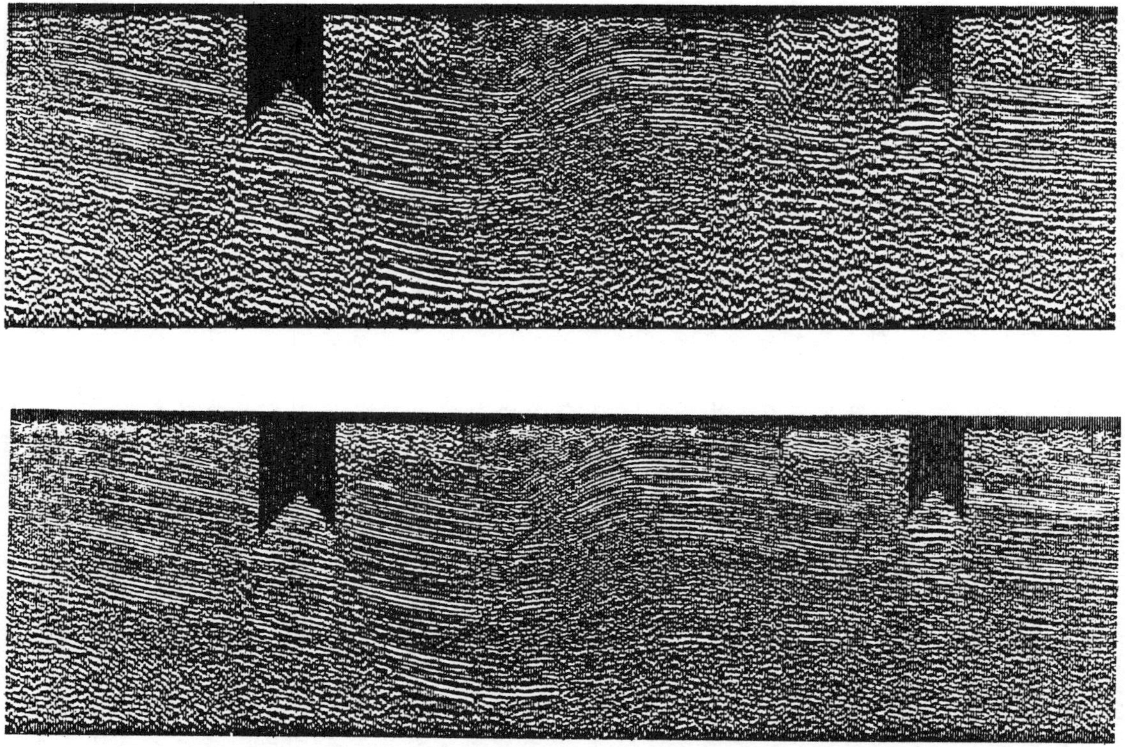

Figure 3.27. Effect of wavelet processing on a land line in an area of appreciable near-surface variation. Undershooting was used to fill in two large gaps on the line: (a) without wavelet processing and (b) with PULSE prestack wavelet processing. The wavelet processing has lessened waveshape variations and increased the frequency and continuity. A high-angle, basement-involved reverse fault occurs in the middle of the line; this fault probably involves strike-slip motion also. Reprinted by permission of Seiscom Delta.

filter, it would be impossible for the wavelet to have any precursor (i.e., a nonzero value) that comes in before its arrival time. It takes some time (a few milliseconds), however, for the seismic wavelet to build up to its peak amplitude, and thus this peak amplitude occurs after the arrival time of the wavelet. Also, some further secondary peaks usually follow the main peak value of the wavelet. The objective of wavelet processing is to convert this actual seismic wavelet into an artificial wavelet that is more pleasing to the eye of the interpreter. Like great art, this artifact does not have to be a wavelet that satisfies physical laws, but it must be esthetically pleasing. The criterion, therefore, is not physics but art. Over the years, seismic interpreters have decided that the most pleasing wavelet to look at is a zero-phase wavelet. A zero-phase wavelet is nothing more than an autocorrelation function. We will see that an autocorrelation is symmetric about time zero, that it has its maximum value at time zero, and that it damps off in both time directions quite rapidly. Thus, in wavelet processing, the actual seismic wavelet, which has it beginning at the arrival time of the reflection, is replaced as well as possible by a zero-phase wavelet that has its center maximum (i.e., time zero) at the arrival time of the reflection (see fig. 3.28). The particular zero-phase wavelet used is chosen to be one that is the sharpest (i.e., that has the greatest bandwidth) consistent with the data. This chosen (artificial) wavelet is often called the *interpreter's wavelet.* Wavelet processing exploits the fact that the sharpness of a wavelet depends on its phase spectrum as well as its amplitude spectrum. For a given amplitude spectrum, the greatest sharpness occurs in the form of a phase spectrum that is zero—that is, a zero-phase wavelet. The wavelet-processing step in the processing sequence brings the actual seismic wavelets into artificial wavelets with zero-phase.

Perhaps at this point we should insert a few comments about nature and nurture. Seismology is concerned with the physics of wave propagation through the earth. The actual recorded seismic waves, before any processing, contain the greatest amount of information about this physical phenomenon. Geologists, however, are interested in a picture of a cross section of the earth. The painting of a good picture requires a significant element of art. The whole purpose of data processing is to throw out the physical picture of wave propagation and to put in its place an artistic picture of the earth. If our artwork is done properly, geologists can look at this picture and see the geology with their own eyes. The ultimate goal of digital processing is to eliminate the need for geologists to know about seismic wave propagation; all these geophysical aspects should be removed in order to give them a good picture of the subsurface strata. Nature gives us the recorded seismic traces, but nurture—in the form of deconvolution, stacking, wavelet processing, and migration—yields the final geologic picture. The discipline known as seismic stratigraphy, introduced in the past decade, attests to the efficacy of seismic processing.

The classical work in the analysis of thin beds was done by Widess (1974). Let us consider a thin bed with a horizontal top interface and a dipping bottom interface so that the bed pinches out. Suppose that the thin bed is soft (i.e., the thin bed is a rock of low impedance embedded in a rock of high impedance). Thus, the top reflection is negative (major peak negative) and the bottom reflection is positive (major peak positive). Let the wavelets be zero-phase wavelets, each centered at the times of the corresponding reflections. The reflection coefficients are assumed to have the same magnitude, so the wavelets have the same

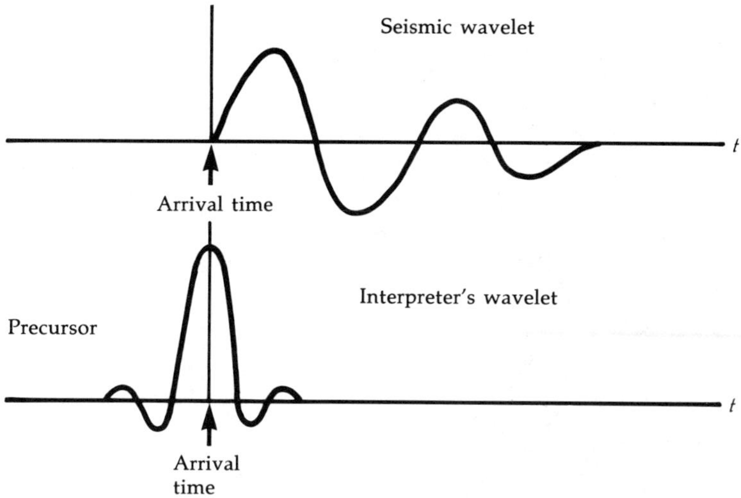

*Figure 3.28. In wavelet processing, a shaping filter converts the actual seismic wavelet (*top*) to a zero-phase interpreter wavelet (*bottom*). The actual seismic wavelet is a physical wavelet, and hence must ocillate only after its arrival time. The zero-phase wavelet is an artifact, and hence can be centered at the arrival time.*

amplitudes but, as we have said, they are of opposite sign. On each side of the main peak, the wavelet has two side peaks (i.e., side lobes) of opposite sign. The time period of the wavelet is taken to be the distance between these side peaks. Let us note the period by T.

The Widess diagram is shown in figure 3.29. The first trace is labeled $2T$, which means that the time thickness of the layer is twice the wavelet period. On examining this first trace, we see a small black circle, which denotes the trough of the top reflection, and a small white circle, which denotes the peak of the bottom reflection. As we see, this time separation is large enough to provide a complete separation of the top and bottom reflections. As we look at the traces plotted under the first trace, we see the effect of the layer becoming thinner. The two reflection wavelets come closer and closer together and, in so doing, they interfere with each other and produce a wavelet complex. Even though the two wavelets are no longer clearly distinguishable, there are attributes by which we can still recognize the complex as due to two interfering wavelets. For a layer thickness of $T/2$, for example, the wavelet complex reaches a maximum amplitude. Thus, for this thickness, we say that the two wavelets in the complex are *tuned.* If we see very large amplitudes on a seismic section, we recognize that the reflections are tuned so as to produce constructive interference between the top and bottom reflections. At this tuning thickness ($T/2$ in this example), the wavelet complex is skew-symmetric about a zero crossing, instead of symmetric about a peak or trough. For thinner layers (corresponding to the traces in the figure below the trace for $T/2$), the wavelet complex loses amplitude, and the period of the complex becomes smaller (i.e., a rise in apparent frequency). A key layer thickness is the case for $T/4$. At this thickness, the amplitude of the wavelet complex is larger than the amplitude of the top reflection alone (as seen on the $2T$ trace). Thus, the form of the wavelet complex

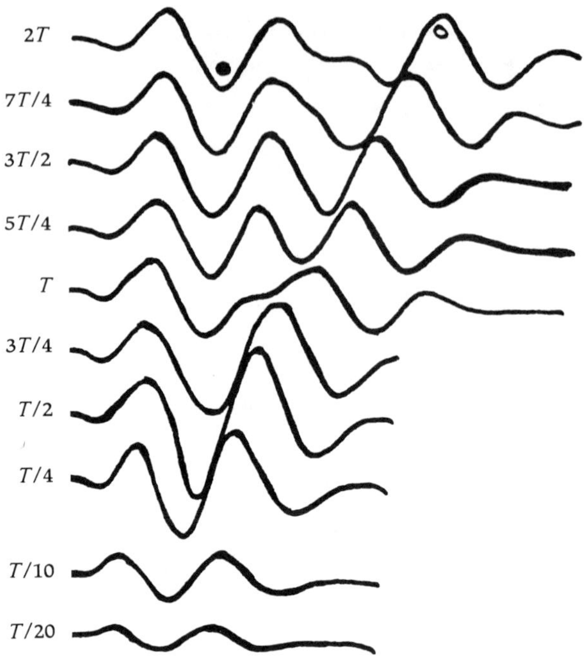

Figure 3.29. The Widess diagram for a thin soft bed that is pinching out in a hard formation. The top trace corresponds to the thick end of the wedge and the bottom trace to the thin edge of the wedge. The period T *is the peak-to-peak period of an individual wavelet. In the top trace, the two wavelets are separated by* 2T*—that is, the distance between the black and white small circles. In the bottom trace, the separation is only* T/20. *Reprinted by permission of IHRDC from Anstey 1980.*

for the $T/4$ case is characterized by a higher frequency than the individual wavelets for the top and bottom reflections, one more half-cycle (four instead of three), and the skew symmetric shape. This form is said to be differentiated. In the case of a soft thin bed to which this Widess diagram applies, the skew symmetry is about an upward-going zero crossing. A soft thin bed is interesting, as it is a likely trap for gas. In the case of a hard thin bed, the skew symmetry is about a downward-going zero crossing. Thus, if we see a bright spot (i.e., an event with high amplitudes on the seismic section), we must check the largest amplitude on each trace. If these largest amplitudes are trough-to-peak, we may have a porous gas sand; if they are peak-to-trough, we might have a tight sand or hard limestone. The last three wavelet complexes on the Widess diagram show virtually no change in wavelet shape, so the amplitude is the only clue to the thickness of the layer. The next to last complex (the one for $T/10$) represents a trough-to-peak amplitude of only about two-thirds of the component individual reflections. The last complex (the one for $T/20$) has only about 40 percent of the individual wavelet amplitude. In the case of a highly porous gas-saturated sand at a shallow depth, this 40 percent amplitude may still be so large that it represents a strong reflection. Thus, at shallow depths it is possible to detect beds that are quite thin (the $T/20$ case). At great depths, however, the thickness necessary for the layer to be detected may be ten times greater (the $T/2$ case), because the reflection amplitudes are smaller, the velocities larger, and the bandwidths smaller. If we actually have a pinch-out, or wedge, like the one illustrated in the Widess diagram, the thick part gives the separated reflections and we can obtain a reference amplitude. From this reference amplitude, we can then calibrate the amplitudes we observe and thus determine the thicknesses as we work our way to the thin end of the wedge. As the wedge becomes thinner, we see on the Widess diagram that a strong reflection alignment of spurious dip appears. This spurious alignment is especially clear on the traces labeled $3T/4$, $T/2$, and $T/4$. We must be careful of the false dips that such spurious alignments give.

In conclusion, the Widess diagram gives us a picture of how to interpret the reflection events from thin beds. The recognition of tuning on the seismic section can yield the thickness of the bed. On the thinning side of the tune, we hope to see the amplitude of the wavelet complex decrease and the frequency increase. On the thickening side of the tune, we hope to see the two separate reflection wavelets emerge as separate entities. The Widess diagram discussed here has the same rock type on each side of the thin bed, so that the two reflections have the same amplitudes but opposite polarity. In many geologic situations, such a model is adequate. If the material above the thin layer is different from the material below, however, then a Widess diagram would have to be drawn for reflection wavelets of the same shape but different amplitudes. Such Widess diagrams may be constructed quickly on the computer (see figs. 3.30, 3.31, and 3.32).

Exercises

1. Find the z-transforms of the causal functions, with values for $k \geq 0$ given by:
   (a) unit spike $\delta(k)$;
   (b) unit-step 1;
   (c) exponential c^k;
   (d) ramp k;
   (e) cosine wave $a^k \cos bkT$;
   (f) sine wave $a^k \sin bkT$.
2. What are the z-transforms of the finite-length wavelet (1, 0, 1/4, 0, 1/16) and the infinite-length wavelet (1, 0, 1/4, 0, 1/16, 0, 1/64, 0, . . .)?
3. Find the z-transforms of the following infinite-length wavelets:

   $$(1, c, c^2, c^3, c^4, \ldots), \quad \text{where } c = 0.5;$$

   $$(1, 2c, 3c^2, 4c^3, \ldots), \quad \text{where } c = 0.5.$$

   Repeat with $c = -0.5$. Plot the wavelets.
4. Let A be the polynomial $1 + z$. Its coefficients are (1, 1). Arrange the coefficients of A, A^2, A^3, A^4, A^5, and A^6. The result is Pascal's triangle, whose entries are the binomial coefficients.
5. Let A be the polynomial with coefficients (1, 1, 1, 1, 1). What are the coefficients of A^2, A^3, and A^4?

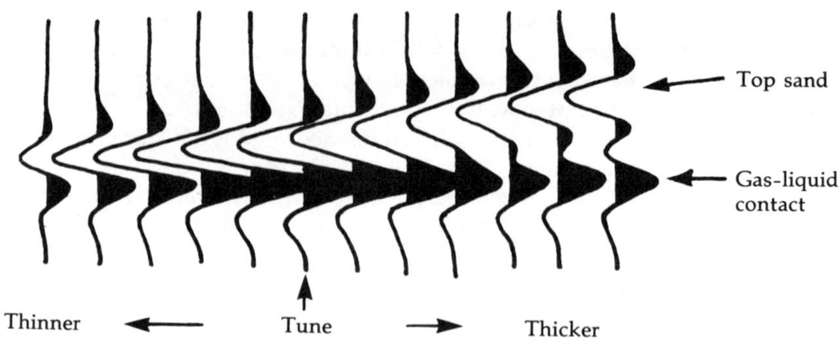

Figure 3.30. The Widess diagram used to explore the details of the reservoir (in this case, the thickness on each side of the tuning thickness). Reprinted by permission of IHRDC from Anstey 1982.

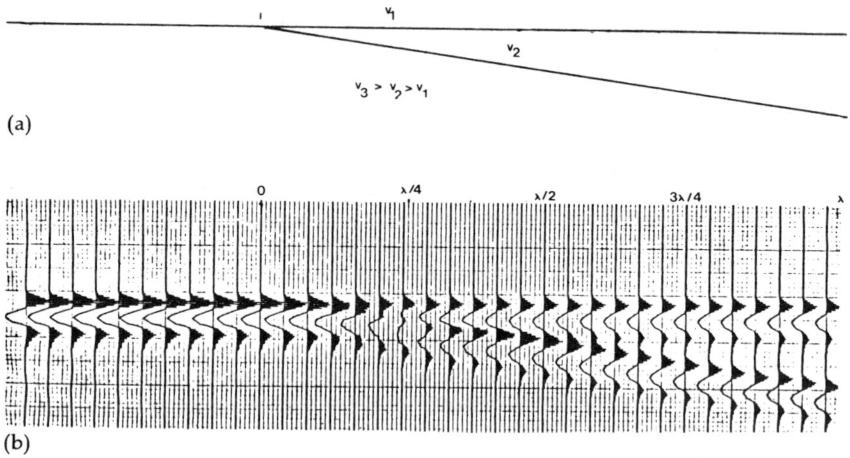

Figure 3.31. Reflection from a wedge. The thickness of the wedge is indicated as a fraction of the dominant wavelength: (a) model; (b) resulting seismic section. Reprinted by permission of IHRDC from Sheriff 1980.

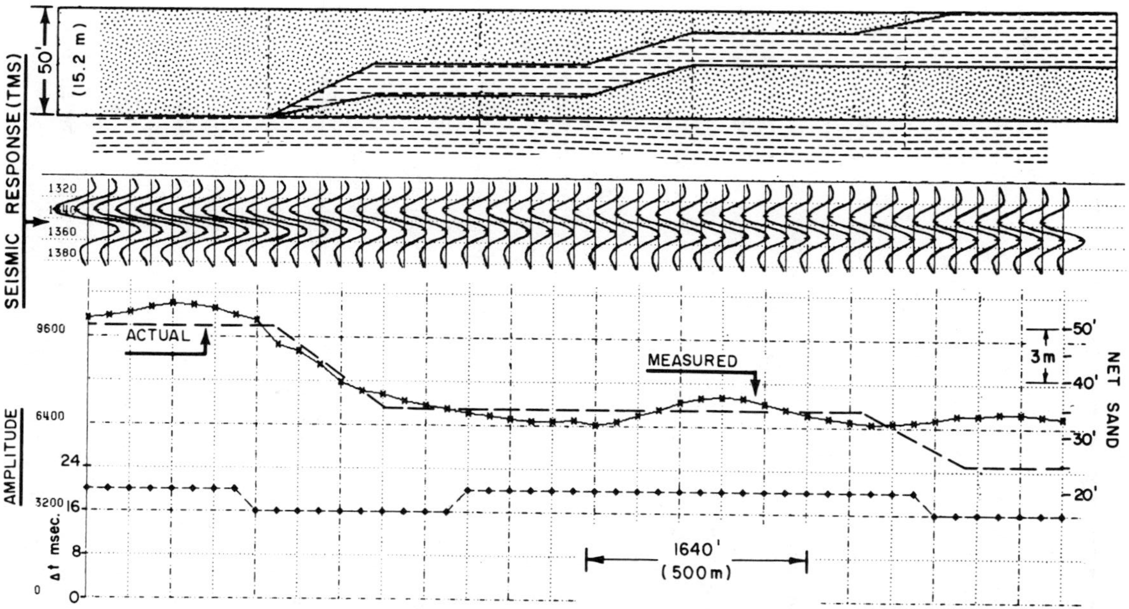

Figure 3.32. A remarkable illustration showing that the amplitude of the seismic reflection obtained from a thin sandstone containing a variable thickness of shale is closely proportional to the net thickness of sand. Reprinted by permission of AAPG from Schramm et al. 1977.

6. Consider two seismic wavelets of the same energy propagating in two materials of different acoustic impedance, such as shale (low impedance) and limestone (high impedance). Can we conclude that the particle velocity will be large in the shale and small in the limestone?
7. Let a single wave propagating in the shale be incident on the interface between the shale and limestone. If all the energy in that packet were to be transmitted across the interface, then it would follow that just on the shale side of the interface the particle velocity is large, while just on the limestone side the particle velocity is small. Why is such total transmission impossible if we insist that the materials should stay in contact across the interface and that no void is to appear between them?
8. In seismic work, do we measure energy or do we measure some other quantity?
9. What is the ratio of the reflected amplitude A_r to the incident amplitude A_i in the case of a pressure wave in material of impedance Z_1 incident on a material of impedance Z_2?
10. What is a free surface and what is a fixed surface?
11. What happens, physically, to cause a compression to be reflected as a rarefaction?
12. Draw a parallel between free and fixed seismic surfaces and the behavior of a rope suspended from the hand.
13. When we work on land with a geophone at the free surface, why is it essential that we use geophones that are sensitive to motion, not to pressure?
14. The proportion of signal amplitude that is transmitted is defined by the transmission coefficient. If the pressure amplitude of the transmitted wave is A_t—and the other symbols are as before—what is the transmission coefficient?
15. By simple algebraic manipulation, show that the pressure transmission coefficient is equal to one *plus* the pressure reflection coefficient. The plus sign means that, if the reflection coefficient is positive, the transmitted signal is larger than the incident signal. How is this possible?
16. If, instead of measuring pressure amplitudes (as we would with a marine hydrophone), we choose to measure particle velocity amplitudes (as with a land geophone), then the reflection and transmission expressions change. Repeat exercises 9, 14, and 15 using particle velocity amplitude in place of pressure amplitude.
17. If we assume a constant source, what happens to the amplitude of the check-shot signal as the borehole geophone passes from a formation of low acoustic impedance to one of higher acoustic impedance?
18. In reflection work, the primary reflection raypaths to any stated reflector always pass twice through every shallower interface. The two-way transmission coefficient for an interface is defined as the product of the downgoing transmission coefficient and the upgoing transmission coefficient. Show that the two-way transmission coefficient T is the same for either pressure amplitudes or particle velocity amplitudes.
19. What is impedance matching?
20. Suppose that the price of a good is proportional to the quantity, so that if supplier 1 has a quantity q, his price is Z_1q, and if supplier 2 has a quantity

q, his price is $Z_2 q$. Suppose that supplier 1 offers supplier 2 a quantity q of which q_r is rejected and q_t is accepted, so that $q = q_r + q_t$. Show that the value before the transaction $Z_1 q^2$ is the sum of the values $Z_1 q_r{}^2$ plus $Z_2 q_t{}^2$ after the transaction. Also show that $Z_1 q = -Z_1 q_r + Z_2 q_t$ (interpret). Show that the rejection coefficient is

$$\frac{q_r}{q} = \frac{Z_2 - Z_1}{Z_2 + Z_1}.$$

21. A vertical iron bar is struck at both ends simultaneously with equal blows from two hammers. Equal compressional disturbances propagate toward one another along the bar. At the instant they cross, the resultant disturbance is the sum of two equal disturbances (see fig. 3.33). Suppose, however, that now the top blow is the same as before but the bottom blow is modified so that the disturbance generated is a rarefaction instead of a compression. This negative disturbance travels upward and meets the downgoing positive disturbance. At the instant of meeting, the two disturbances cancel. At this instant, what happens to the energy in the bar?
22. Let three geophones be oriented such that one geophone records the vertical component of a seismic wave and the second and third geophones record only the horizontal components, the second in the direction of the shot and the third at right angles to this. What are the responses of the geophones to (a) the direct compressional wave traveling along the surface from shot to geophones, (b) a compressional wave reflected from a deep horizon, and (c) a shear wave converted by reflection at a deep interface.
23. Show that normal-incident particle velocity reflection and transmission coefficients c and t satisfy $t = 1 + c$ and that those for pressure satisfy $t_p = 1 + c_p$.
24. Let layer 1 be above layer 2. If c and $t = 1 + c$ are the particle velocity reflection and transmission coefficients for a normally incident wave in layer 1 striking from above the interface with layer 2, show that the corresponding coefficients C and T for a normal-incident wave in layer 2 striking from below the interface with layer 1 are given by $C = -c$ and $T = 1 - c$. Also show that in the case of pressure coefficients, the corresponding results are $C_p = -c_p$ and $T_p = 1 - c_p$. Thus, show that for the eight coefficients $c, t, C, T, c_p, t_p, C_p, T_p$, there is only one independent coefficient, which we may take to be c. Show that $-1 \leq c \leq 1$, and find corresponding inequalities for the other coefficients.
25. If the acoustic impedance across an interface doubles, so that $Z_2 = 2Z_1$, show that $c = -1/3$, $t = 2/3$, $c_p = 1/3$, and $t_p = 4/3$. A geophone measures particle velocity. Because $c = -1/3$, the reflected wave recorded by a geophone would have a smaller amplitude than the incident wave but would have opposite polarity. Explain physically what is meant by this change in polarity. A hydrophone measures pressure. Because $c_p = 1/3$, the reflected wave recorded by a hydrophone has the same smaller amplitude, but now the same polarity. Explain physically why the polarity is the same. Explain why the transmitted wave always has the same polarity as the incident wave but why, in this case, the amplitude of the trans-

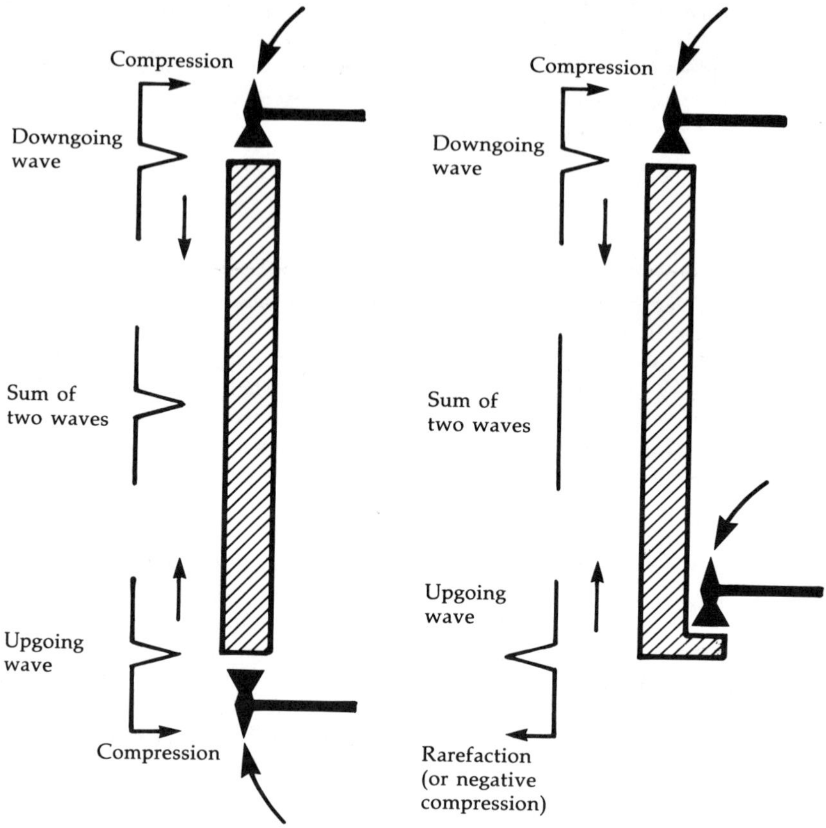

*Figure 3.33. (*left*) Upgoing and downgoing compressional waves are generated. (*right*) A downgoing compressional wave and an upgoing rarefaction wave are generated.*

mitted wave is smaller in particle velocity but larger in pressure than the incident wave.

26. If an upcoming wave in water strikes the surface, find the reflection and transmission coefficients for both pressure and particle velocity. How much energy is transmitted across the surface? (Assume that the impedance of air is zero.)
27. An impulsive shot is set off below the water surface, generating an upgoing and a downgoing wave. The downgoing wave is called the primary pulse. The upgoing wave is reflected at the surface, thus creating another downgoing pulse, called the ghost. Show that the primary and the ghost are always 180° out of phase, either for pressure or for particle velocity.

4. *Correlation and convolution*

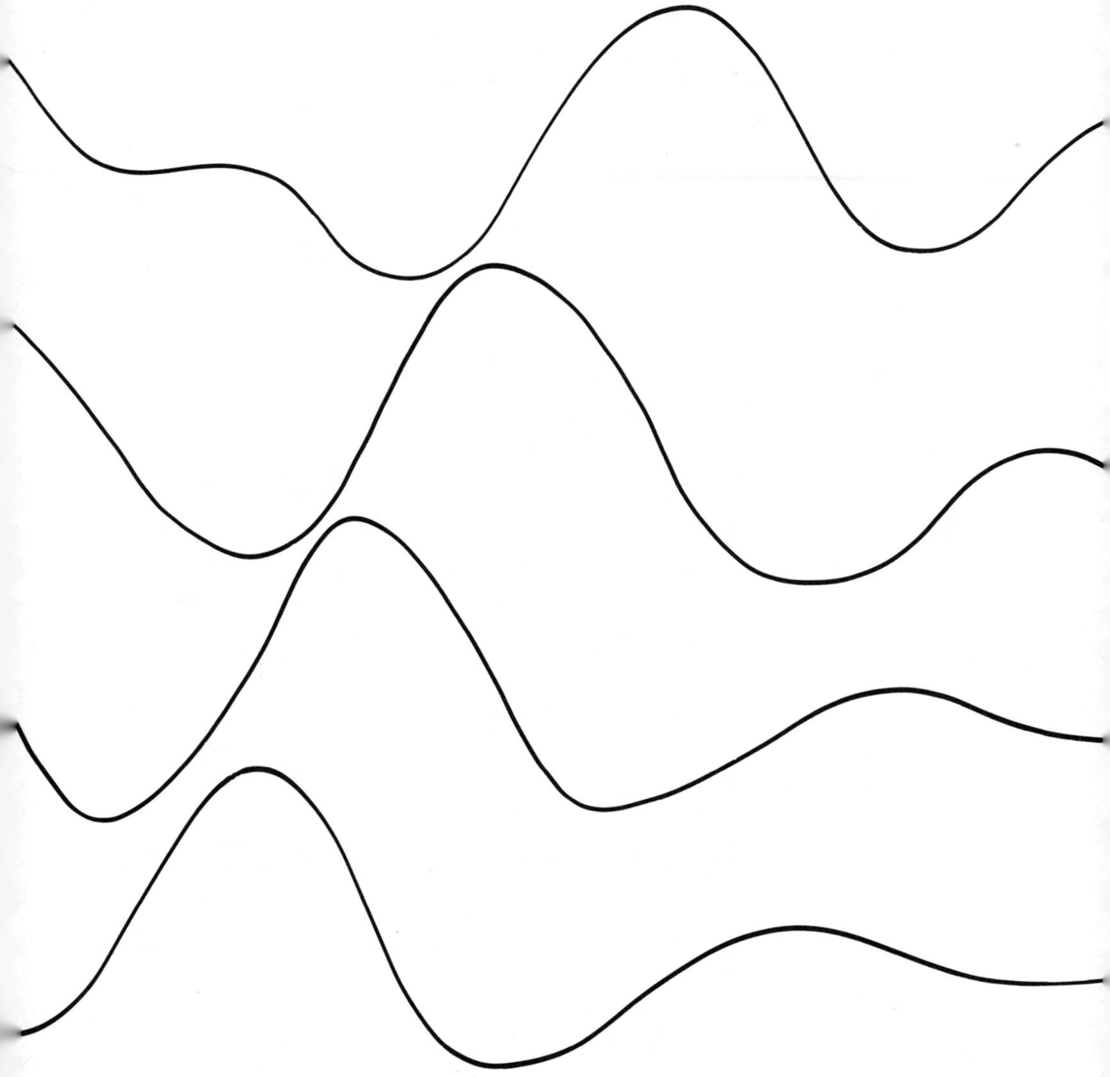

Correlation

The concept of correlation is basic to seismology. Anything so fundamental is bound to have different interpretations and uses, but all of them have the same general meaning. If we identify a waveform on one seismic trace as being similar to a waveform on another trace, we say that the two waveforms *correlate.* In a mathematical sense, correlation measures the degree of linear relationship between a pair of waveforms. Correlation is a measure of how much two waveforms look alike, or, alternatively, a measure of the extent to which one waveform can be considered a linear function of the other.

If we take the wiggly line that represents the waveform, such as a wavelet, we can convert it to a list of numbers. These numbers are the digitized values of the waveform at equal intervals of time (see fig. 4.1). Each number represents the amplitude of the waveform at a given instant of discrete time. Let us consider two different waveforms: $A = (2, -2, 1)$ and $B = (2, 1, -1)$ (see fig. 4.2). We put each list of numbers in a horizontal row. We hold the row for B in place, but we consider different positions for A by shifting it various amounts. Let us consider the case when neither is shifted:

$$A\colon (2,\ -2,\ 1),$$

$$B\colon (2,\ 1,\ -1).$$

The first number of the A row (2) is multiplied by the corresponding number of the B row (2), then the next in A (-2) by the next in B (1), and again the next (1) by the next (-1). We thus obtain a row of products: $2(2)$, $(-2)(1)$, $(1)(-1)$. We add the products together and obtain a single number: $4 - 2 - 1 = 1$. This number is a measure of the similarity of the two waveforms when they are not shifted relative to each other. We call this number the cross-correlation value for time shift 0.

At this point, we should define what we mean by a time advance. If a meeting was to be held at 4 o'clock but we advanced the meeting to 3 o'clock, it means that we shifted the meeting against the direction of time. By convention, time runs to the right, so a time advance is represented by a shift to the left. We should also define a time delay. If we delay a 4 o'clock meeting to 5 o'clock, we have moved it to the right on the time axis by one hour. A delay, of course, is a negative time advance.

Let us advance waveform A by minus one time unit; that is, we shift it to the right by one notch. We have

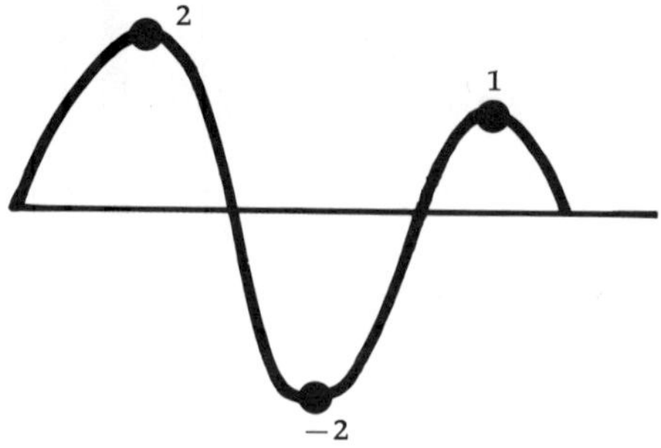

Digitize → (2, −2, 1) Wavelet A
$k = 0$

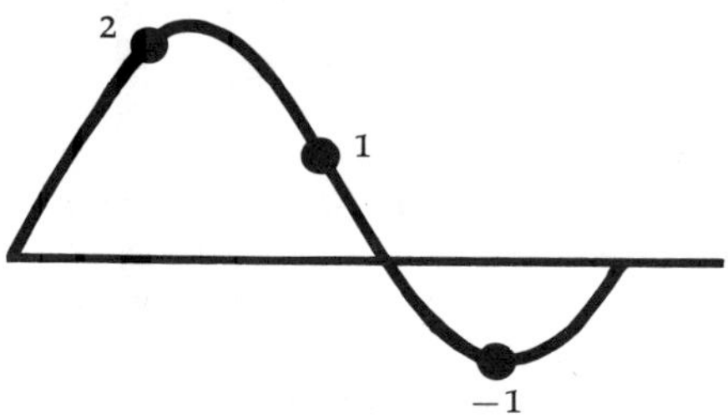

Digitize → (2, 1, −1) Wavelet B
$k = 0$

Figure 4.1. The digitization of two wavelets. The leading coefficient of each wavelet is at time 0.

Wavelet A advanced −3:	0	0	0	0	0	0	2	−2	1	0
Wavelet A advanced −2:	0	0	0	0	0	2	−2	1	0	0
Wavelet A advanced −1:	0	0	0	0	2	−2	1	0	0	0
Wavelet A in place:	0	0	0	2	−2	1	0	0	0	0
Wavelet A advanced 1:	0	0	2	−2	1	0	0	0	0	0
Wavelet A advanced 2:	0	2	−2	1	0	0	0	0	0	0
Wavelet A advanced 3:	2	−2	1	0	0	0	0	0	0	0
Wavelet B in place:	0	0	0	2	1	−1	0	0	0	0
Time index n:	−3	−2	−1	0	1	2	3	4	5	6

Figure 4.2. Wavelet B *is held in place, while Wavelet* A *is shown in positions from an advance of −3 to an advance of +3. An advance is defined as a shift against the direction of time. Since time runs to the right, a positive advance represents a shift to the left, whereas a negative advance represents a shift to the right.*

A (advanced -1 time unit): (2, -2, 1),

B: (2, 1, -1).

The sum of cross products is $(2)(1) + (-2)(-1) = 4$. Since the product of two positive numbers is positive, and the product of two negative numbers is also positive, we see that both matching peaks and matching troughs yield positive numbers that increase the value of the sum. The sum is the cross correlation for the time shift (advance) of -1, and in this case is a large value (4).

Let us now advance waveform A by plus one time unit; that is, we shift it to the left by one notch. We have

A (advanced by 1 unit): (2, -2, 1)

B: (2, 1, -1).

The cross correlation for the time shift (advance) of 1 is $(-2)(2) + (1)(1) = -3$. The lining up of the trough (-2) on A with the peak (2) on B produces the negative number -4, which decreases the sum, and thus the waveforms in this position are negatively correlated.

In summary, a positive match (peak-to-peak and trough-to-trough) produces a positive cross-correlation value, and a negative match (peak-to-trough and trough-to-peak) produces a negative cross-correlation value.

Let us now shift waveform A one more notch; that is, we shift waveform A a total of two time units to the left. We have

A (advanced by 2 units): (2, -2, 1),

B: (2, 1, -1).

We now repeat the process of multiplying and adding. We multiply the first number on the top row by the first number on the bottom row. We next multiply the second numbers together, then the third, and so on. Again we add all the products. The sum is the cross-correlation value for a time shift of 2. In this case, there is only one nonzero product—namely, (1)(2)—so the cross-correlation value for this time shift is 2.

We now shift the top row one unit farther to the left and calculate the products and their sum. This sum is the cross-correlation value for a time shift of three. We keep repeating the process and thus compute the cross-correlation function for all time shifts, both to the left and to the right (i.e., both positive and negative time shifts) (see fig. 4.3). We can plot the values of the cross correlation and connect the points by a curve (see fig. 4.4). The result is a wiggly line, which is a visible display of how alike the two waveforms are with respect to each position relative to each other. There will be one shift (-1) that makes the best fit, and this is the shift that has the largest correlation value. A shift either way from this one produces poorer fits and smaller correlations. Because typical seismic waveforms are made up of alternating peaks and troughs, the highest correlation will be at the time shift that best places the waveforms peak to peak and trough to trough. A shift such that peaks on one waveform line up with troughs on the other yield a negative cross-correlation value. As we keep

Relative positions	Sum of cross products (or cross correlation)
A advanced -3 and B in place	$2(0) + (-2)(0) + (1)(0) = 0$
A advanced -2 and B in place	$2(-1) + (-2)(0) + (1)(0) = -2$
A advanced -1 and B in place	$2(1) + (-2)(-1) + (1)(0) = 4$
A in place and B in place	$2(2) + (-2)(1) + (1)(-1) = 1$
A advanced 1 and B in place	$2(0) + (-2)(2) + (1)(1) = -3$
A advanced 2 and B in place	$2(0) + (-2)(0) + (1)(2) = 2$
A advanced 3 and B in place	$2(0) + (-2)(0) + (1)(0) = 0$

Figure 4.3. Computation of the cross-correlation value for each time shift.

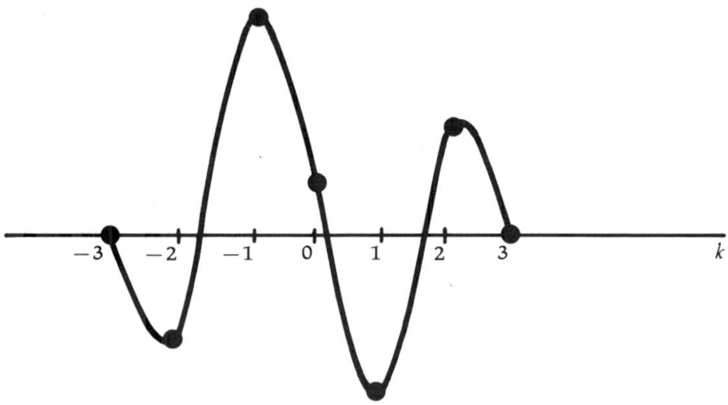

Figure 4.4. The cross correlation of A *and* B *is the function made up of the sum of cross products of wavelet* A *advanced and wavelet* B *in place.*

shifting, we will obtain a positive cross-correlation value as we come into a peak-to-peak position again, but this cross-correlation value will be smaller than the case of best fit. We thus see that the plot of cross correlation versus time shift will be a curve of peaks and troughs, with some maximum value corresponding to the best-fit position of the waveforms.

To summarize, the cross-correlation function is a measure of the similarity between two waveforms. Corresponding values of the two waveforms are multiplied together, and the products are summed to give the value of the cross correlation. Wherever the waveforms are nearly the same, the products will usually be positive, and hence the cross correlation is large; wherever the waveforms are unlike, some of the products will be positive and some negative, and hence the sum will be small. If the cross correlation should have a large negative value, it means that the two waveforms would be similar if one were inverted (that is, they are similar except that they are out of phase). The two waveforms might be dissimilar when lined up in one fashion and yet similar when one waveform is shifted with respect to the other; thus, the cross correlation is a function of the relative shift between the waveforms. By convention, we call a shift positive if it involves moving the first waveform to the left with respect to the second waveform. Thus, a positive shift represents a time advance of the first waveform with respect to the second waveform. A negative shift represents a time delay (i.e., a negative time advance) of the first waveform with respect to the second. We express the cross correlation of two waveforms $x(k)$ and $y(k)$ by

$$\phi_{xy}(k) = \sum_{n} x(n + k)y(n),$$

where k is the advance of x relative to y. Note that $\phi_{xy}(k)$ is a discrete time function rather than a continuous function, because x and y are discrete time functions.

Suppose that we reverse the roles of x and y. We get another cross-correlation function, which is

$$\phi_{yx}(k) = \sum_{j} y(j + k)x(j).$$

How are these two cross-correlation functions related? Let us consider the second one for a shift of $-k$; the result is

$$\phi_{yx}(-k) = \sum_{j} y(j - k)x(j).$$

If we define $n = j - k$, then $j = n + k$. Thus, this equation becomes

$$\phi_{yx}(-k) = \sum_{n} y(n)x(n + k)$$

or

$$\phi_{yx}(-k) = \sum_{n} x(n + k)y(n).$$

The expression on the right, however, is the expression for $\phi_{xy}(k)$. Thus, we have the important result

$$\phi_{xy}(k) = \phi_{yx}(-k),$$

which says that either one of the cross-correlation functions is the time reverse of the other. The following two operations give the same sum of cross products:

(1) Shift x to the left and hold y in place.
(2) Shift y to the right and hold x in place.

There are many uses for the operation of cross-correlation seismic processing. One important use is in automatic static programs. Cross correlation provides a means of determining the amount of time shift that will result in the optimum alignment of two seismic traces. If one trace has been delayed with respect to another—for example, in passing through the near-surface layers—the shift that maximizes the cross correlation is the one that produces the optimum alignment (match) of the two traces. The magnitude of the cross correlation indicates quantitatively how much improvement the shift produces. Cross correlation is a powerful method and is especially useful when the data quality is poor. Another use of cross correlation is in stacking programs. It can be used to determine the amount of normal moveout to introduce in order to align traces from different offsets before stacking. Various criteria are used to make possible the automatic determination and application of the optimum time shifts. In all cases, tests must be incorporated to ensure that the shifts so introduced are consistent. In an automatic statics program, for example, a criterion must be incorporated in the program to ensure that the same corrections are always assumed for the weathering beneath any particular location.

Convolution

Convolution is a process similar to correlation, with one important difference. When we correlate two digital waveforms, we place one above the other, shift one to the right or left, multiply vertical entries, and then add the products. The resulting sum is the correlation value for the time shift in question.

To convolve two digital waveforms, however, we first reverse the direction of one, so that its values are in the opposite order with respect to time. We then carry out the correlation process as described in the preceding section, and the resulting sum is the convolution value for the particular time shift. The curve of convolutional values for all the time shifts gives the convolved waveform of the two given waveforms.

In correlation, we keep the two waveforms in their natural order and then shift, multiply, and sum to obtain each correlation value. In convolution, we reverse one waveform, and then shift, multiply, and sum to obtain each convolution value.

It is easier for people to think about correlation, since correlation is a mental process we undertake as we try to compare similar things that we see with our own eyes. We can usually tell, to an astounding degree of accuracy, how well two things compare in shape. Two different but seemingly identical paintings, for example, can be examined by many people, who will all agree on which one is the forgery. Most natural systems involve convolution and not correlation, however, so convolution becomes the more important operation when we wish to describe how natural things are generated. Each natural system has a memory; this memory can be expressed in terms of a waveform that we call the *memory function* or, more commonly, the *impulse response function.* When a given input signal is fed into a natural system, the memory operates on the input, and the result is the corresponding output signal. The mathematical relationship is that the output is equal to the convolution of the input with the memory function. From our human experience, however, it is more natural to decide to reverse the time direction of the memory function so that it looks into the past. Then the output is the correlation of the input with this reversed memory function.

Let us now consider the natural system represented by the watershed of a lake. Suppose that for each extra inch of rain in any month (i.e., the amount of rain above the average rainfall), the water level of the lake increases by 2 in. in the same month and by another inch the following month because of runoff from nearby fields and swamps. The memory function is thus made up of two spikes; the first spike is the direct effect of 2 in. for the same month, and the second spike is the delayed effect of 1 in. for the following month. We thus represent the memory function as the wavelet (2, 1). Suppose that the rainfall for five successive months is (2, 3, 1, 4, 3). The extra height of the lake is the output of the natural watershed system. The memory function is (2, 1), and the input is the rainfall (2, 3, 1, 4, 3). We can obtain the output by convolving these two waveforms. Alternatively, the output can be found by correlating the input (2, 3, 1, 4, 3) with the reverse of the memory function (1, 2). We thus write

Input		2	3	1	4	3	
Reversed memory	1	2					
Product	0	4	0	0	0	0	Sum = 4

Input	2	3	1	4	3	
Reversed memory	1	2				
Product	2	6	0	0	0	Sum = 8

Input	2	3	1	4	3	
Reversed memory		1	2			
Product	0	3	2	0	0	Sum = 5

Similarly, the remaining sums are 9, 10, 3. Thus, the output (or convolution) is (4, 8, 5, 9, 10, 3).

Instead of this approach, we can use another approach that more nearly describes the action of the system. The extra 2 in. rainfall in the first month increases the level of the lake by 4 in. the first month and 2 in. the second month. The extra 3 in. the second month increases the lake by 6 in. the second month and by 3 in. the third month. The extra 1 in. the third month increases the lake by 2 in. the third month and 1 in. the fourth month. In the first month, the lake is increased by 4 in. In the second month, the lake is increased by 2 plus 6, or 8 in. In the third month, the lake is increased by 3 plus 2, or 5 in. We can summarize this computational process as follows:

Input (across): 2 3 1 4 3

Memory (downward):

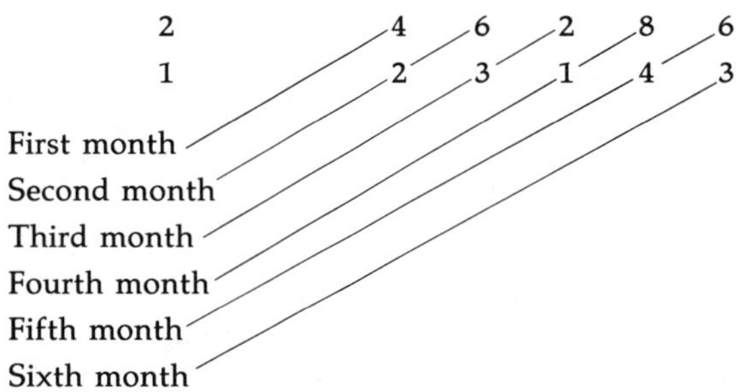

We see that the output obtained,

Output: (4, 2+6, 3+2, 1+8, 4+6, 3) = (4, 8, 5, 9, 10, 3),

is the same as that obtained before.

Every signal that is recorded in a scientific experiment is changed in some manner both by natural physical processes and by the instruments used to measure it. This modification may be thought of as a filtering process, and its effect may be described mathematically by a convolution. Since the convolution operation is basic to all time-sequence analysis, we will describe it in some detail. In the words of Sir Isaac Newton, however, "In learning the sciences examples are of more use than precepts." Interested readers can think of examples of natural and man-made processes around them and see the process of convolution at work.

The physical significance of a convolution is particularly simple in its digital form. Let us assume that we feed into a system data sampled at regular intervals —for example, a digital seismic trace. The output of the system can be computed if we know the impulse response of the system—that is, the response of the system when the input is a unit impulse. The *unit impulse* is defined as the function that has zero values everywhere except at the origin. We write the unit impulse δ_k or $\delta(t)$, depending on whether we are dealing with sampled data or continuous functions. In the discrete case, the value of the unit impulse at the origin is one. The impulse response of a system will be zero prior to time zero and then will have the values $b_0, b_1, b_2, \ldots$ at successive sampling intervals. We represent this process diagramatically as

$$\delta_k \rightarrow \boxed{\text{System}} \rightarrow (b_0, b_1, b_2, \ldots).$$

We can denote the impulse response $(b_0, b_1, b_2, \ldots)$ by b_k. Most systems with which we deal are linear and time-invariant (or very nearly so). A linear system is one in which the output is directly proportional to the input, whereas a time-invariant system is one in which the output is independent of the time when the input occurred. We use the notation δ_{k-n} for a unit impulse that occurs at $k = n$ rather than at $k = 0$. We can then illustrate linear and time-invariant systems as follows (here a_0 is an arbitrary constant). For the linear system:

$$a_0\delta_k \rightarrow \boxed{\text{System}} \rightarrow a_0 b_k = (a_0 b_0, a_0 b_1, a_0 b_2, \ldots);$$

and for the time-invariant system:

$$\delta_{k-n} \rightarrow \boxed{\text{System}} \rightarrow b_{k-n} = (0, 0, 0, \ldots, 0, b_0, b_1, b_2, \ldots).$$

In the parentheses on the right, the first output different from zero is b_0, which occurs at the instant $k = n$. If a_n is an arbitrary constant, then the foregoing two properties can be summarized as

$$a_n \delta_{k-n} \rightarrow \boxed{\text{System}} \rightarrow a_n b_{k-n} = (0, 0, 0, \ldots, 0, a_n b_0, a_n b_1, a_n b_2, \ldots).$$

We now make an important observation. Any input that consists of a series of sampled values can be represented by a series of unit impulses multiplied by appropriate amplitude factors. We can then use the foregoing two properties to find the output for each impulse, and by superimposing these we get the output for the arbitrary input. This is the process of convolution.

We shall illustrate convolution by considering the output for the watershed example. The impulse response is (2, 1), and the input is (2, 3, 1, 4, 3). We apply to the input the series of impulses $2\delta_k$, δ_{k-1}. The subscript of δ_{k-1} means that the impulse has been delayed by one sampling interval; this represents the runoff from nearby swamps and meadows in the month following the rainfall. We obtain the outputs:

$$2\delta_k \rightarrow (4, 6, 2, 8, 6),$$

$$\delta_{k-1} \rightarrow (0, 2, 3, 1, 4, 3).$$

Summing, we obtain the output

$$(2\delta_k + \delta_{k-1}) \rightarrow (4, 8, 5, 9, 10, 3).$$

This is the same output as before.

Let us now consider an arbitrary input $(a_0, a_1, a_2, \ldots)$ to the linear time-invariant system with impulse response $(b_0, b_1, b_2, \ldots)$. This input corresponds to the sequence of impulse inputs given by

$$a_0 \delta_k = (a_0, 0, 0, \ldots),$$

$$a_1 \delta_{k-1} = (0, a_1, 0, \ldots),$$

$$a_2 \delta_{k-2} = (0, 0, a_2, \ldots),$$

$$\ldots$$

The corresponding outputs are

$$a_0 b_k = (a_0 b_0, a_0 b_1, a_0 b_2, \ldots),$$

$$a_1 b_{k-1} = (0, a_1 b_0, a_1 b_1, \ldots),$$

$$a_2 b_{k-2} = (0, 0, a_2 b_0, \ldots),$$

$$\ldots$$

The entire input is the sum of the component inputs; that is,

$$a_k = \sum_{n=0}^{\infty} a_n \delta_{k-n},$$

where k takes on all integer values from zero to infinity. Likewise, the entire output is the sum of the component outputs; that is,

$$c_k = \sum_{n=0}^{\infty} a_n b_{k-n},$$

where k takes on all integer values from zero to infinity. This equation is the *convolutional formula.* It states that the output $(c_0, c_1, c_2, \ldots)$ is equal to the convolution of the input $(a_0, a_1, a_2, \ldots)$ with the impulse response $(b_0, b_1, b_2, \ldots)$.

The expressions for the individual output values can be obtained simply by adding the component outputs. The component outputs are as follows:

$$\begin{array}{l}
(a_0 b_0, \; a_0 b_1, \; a_0 b_2, \; a_0 b_3, \; \ldots), \\
(\quad 0, \; a_1 b_0, \; a_1 b_1, \; a_1 b_2, \; \ldots), \\
(\quad 0, \quad\;\; 0, \; a_2 b_0, \; a_2 b_1, \; \ldots), \\
(\quad 0, \quad\;\; 0, \quad\;\; 0, \; a_3 b_0, \; \ldots).
\end{array}$$

Summing of the individual columns gives, respectively,

$$c_0 = a_0 b_0,$$

$$c_1 = a_0 b_1 + a_1 b_0,$$

$$c_2 = a_0 b_2 + a_1 b_1 + a_2 b_0,$$

$$c_3 = a_0 b_3 + a_1 b_2 + a_2 b_1, + a_3 b_0.$$

Alternatively, let us rearrange the table of component outputs by shifting all nonzero entries left to eliminate the zeros. The resulting table is

$$\begin{array}{cccc}
a_0 b_0 & a_0 b_1 & a_0 b_2 & a_0 b_3 \\
a_1 b_0 & a_1 b_1 & a_1 b_2 & a_1 b_3 \\
a_2 b_0 & a_2 b_1 & a_2 b_2 & a_2 b_3 \\
a_3 b_0 & a_3 b_1 & a_3 b_2 & a_3 b_3
\end{array}$$

Now, to obtain $c_0, c_1, c_2, \ldots$, we must add along the dashed diagonal lines, as shown in the table.

If the signals a and b were two-sided, then we must let the running index n appearing in the convolutional formula run over negative values as well as positive values; that is, the convolutional formula becomes

$$c_k = \sum_{n=-\infty}^{\infty} a_n b_{k-n} \quad (\text{for } k = \ldots, -3, -2, -1, 0, 1, 2, 3, \ldots).$$

The *convolution* of the two wavelets

$$a = (a_0, a_1, a_2, \ldots)$$

and

$$b = (b_0, b_1, b_2, \ldots)$$

is defined to be

$$c = a * b = (c_0, c_1, c_2, \ldots),$$

where the coefficients of c are given by the formula

$$c_k = \sum_{n=0}^{k} a_n b_{k-n}.$$

The important point here is that the running index n ranges only over a finite number of integers (i.e., $n = 0, 1, 2, \ldots, k$). The asterisk used as a binary operation between two time functions denotes convolution. We introduced the asterisk as a symbol for convolution in Chapter 3, and this usage is well accepted and very helpful.

To illustrate the use of this formula, suppose that

$$a = (a_0, a_1) = (2, 1)$$

and

$$b = (b_0, b_1) = (3, 4).$$

Each of these wavelets has two coefficients, or has length 2. Using the convolution formula, we see that the convolution of a and b is obtained by

$$c_0 = a_0 b_0 = 2 \cdot 3 = 6,$$

$$c_1 = a_0 b_1 + a_1 b_0 = 2 \cdot 4 + 1 \cdot 3 = 11,$$

$$c_2 = a_1 b_1 = 1 \cdot 4 = 4.$$

Hence, the convolution is

$$c = a * b = (6, 11, 4).$$

From the nature of the convolution process, we may deduce this useful fact: the duration of the output wavelet is always equal to the sum of the durations of the input wavelet and the impulse-response wavelet, minus one. In this example, both a and b had length 2, so the output c has length $2 + 2 - 1 = 3$.

Convolution can be performed as a geometric operation of folding (see fig. 4.5). The two signals to be convolved are written along the vertical and horizontal axes of a graph, and the products of the individual terms formed in the

	b_0	b_1	b_2	b_3	...
a_0	a_0b_0	a_0b_1	a_0b_2	a_0b_3	
a_1	a_1b_0	a_1b_1	a_1b_2	a_1b_3	
a_2	a_2b_0	a_2b_1	a_2b_2	a_2b_3	
a_3	a_3b_0	a_3b_1	a_3b_2	a_3b_3	
...					

$$c_0 = a_0b_0$$

$$c_1 = a_0b_1 + a_1b_0$$

$$c_3 = a_0b_2 + a_1b_1 + a_2b_0$$

$$\ldots$$

Figure 4.5. Convolution as folding. Place the first signal down the vertical margin, and place the second signal across the horizontal margin. The respective cross products become the entries in the table. The convolution is obtained by taking the sum of the entries in each northeast-southwest diagonal.

interior. The convolution is obtained by summing the products lying along diagonal lines, as illustrated below:

	3	4
2	6	8
1	3	4

$$c_0 = 6,$$

$$c_1 = 3 + 8 = 11,$$

$$c_2 = 4.$$

Convolution can also be performed as the geometric operation of sliding (see fig. 4.6). One wavelet is held fixed, and the other is reversed. In each relative position, the adjacent terms are multiplied and the resulting products summed, as illustrated here:

```
  (2  1)
(4  3)
```

$$c_0 = 2 \cdot 3 = 6,$$

```
  (2  1)
  (4  3)
```

$$c_1 = 2 \cdot 4 + 1 \cdot 3 = 11,$$

```
  (2  1)
     (4  3)
```

$$c_2 = 1 \cdot 4 = 4.$$

Convolution may be performed by *multiplication of polynomials.* Thus, we write the polynomials:

$$A(z) = a_0 + a_1 z,$$

$$B(z) = b_0 + b_1 z,$$

or

$$(a_0, a_1, a_2, \ldots) \qquad\qquad c_0 = a_0 b_0$$
$$(\ldots, b_2, b_1, b_0)$$

$$(a_0, a_1, a_2, \ldots) \qquad\qquad c_1 = a_0 b_1 + a_1 b_0$$
$$(\ldots, b_2, b_1, b_0)$$

$$(a_0, a_1, a_2, \ldots) \qquad\qquad c_2 = a_0 b_2 + a_1 b_1 + a_2 b_0$$
$$(\ldots, b_2, b_1, b_0)$$

Figure 4.6. Convolution as sliding. Write down one signal and hold it in place. Reverse the other signal and slide it to the right. In each position, the sum of the cross products gives the convolution.

$$A(z) = 2 + z,$$

$$B(z) = 3 + 4z.$$

These polynomials are the z-transforms of the respective wavelets. Multiplying these polynomials, we obtain

$$C(z) = (a_0 + a_1 z)(b_0 + b_1 z) = (2 + z)(3 + 4z).$$

The resulting polynomial

$$C(z) = a_0 b_0 + (a_0 b_1 + a_1 b_0)z + a_1 b_1 z^2 = 6 + 11z + 4z^2$$

has coefficients that are equal to the convolution

$$c_0 = a_0 b_0 = 6,$$

$$c_1 = a_0 b_1 + a_1 b_0 = 11,$$

$$c_2 = a_1 b_1 = 4.$$

Thus, multiplication of polynomials corresponds to convolution of their coefficients.

One implication is that convolutions can be taken in any order (that is, that convolution is *commutative*), since polynomial products can be taken in any order. For example,

$$a * b = b * a,$$

since

$$A(z)B(z) = B(z)A(z).$$

By the same reasoning, we see that convolution is *associative;* that is,

$$(a * b) * c = a * (b * c).$$

Convolution is also *distributive* with respect to addition; that is,

$$a * (b + c) = (a * b) + (a * c).$$

To cross correlate two waveforms, we simply convolve one with the reverse of the other. We must be careful which one we reverse, however, as the cross correlation is not symmetric. The cross correlation of a with b is

$$\phi_{ab}(k) = \sum_n a_{n+k} b_n,$$

which is the same as the convolution of a with the reverse of b. The cross correlation of b with a is

$$\phi_{ba}(k) = \sum_n b_{n+k} a_n,$$

which is the same as the convolution of b with the reverse of a. Suppose that $a = (a_0, a_1)$ and $b = (b_0, b_1)$. Then $\phi_{ab}(k)$ is found by convolving (a_0, a_1) with (b_1, b_0). The folding table is

	b_1	b_0
a_0	$a_0 b_1$	$a_0 b_0$
a_1	$a_1 b_1$	$a_1 b_0$

so that

$$\phi_{ab}(-1) = a_0 b_1,$$

$$\phi_{ab}(0) = a_0 b_0 + a_1 b_1,$$

$$\phi_{ab}(1) = a_1 b_0.$$

Similarly, $\phi_{ba}(k)$ is the convolution of (b_0, b_1) with (a_1, a_0). We have the table

	a_1	a_0
b_0	$b_0 a_1$	$b_0 a_0$
b_1	$b_1 a_1$	$b_1 a_0$

so that

$$\phi_{ba}(-1) = b_0 a_1,$$

$$\phi_{ba}(0) = b_0 a_0 + b_1 a_1,$$

$$\phi_{ba}(1) = b_1 a_0.$$

As a check, we see that one cross correlation is the time reverse of the other:

$$\phi_{ab}(k) = \phi_{ba}(-k).$$

Convolution, cross correlation, and summing (i.e., stacking) are the three basic processes that constitute practically the entire range of digital signal processing in the time domain. Summing (or stacking) of traces is a simple, well-known process of adding corresponding samples from two or more time series. Cross correlation and convolution both follow the same operational procedure of cross-multiplying corresponding samples of two shifted time series and then summing the products to obtain a single output point. One time series is next shifted in time by one sample increment with reference to the other. Then the process is repeated to obtain the second output point, and so on for all points. The only operational difference between cross correlation and convolution is that in convolution, one time series is reversed end for end before any arithmetic is done. Because convolution corresponds to polynomial multiplication, convolution has much in common with the mathematical process of multiplication, including the commutative property. Thus, the order in which any series of convolution operations is done may be changed without affecting the final result. This property of commutation is not true for cross correlation, however, with one exception. The exception, of course, is the autocorrelation of a (real-valued) signal. In the next section we will see that the autocorrelation is a special case of cross correlation in which a signal is cross correlated with itself. The autocorrelation function is necessarily symmetric, whereas the cross-correlation function generally is not symmetric (see figs. 4.7 and 4.8).

Autocorrelation

Autocorrelation is a special case of cross correlation. Instead of correlating two different waveforms, autocorrelation involves correlating a waveform with itself; that is, the two rows used in the correlation process are made up of the same digital waveform. The same procedure is used. For each time shift, the corresponding numbers in each row are multiplied and the products are added to produce the autocorrelation value for that time shift. When there is no time shift, there is perfect match, so the highest autocorrelation occurs at zero time shift. As we shift to the right, or to the left, the autocorrelation values fall off but then can build up again, so that we can get a wiggly curve for the autocorrelation function. Since a shift in either direction by the same amount produces the same set of products, we see that the autocorrelation curve is symmetric; that is, the value of the autocorrelation for time shift k is the same as the autocorrelation for time shift $-k$. Because there is no need to show the two identical halves of the autocorrelation curve, we often display only the right-hand side, which is called the half-autocorrelation. A signal often has some repeating feature; for example, a seismic trace may have a multiple reflection that repeats itself every 100 msec. Then, when the trace is shifted 100 msec with respect to itself, the two will have a good fit, and so the autocorrelation will have a subsidiary peak at 100 msec.

The special case in which a time series is being correlated with itself is called autocorrelation. In this case, the cross-correlation equation becomes the equation for autocorrelation:

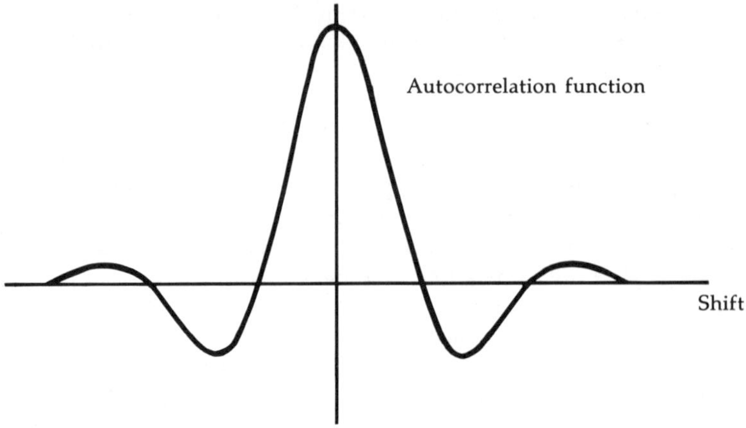

Figure 4.7. The autocorrelation is a symmetric zero-phase function. It is the cross correlation of a waveform with itself.

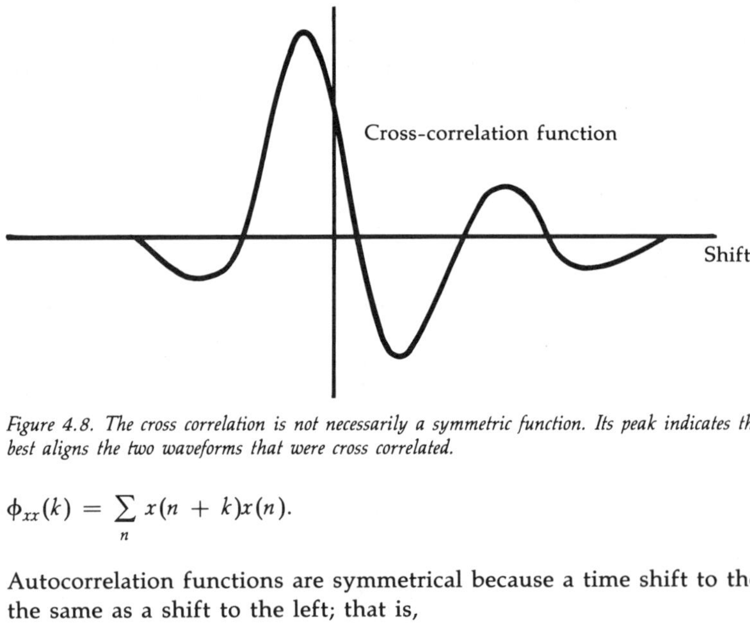

Figure 4.8. The cross correlation is not necessarily a symmetric function. Its peak indicates the shift that best aligns the two waveforms that were cross correlated.

$$\phi_{xx}(k) = \sum_n x(n + k)x(n).$$

Autocorrelation functions are symmetrical because a time shift to the right is the same as a shift to the left; that is,

$$\phi_{xx}(k) = \phi_{xx}(-k).$$

The autocorrelation has its peak value at zero time shift; that is, a time series is most like itself without any time shift. If the autocorrelation should also have a large value at some non-zero time shift k_0, it indicates that the time series tends to be periodic with the period k_0. Thus the autocorrelation function may be thought of as a measure of the repetitiveness of a time series.

Using the above formula, the autocorrelation of wavelet $(b_0, b_1, b_2, \ldots)$ is

$$r_k = \sum_{n=0}^{\infty} b_{n+k} b_n \qquad \text{(for all integers } k\text{)}.$$

For a finite-length wavelet $(b_0, b_1, \ldots, b_n)$, this equation becomes

$$r_k = \begin{cases} \sum_{n=0}^{N-k} b_{n+k} b_n & \text{for } k = 0, 1, 2, \ldots, N-1, N, \\ r_{-k} & \text{for } k = -N, -N+1, \ldots, -2, -1, \\ 0 & \text{for } k < -N \text{ and } k > N. \end{cases}$$

Let us now define the reverse wavelet. If $(b_0, b_1, \ldots, b_N)$ is a finite wavelet (with time origin at b_0), then $(b_N, b_{N-1}, \ldots, b_1, b_0)$ is the reverse wavelet (where the time origin is also at b_0, so b_N now occurs at time index $-N$). For example, (4, -3, 2) is the reverse of the wavelet (2, -3, 4). Another example is (1, 2, 5) which is the reverse wavelet of (5, 2, 1). Also we may say that the wavelets (4, -3, 2) and (2, -3, 4) are the reverses of each other. So also are (2.5, -1) and (-1, 2.5).

To autocorrelate a finite-length wavelet, we merely convolve the wavelet with its reverse. As an example, let us find the autocorrelation of the wavelet (b_0, b_1, b_2). Its reverse wavelet is (b_2, b_1, b_0). The center term is

$$r_0 = b_0 b_0 + b_1 b_1 + b_2 b_2.$$

For each summand in r_0, the first index minus the second index is 0. We see that the autocorrelation r_0 for time index 0 is equal to the energy of the wavelet (where the energy is defined as the sum of squares of the coefficients of the wavelet). The r_k's for the negative k's are found as follows. For r_{-1} we have

$$r_{-1} = b_0 b_1 + b_1 b_2.$$

For each summand in r_{-1}, the first index minus the second index is -1. For r_{-2} we have

$$r_{-2} = b_0 b_2.$$

For each summand in r_{-2}, the first index minus the second index is -2. The r_k's for positive k's are as follows. The expression for r_1 is

$$r_1 = b_1 b_0 + b_2 b_1.$$

For each summand in r_1, the first index minus the second is 1. For r_2 we have

$$r_2 = b_2 b_0,$$

where for each summand the first index minus the second index is 2. (Note that we now us the symbol r for autocorrelation, whereas in previous chapters we used r for the receiver coordinate.)

Let us multiply each coefficient of a wavelet $(b_0, b_1, b_2, \ldots)$ by -1. The resulting wavelet is $(-b_0, -b_1, -b_2, \ldots)$. It is easy to see that the new wavelet has the same autocorrelation as the original wavelet. As a result, it is convenient to introduce the convention that two wavelets are equivalent whenever their corresponding coefficients are equal, except for a factor of -1. Thus, the two wavelets (5, 1) and (-5, -1) are equivalent, since their corresponding coefficients are equal, except for the factor of -1.

Let us now say a word about normalized correlation functions. The autocorrelation value at zero shift is called the energy of the time series; the energy is

$$\phi_{xx}(0) = \sum_{n} x^2(n).$$

This terminology is justified on the basis that a physical time series is usually a voltage, current, pressure, or particle velocity, and hence $x^2(n)$ is proportional to energy. For infinitely long time series, of course, we must normalize the correlation function. In such cases, we define the cross correlation as

$$\phi_{xy}(k) = \lim_{N \to \infty} \frac{1}{2N + 1} \sum_{k=-N}^{N} x(n + k)y(n)$$

and similarly for the autocorrelation. In such cases, $\phi_{xy}(0)$ represents power. Often the autocorrelation function is normalized by dividing by the energy:

$$\phi_{xx}(k)_{\text{norm}} = \frac{\phi_{xx}(k)}{\phi_{xx}(0)}.$$

The cross-correlation function is normalized in a similar manner by dividing by the geometric mean of the energy of the two time series:

$$\phi_{xy}(\mathrm{k})_{\text{norm}} = \frac{\phi_{xy}(k)}{[\phi_{xx}(0)\phi_{yy}(0)]^{1/2}}.$$

Normalized correlation values must lie between ± 1. A value of $+1$ indicates perfect copy. A value of -1 indicates perfect copy if one of the time series is inverted (i.e., multiplied by -1).

The concept of cross correlation and autocorrelation can be expressed in integral form applicable to continuous functions. The definitions now take the forms

$$\phi_{xy}(\tau) = \int_{-\infty}^{\infty} x(t + \tau)y(t)\,dt,$$

$$\phi_{xx}(\tau) = \int_{-\infty}^{\infty} x(t + \tau)x(t)\,dt.$$

Also, convolution takes the form of the integral

$$y(t) = \int_{-\infty}^{\infty} x(\tau)f(t - \tau)\,dt.$$

Zero-phase wavelets

The autocorrelation function is a symmetric two-sided function that has a central positive maximum. Except in the case of periodicity, the autocorrelation will damp out symmetrically in both directions from this central maximum. Moreover, it will damp out with a certain degree of rapidity. If we perform a Fourier analysis on an autocorrelation, we will find the following. The autocorrelation curve can be represented as a sum of cosine curves of different frequencies and different amplitudes. At the central point of the autocorrelation, all of these cosine curves will be in phase; that is, the crest of each cosine wave will occur at the maximum (or central) value of the autocorrelation. No negative cosine curves (i.e., those that are 180° out of phase) occur, nor do any sine curves occur. Thus, there are no phase differences; that is, the phase is zero everywhere. As a result, any autocorrelation function is a *zero-phase waveform.* Conversely, any zero-phase waveform must be in the form of an autocorrelation. In other words, a zero-phase waveform and an autocorrelation function are the same thing (see fig. 4.9).

The resolution of reflected events on a seismic trace (i.e., their distinct separation in time so that they stand out and can be visually identified as individual events) is determined by the length properties of the seismic wavelet $w(k)$ attached to each event. A quantitative measure of the effective length of a wavelet $w(k)$ is the second-order moment (or spread parameter):

$$\sigma^2 = \frac{\sum_k k^2 w^2(k)}{\sum_k w^2(k)}.$$

This moment is zero for a unit spike δ_k. The Fourier analysis of a spike yields cosine curves of all frequencies. All the cosine curves have the same amplitude and all are in phase with their crests at time zero. A spike is the sharpest of all the zero-phase wavelets. In seismic work, we cannot attain an infinitely large range of frequencies but must work in a narrower range, as governed by the frequency characteristics of the earth and our instruments. We also must take into account our ultimate objectives and the economic costs involved. Within these limitations, we find ourselves working in a relatively narrow frequency band, such as the band from 5 Hz to 50 Hz. Within such a band, what sort of wavelet will have the smallest spread? We want to retain the sharpness of the spike as much as possible, so we will retain the cosine waves in this range of frequencies. Because all of the retained cosine waves are still in phase with their crests at time zero, the resulting wavelet $w(k)$ will be symmetric and zero-phase. Because all of the retained cosine waves will have the same amplitude, the amplitude spectrum will be level (flat) and will cover the entire available frequency range. The resulting wavelet will have the smallest value of σ^2 for the given frequency band. It will have a sharp central peak at $k = 0$. It will oscillate, however, and these secondary oscillations (side lobes) are undesirable (see fig. 4.10). We can reduce these secondary oscillations (side lobes) at the expense of increasing the width of the central peak. The result will be a zero-phase wavelet

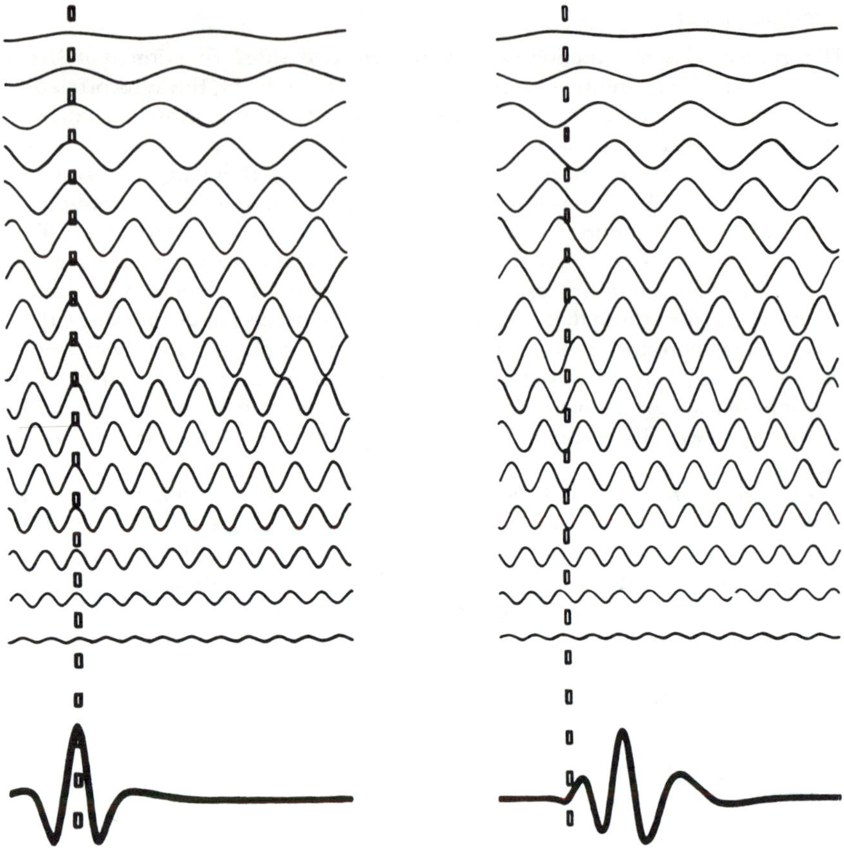

*Figure 4.9. (*left*) Fourier analysis of a zero-phase waveform. (*right*) Fourier analysis of a one-sided waveform. Both waveforms are made up of sinusoidal curves with the same amplitude spectrum. In the case of the zero-phase waveform, however, all the sinusoidal components are cosine waves with positive crests all occurring at the central time point. In the case of the one-sided waveform, all the sinusoidal waves must have a phase relationship such as to cancel out on the left-hand side. Reprinted by permission of Gebrüder Borntraeger from Anstey 1970.*

whose amplitude spectrum will no longer be absolutely flat in the available frequency range but still will be smooth and broad.

We can therefore conclude that zero-phase wavelets (i.e., autocorrelation functions) are the best choice for interpreter wavelets, especially those with smooth and broad amplitude spectra. An additional argument for broad-band zero-phase wavelets is that for a given amplitude spectrum, the zero-phase wavelet has the largest central amplitude. The reason is that a zero-phase wavelet, in its Fourier analysis, is made up of only cosine waves in phase at its central point. There are no negative cosine waves and certainly no sine waves, both of which would reduce the peak amplitude. Hence, in wavelet processing we transform the one-sided reflection wavelets (i.e., one-sided with respect to their arrival times) on the received seismic trace to their zero-phase counterparts (i.e., zero-phase with respect to their arrival times). These zero-phase interpreter

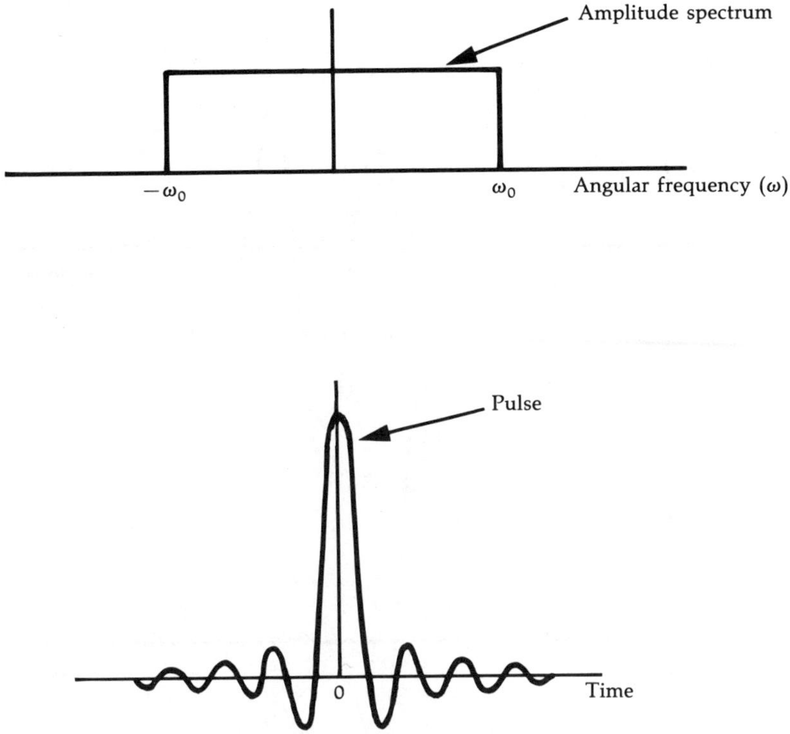

Figure 4.10. An approximation to the delta function, which shows a pulse having a constant amplitude spectrum up to the cutoff frequency ω_0.

wavelets maximize the detectability of the arrival times of the reflected events (see fig. 4.11).

Vibroseismic recording and processing

In the late 1950s, the chirped radar method was introduced in seismology as a vibroseismic technique called Vibroseis (a registered trademark of CONOCO), a major development in seismic exploration. In the vibroseismic method, a vibrator is used as an energy source to generate a wavetrain (the chirp signal or sweep signal) of controlled frequencies. The chirped waveform is a sinusoidal vibration of continuously varying frequency. The waveform is applied by the vibrator to the earth over a period typically lasting 6 sec or longer (see fig. 4.12). This period of time is called the *sweep period.* In downsweeping, the frequency begins high and decreases to a low frequency at the end of the sweep period. In upsweeping, the low frequency occurs first. Thus, unlike other energy sources, which deliver energy into the ground in the shortest possible time, the chirp source (the vibrator) passes energy into the ground for 6 sec or more. The hydraulic vibrator drives a steel plate firmly pressed against the ground. The chirp signal imparted to the ground consists of a sine wave whose frequency increases continuously from 10 Hz to 40 Hz during the 6 sec sweep. As a result,

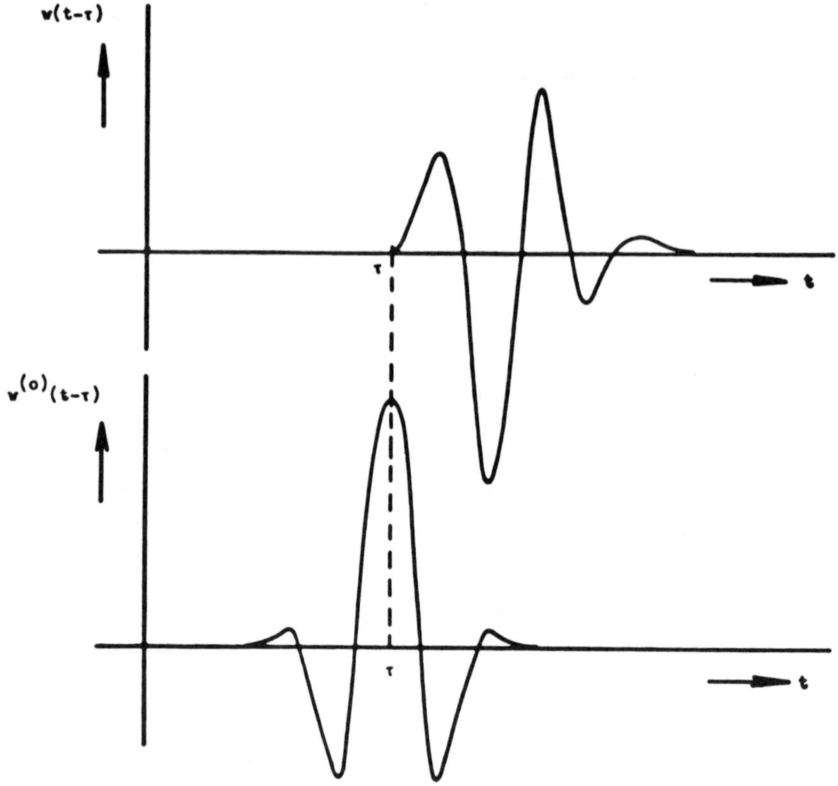

Figure 4.11. (top) *A one-sided recorded seismic wavelet (the physical wavelet).* (bottom) *Its zero-phase counterpart (i.e., the interpreter wavelet). Reprinted by permission of Geophysics from Berkhout 1977.*

each returning reflection event is a similar wavetrain (chirp signal) of about 6 sec duration. Since reflections occur at much closer intervals, measured in milliseconds, the result is an uninterpretable trace consisting of the superposition of the many reflected chirp signals. It is absolutely necessary, therefore, to process the data in order to resolve the arrival times of the reflected events. The processing must compress each returning chirp signal in a short-duration wavelet and thus remove much of the overlap.

The use of chirp signals in exploration removes a dependence on explosive sources with their often relatively unpredictable characteristics. Instead of a sharp impulse, the vibrator source is a prolonged signal of much lower amplitude. It is 6 sec or more in length, as compared to a fraction of a millisecond for dynamite or a few milliseconds for air guns. Its decreased amplitude and increased length make the chirp signal easier to control when initiated. Moreover, the total power emitted by the vibrator source can be increased arbitrarily by increasing the length of time the signal is emitted. Very long chirp signals are used in deep crustal reflection exploration.

Facsimiles of the chirp waveform are reflected at all the subsurface interfaces

Figure 4.12. Truck-mounted vibrators. Reprinted by permission of John Wiley & Sons, Inc. from Reflection Seismology by K.H. Waters 1978. Copyright © 1978 John Wiley & Sons, Inc.

in the travelpath. There are some modifications of the amplitude spectrum by short-period multiple reflections and by absorption in the earth. The seismic trace received by the geophone group at the surface of the earth is the summation of all the reflected chirp waveforms. These reflected waveforms are different in amplitudes and arrive at different times (the reflection times). Whereas a seismic trace recorded from a conventional source is about 6 sec long, the vibroseismic trace will be at least twice as long (6 sec of the conventional trace plus 6 sec or more because of the length of the sweep). The vibroseismic trace will not be recognizable as a reflection record at the recording stage because of the complexity of the long, low-amplitude chirp signals.

Cross correlation forms the basis of vibroseismic processing. The trace that the geophone group records when we use a vibroseismic source bears little resemblance to a dynamite seismic trace (dynoseismic trace). To convert the recorded vibroseismic trace to a trace that corresponds to a dynamite trace, we cross correlate the vibroseismic trace with the sweep (chirp) signal. The overall effect is a trace in which the basic wavelet is a noncausal interpreter wavelet given by the autocorrelation of the sweep signal. This autocorrelation function is quite sharp and has sizable values over only a very narrow range of time shifts. Therefore, the overlap produced by the passage of a long sweep through the earth has been almost entirely eliminated.

Let the reflected events have strength c_n and arrival time τ_n. Let the sweep signal be denoted by $s(t)$ and let the received vibroseismic trace be $x(t)$. The convolutional model for N reflection events is

$$x(t) = \sum_{n=1}^{N} c_n s(t - \tau_n).$$

In vibroseismic data processing, the received trace is cross correlated with the original signal $s(t)$. We must initially record the shape of the received signal in order to have it available for this data-processing step. Both the received trace $x(t)$ and the sweep signal $s(t)$ are recorded in the same medium so that a time reference for them can be established. The cross-correlation process represents a detection filter that seeks out and marks the individual reflected events in the received trace. As we know, cross correlation is the same as convolution, with one signal turned around in time. The processed trace is therefore

$$\begin{aligned} x(t) * s(-t) &= \sum_{n=1}^{N} c_n s(t - \tau_n) * s(-t) \\ &= \sum_{n=1}^{n} c_n \phi_{ss}(t - \tau_n). \end{aligned}$$

Here $\phi_{ss}(t)$ is the autocorrelation of the sweep signal, and hence $\phi_{ss}(t - \tau_n)$ is the delayed autocorrelation pulse with its central (or peak) value at the reflection time τ_n. Thus, the cross-correlation process causes each of the long reflection signals $s(t - \tau_n)$ to be compressed to zero-phase pulses $\phi_{ss}(t - \tau_n)$. The arrival times τ_n as well as the amplitudes c_n of the reflected signals are preserved after cross correlation (see fig. 4.13).

Regarding the choice of the vibroseismic input signal $s(t)$, we want the autocorrelation of the chosen signal to be as clean as possible—that is, to consist of a narrow central pulse that dies away on each side as rapidly as possible. Some side lobes will be present, but we want their magnitudes to be small. A very ringy shape (i.e., a sequence of high side lobes) can hide several reflections and thus reduce the resolution obtainable. A fast-decaying shape around the central peak is most desirable. The standard linear swept-frequency signal (chirp signal) as used in radar has been adopted in most vibroseismic operations, and over the years has proved to be very successful. Other types of signals have been tried, such as the pseudorandom sweep (as discussed in the next section). Such waveforms have a variable envelope, however, not the constant-amplitude envelope, as does the swept-frequency signal. For practical reasons, then, the pseudorandom sweep with its variable envelope is not desirable, as it is better to operate the vibrators at maximum amplitude all the time in order to impart maximum energy to the earth.

Conventional seismic techniques are of the pulse type; that is, a very large burst of energy of short time duration is transmitted into the earth. The vibroseismic method, on the other hand, transmits a low amplitude signal over a long time duration. Time resolution is accomplished by compressing the reflected signals by the data-processing techniques of cross correlation. We will now discuss the transmitted signal and explain why cross correlation can compress it into a much sharper pulse.

The transmitted signal produced by the vibrators is a linear frequency modulation (FM) sweep. This signal is straightforward to implement and has desirable overall properties. The mathematical form of the swept signal is

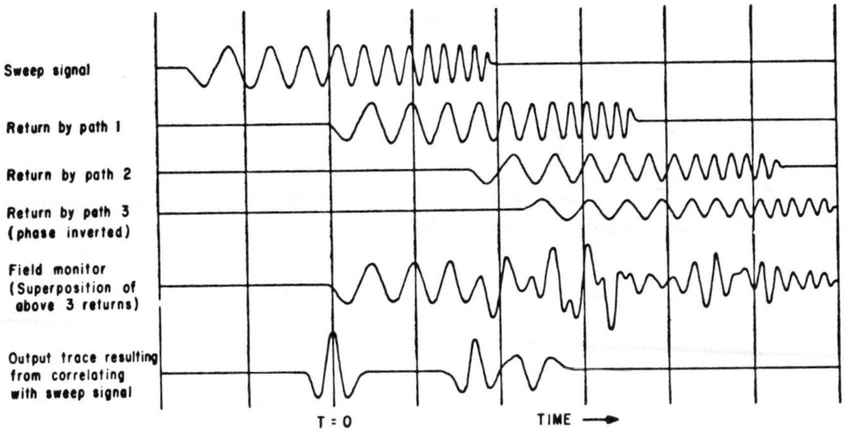

Figure 4.13. Illustration of the vibroseismic technique. Reprinted by permission of the Society of Exploration Geophysicists from Sheriff 1973.

$$s(t) = \begin{cases} \cos\left[2\pi\left(f_0 + \frac{kt}{2}\right)t\right] & \text{for } -\frac{T}{2} \le t \le \frac{T}{2}, \\ 0 & \text{elsewhere,} \end{cases}$$

where t = time,
f_0 = carrier frequency,
k = rate of frequency sweep,
T = signal width,
f = swept spectrum bandwidth = kT.

Note that the signal has been arbitrarily centered at time $t = 0$; the signal runs from $-T/2$ to $T/2$. Since frequency is the rate of change of phase, the frequency at time t is

$$\frac{d}{dt}\left[\left(f_0 + \frac{kt}{2}\right)t\right] = f_0 + kt.$$

Thus, instantaneous frequency runs from $f_0 - (kT/2)$ at time $t = -T/2$ to $f_0 + (kT/2)$ at time $t = T/2$. The amplitude of the sweep is unity for its entire duration. In vibroseismic work, typical values of the parameters would be

$$f_0 = 25 \text{ Hz},$$

$$k = 5 \text{ Hz/sec},$$

$$T = 6 \text{ sec},$$

$$\Delta f = kT = 30 \text{ Hz},$$

so that the lower frequency is 10 Hz and the upper frequency 40 Hz.

The filter required in order to compress the linear FM sweep is called a *matched filter* or a *correlation filter.* It has an impulse response $h(t)$ that is the time reverse of the signal; that is,

$$h(t) = \cos\left[2\pi\left(f_0 - \frac{kt}{2}\right)t\right] = s(-t)$$

for $-T/2 \leq t \leq T/2$, and $h(t) = 0$ elsewhere. The output of the matched filter resulting from the sweep $s(t)$ as input is

$$\begin{aligned} g(\tau) &= \int_{-\infty}^{\infty} s(t)h(\tau - t)\,dt \\ &= \int_{-\infty}^{\infty} s(t)s(t - \tau)\,dt, \end{aligned}$$

which is the autocorrelation function of the sweep. Upon evaluation, this autocorrelation function is

$$g(\tau) = \frac{T}{2}\frac{\sin \pi k\tau(T - |\tau|)}{\pi k\tau T}\cos 2\pi f_0\tau$$

for $-T \leq \tau \leq T$, and $g(\tau) = 0$ elsewhere. This autocorrelation function has been named the Klauder wavelet by Geyer (1969). This expression shows that the autocorrelation is approximately $(\sin x)/x$ in shape. The pulse width of the Klauder wavelet at the -4 dB point is that value τ_0 for which $k\tau_0 T = 1$; that is, $\tau_0 = 1/kT$. Thus, the pulse width has been reduced by a factor of $(1/kT)/T = 1/kT^2 = 1/(T\,\Delta f)$. As an example, the pulse width of a 10-40 6 sec sweep is reduced by the factor $1/(6 \cdot 30) = 1/180$. The maximum amplitude of the Klauder wavelet is $T/2$, compared with unity for the sweep. An excellent discussion of the properties of the Klauder wavelet is given by Geyer (1969).

We will now explain why the cross correlation process can compress the swept-frequency signal. A linear frequency sweep is intrinsically indistinguishable from a signal swept in a regular succession of quantum frequency steps. Consider the system illustrated in figure 4.14. In the figure, four discrete frequencies are transmitted in turn, forming the overall transmitted signal. The first frequency f_1 is transmitted for a time T, then frequency f_2 for time T, then f_3 for time T, and finally f_4 for time T. We assume that the time length T for each frequency is identical. To process the received signal, each segment is time-delayed in such a manner that all the segments pile up or coincide in time. Thus, f_1 is delayed for $3T$ seconds, f_2 for $2T$ seconds, f_3 for T seconds, and f_4 is not delayed. The summed output pulse width is T seconds. Because the original transmitted signal was $4T$ seconds long, the output resolution has been increased by a factor of four with no decrease in transmitted energy. The more frequencies used and the less time spent at each frequency, the better is the result. The limit of more and more frequencies is a *linearly swept signal.* The delay required in the processing would then be a linearly decreasing delay versus frequency method. This is the foundation of the vibroseismic method. The method transmits a swept-frequency signal, receives the resulting reflections, and then processes the received data in such a manner that the reflections are

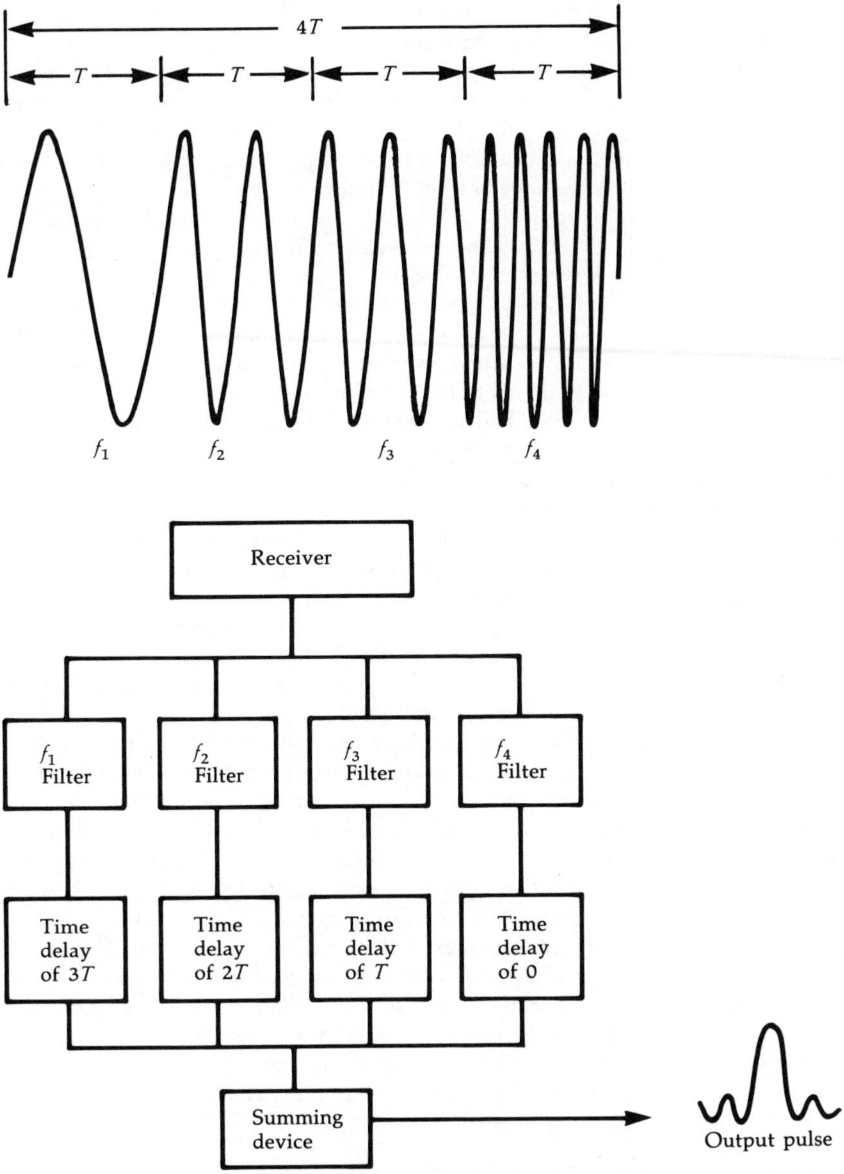

Figure 4.14. Explanation of why cross correlation can compress the swept-frequency signal. (top) *Four different frequencies are transmitted sequentially, starting with* f_1. *The total pulse width is* 4T. (bottom) *The four return frequencies are separated and then delayed in such a manner that they pile up in time. The output pulse width is* T.

compressed in time to give short, intense pulses. The final narrow pulses are called the correlated or compressed record.

When a linearly swept signal is processed by a linear delay versus frequency network, as in figure 4.15, the various frequencies are delayed so that they pile up in time at the output. This piling up does not result in a perfectly rectangular pulse; instead, the output pulse assumes the shape shown in the figure. This pulse has the approximate shape of a sin t/t pulse. If the side lobes of this pulse are eliminated, a good approximation to a conventional rectangular pulse results. By controlling, or shaping, the amplitude of the swept-frequency signal, the side lobes can be very greatly reduced in magnitude (fig. 4.16).

There is another significant advantage to the vibroseismic technique. The processing of the returned data only piles up, or compresses, one particular swept frequency whose slope and bandwidth exactly match the transmitted signal. Random noise signals will not pile up but will come out of the network with the same amplitude they had when they went in, whereas the real signal will pile up and thus increase in amplitude. Thus, the signal-to-noise ratio of the received data is considerably improved by the processing.

In summary, by increasing the transmitted pulse length, the peak power requirements are decreased without degrading time resolution, and the signal-to-noise ratio is improved. The transmitted sweep signal can be chosen so that the band that is emitted corresponds closely with the natural frequency response of the area. Thus, the frequencies that would be attenuated by the earth need not be transmitted. If frequencies above 60 Hz are severely attenuated by the earth in that area, for example, the sweep frequency can be stopped at 60 Hz. The cross-correlation process is done before applying normal moveout (NMO), and then the rest of the processing steps are carried out in more or less the same manner as for conventional records. Account must be made, however, of the fact that the basic wavelet on the correlated vibroseismic trace is zero-phase, not one-sided as on a recorded conventional trace. The vibroseismic method employs an energy source that is almost ideal in seismic exploration. Furthermore, it has several advantages over explosive techniques: shot holes do not have to be drilled, thus eliminating a large expense, and exploration can take place in areas (such as cities) where explosives are not feasible. The signal-to-noise ratio can be effectively improved by direct summation of a number of vibrator sweeps. There are practical limits to the number of vibrator sweeps that can be used, however, and therefore to the improvement of signal quantity. Explosive sources are still the most commonly used type of source, and they are especially needed under conditions that require large source energies.

Comparison of the vibroseismic method with radar

The linear FM sweep is also used as the transmitted signal in chirp radar systems. Radar and seismic detection systems share certain fundamental features, but there are also important differences.

In radar, one is concerned with certain universal limitations and trade-off relationships between resolving power and discriminating power. Consider the transmission of a single simple pulse with pulse length τ. The resolvable range increment, or cell, is then equal to τ—namely, the pulse length resolvable at the detector. Thus, the *range discriminating power* is $1/\tau$ resolvable range elements per

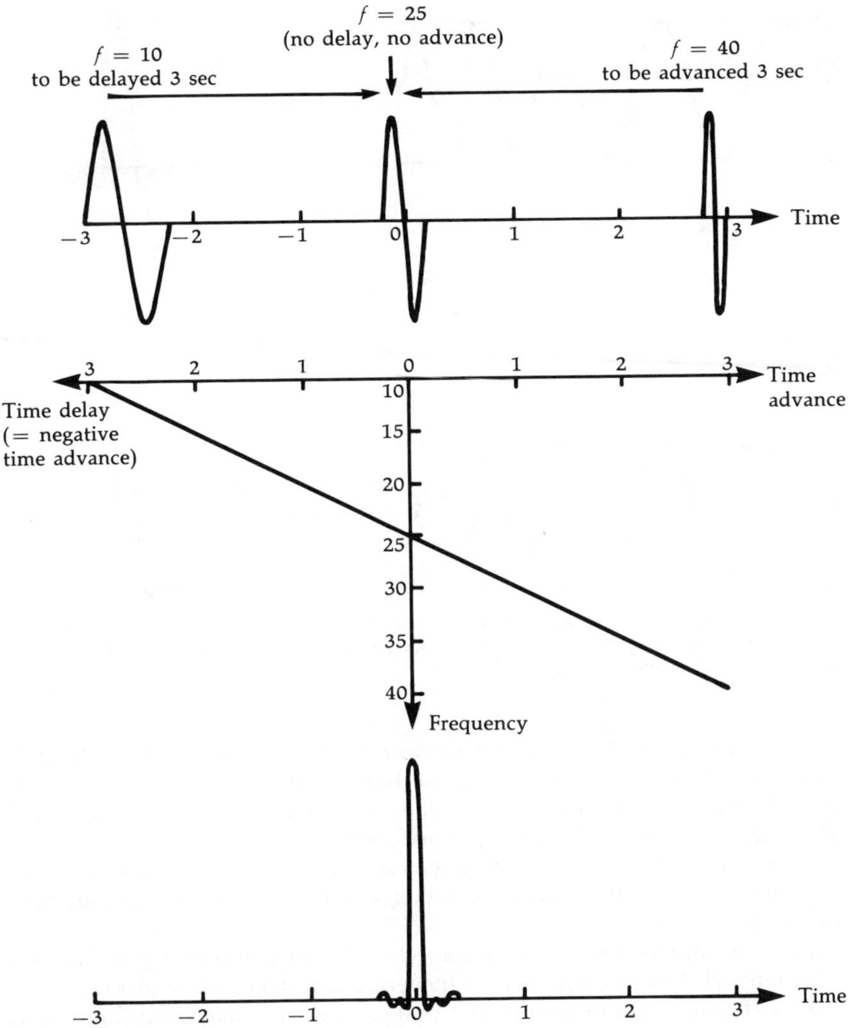

Figure 4.15. (top) *The input to the filter is the swept-frequency signal. Illustrated is a schematic representation of a swept-frequency signal, showing lower, intermediate, and upper frequency components.* (middle) *The filter is a linear delay versus frequency network.* (bottom) *The output is a narrow pulse resulting from the pileup of all frequency components.*

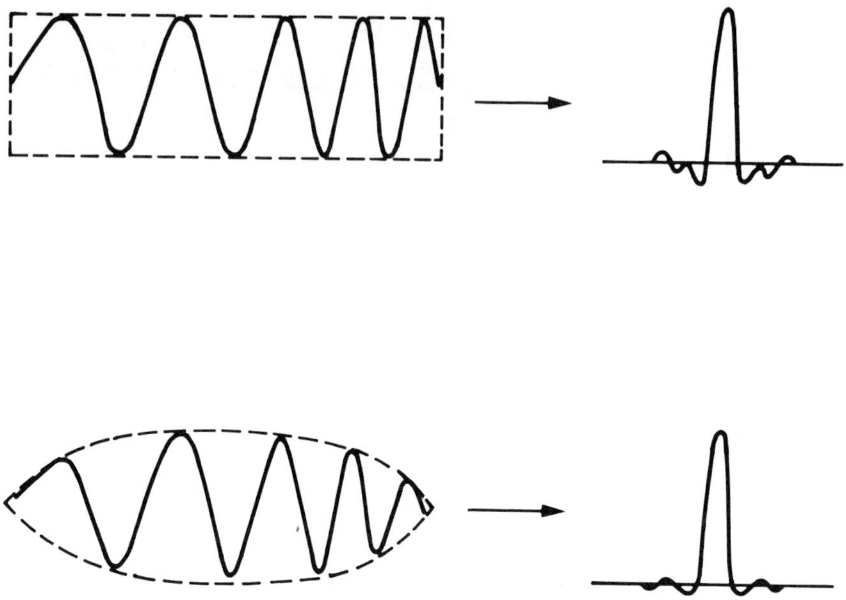

Figure 4.16. (top) *A swept-frequency signal with rectangular envelope collapses to a Klauder wavelet with prominent side lobes.* (bottom) *A swept-frequency signal with tapered envelope collapses to a broader wavelet with reduced side lobes.*

second. Now $1/\tau$ is equal to B, the bandwidth at the detector of the echo from a point target. Let T be the observation time—that is, the time for which the receiver output is observed and in which an echo might reasonably be expected. The *range resolving power* of this single-pulse system is the number of resolvable range elements (each of length τ) in the total detector observation time T, and is therefore T/τ, which is TB, the product of observation time and detector bandwidth.

The effect of target motion with respect to the radar system is approximately a shift in each Fourier component in the received signal by a Doppler frequency f. If we consider a continuous wave (CW) radar with coherent integration over the observation time T, then the frequency misalignment f can be resolved for frequencies f for which fT is spaced by integers. Hence, at the receiver, we are able to discriminate Doppler shifts spaced by $1/T$ Hz. Thus, the Doppler discriminating cell is $1/T$—that is, $1/T$ Hz per cell. The *Doppler discriminating power* is then equal to the number of cells per hertz, which is T, the detector observation time. The *Doppler resolving power* is TB, the product of observation time and detector bandwidth.

In general, radar is concerned with two dimensions: observed range and Doppler. In pulse radar, a short pulse is transmitted. This short pulse contains a wide band of frequencies, resulting in good range resolution but poor Doppler resolution. In CW radar, a long signal is transmitted. This pulse contains a narrow band of frequencies; the result is good Doppler resolution but poor range resolution. In chirp radar, a long FM pulse is transmitted. This pulse contains a wide band of frequencies. With pulse-compression techniques, the long FM pulse yields both good range resolution and good Doppler resolution (see fig. 4.17).

There is no counterpart in seismology to the Doppler effect in radar. Because the underground strata are fixed, there can be no frequency shift due to a moving target. For many years, only short pulses generated by explosives were used in seismic prospecting, comparable to the short pulses used in pulse radar. With the introduction of Vibroseis, however, the counterpart to chirp radar was established in seismology. Whereas much of the theoretical justification of long FM signals in radar is based on increasing Doppler resolution, the corresponding justification in seismology rests on other grounds.

The linear FM signal has found broad application in a large number of radar applications as well as in seismology. It is worthwhile, however, to consider other specific types of signals, for they shed insight onto the problem, even though their practical implementation may never be desired. It is often the case that, after careful consideration, the engineer will decide to rely on a less exotic but more reliably instrumented type of signal. In this event, the basic theory associated with waveform design will indicate the loss of information-gathering capacity resulting from the choice of a less-than-optimal signal. In addition, if it is necessary to push the waveform design to the limit, despite the cost, then the theory indicates the ultimate restrictions that apply to all classes of signals.

The linear V-FM waveform is made up of two parts. The part from $-T$ to 0 is a linear sweep of decreasing frequency, whereas the part from 0 to T is a linear sweep of increasing frequency. Thus, the plot of instantaneous frequency versus time is a V-shaped function, which explains the terminology V-FM. The autocorrelation of the linear V-FM waveform is basically a $\sin x/x$ function on top of a baseline pedestal.

The quadratic FM waveform is made up of two parts. The part from $-T$ to 0 is a quadratic sweep of decreasing frequency, whereas the part from 0 to T is a quadratic sweep of increasing frequency. Thus, the plot of instantaneous frequency versus time is a U-shaped function, where the U is a parabola. The autocorrelation of the quadratic FM waveform has a baseline pedestal, similar to the pedestal of the autocorrelation of the linear V-FM waveform. These baseline pedestals are associated with the autocorrelations of even-function FM waveforms. Because of these high side lobes, signals of this type are not suitable when the objects being observed are densely packed, as in seismology. The importance of these signals in radar is that they yield uncorrelated measurements of range and velocity.

In signal design for seismic applications, the signal autocorrelation is of paramount importance. It is desired that this function exhibit very low-range side lobes in order to reduce the masking effects of large signals on smaller signals. In order to achieve this side lobe reduction with a linear FM signal, the signal may be tapered. However, tapering leads to a signal-to-noise ratio loss relative to that obtained with the ideal matched filter. Nonlinear FM functions with odd symmetry represent a means for obtaining waveforms with low side lobes.

Other methods can be employed in many radar applications as alternatives to the FM techniques just discussed; for example, discrete signals can be used. The simplest discrete signals consist of binary sequences—that is, sequences made up of $+1$'s and -1's, such as the sequence $(1, 1, 1, -1, 1, -1, -1, 1)$. Binary signals can be transmitted as a sequence of pulses or can be imposed on

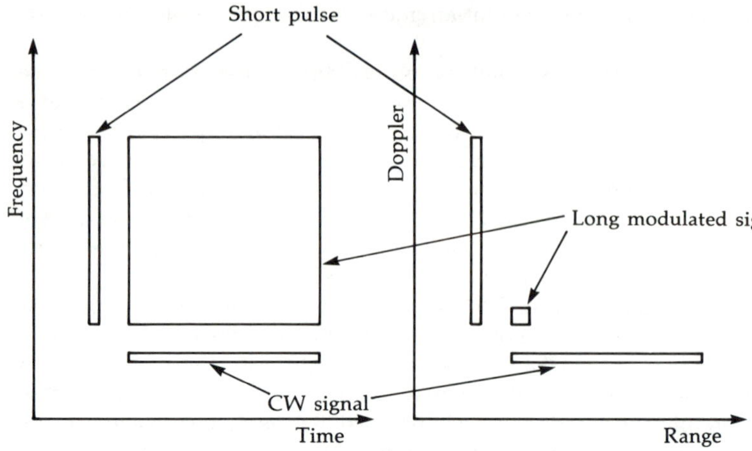

Figure 4.17. (left) *Signal space depicting received signals.* (right) *Target space depicting discrimination cells. The smaller the cell, the better the discrimination.*

a CW carrier signal, where the coding function results in phase reversals of the carrier signal. The autocorrelation of the foregoing sequence is

$$(1, 0, -1, -2, 2, 0, -1, 8, -1, 0, 2, -2, -1, 0, 1).$$

It is seen that the highest level of the side lobes is ± 2, or one-quarter of the peak amplitude of 8. Much effort has been directed to the study of discrete signals that have spike-like autocorrelations. One such class of signals that exhibit an optimum side lobe level is the Barker binary phase codes. These codes have the property that the peak side lobe level never exceeds ± 1; however, no code of this class with length greater than 13 is known to exist. The known Barker codes are

$$(1, 1, -1)$$
$$(1, 1, -1, 1)$$
$$(1, 1, 1, -1, 1)$$
$$(1, 1, 1, -1, -1, 1, -1)$$
$$(1, 1, 1, -1, -1, -1, 1, -1, -1, 1, -1)$$
$$(1, 1, 1, 1, 1, -1, -1, 1, 1, -1, 1, -1, 1)$$

Another well-known class of binary codes is pseudorandom sequences. As such a sequence is made infinitely long, the side lobes of the autocorrelation become negligible in relation to the main lobe. In applications, however, a finite-length sequence is required, and truncation of the infinite-length signal necessarily increases the side lobe level.

Finally, a general class of discrete infinite sequences that have spike autocorrelations consists of the all-pass functions (Robinson, 1980). An example of such a sequence is

$$c, (1 - c^2), -c(1 - c^2), c^2(1 - c^2), -c^3(1 - c^2), \ldots$$

where c is a number less than one in magnitude. If $c = 0.5$, for example, then the sequence is 0.5, 0.75, −0.375, 0.1875, −0.09375, ...

At the Geophysical Analysis Group (GAG) meetings in the early 1950s, especially the one in Tulsa in 1954, there were lively discussions about the possibility of transmitting a controlled-source wavelet of some length into the ground instead of using the time-honored dynamite source. Although J. M. Crawford and W. E. Doty of Continental Oil Company had always been active participants in all GAG meetings, they were remarkably quiet on this issue, and the general consensus of many of the oil company members present was that such a possibility would be beyond technical capabilities. A few years later, however, Continental Oil disclosed the Vibroseis method, invented by Crawford, Doty, and Lee (1960). The principles of the method are straightforward, but its implementation in the field represented a major breakthrough, even as did the implementation of the first radar transmitters in the late 1930s and early 1940s. Vibroseis, a registered trademark and service mark of Continental Oil Company (now CONOCO), was one of the major innovations in seismic prospecting. Much of the mystique of the Vibroseis system had to do with ingenious pieces of equipment, as the field vibrators had to work together precisely and reliably. According to Waters (1978), their completion represented a tour de force in experimental geophysics.

Exercises

1. Given the wavelets (1, −1, 1, −1), (1, 1, 1, 1), and (1, 2, 3, 4), find the autocorrelation of each wavelet and the cross correlation of each pair of wavelets.
2. Given the wavelets (3, 2, 1), (3, 2, −1), (3, −2, 1), and (3, −2, −1), find the autocorrelation of each wavelet and the cross correlation of each pair of wavelets. Plot the results in a 4 × 4 table, with the autocorrelations on the diagonal and the cross correlations off the diagonal. Show that only the right-hand side of each correlation function is needed in such a table.
3. Find the autocorrelations of each of the following wavelets by multiplying the wavelet's z-transform by the z-transform of its time reverse:

   (1, −4, 3, 2),
   (5, −3, 2, −1),
   (6, 2, −1, 1).

   Find the cross correlation between each pair of wavelets by means of the z-transform.
4. Given the wavelet $a = (4, -2, 1)$, according to our convention the first coefficient (4) occurs at time $k = 0$. The time reverse of this wavelet would be the noncausal function (. . ., 0, 0, 1, −2, 4, 0, 0, . . .), where the last coefficient (4) occurs at $k = 0$. Compute the autocorrelation by convolving the wavelet with its time reverse. If we shift the time-reverse function forward by two time units, we obtain the (causal) wavelet $a^R = (1, -2, 4)$. Its first coefficient (1) occurs at $k = 0$. Compute $a * a^R$ and show that it is the same as the shifted autocorrelation of a (where the time shift is two units to the right). Explain this in terms of z-transforms.

5. Given the following two-length (causal) wavelets:

$$a = (2, -1), \qquad a^R = (-1, 2),$$
$$b = (3, -2), \qquad b^R = (-2, 3),$$
$$c = (3, 1), \qquad c^R = (1, 3),$$
$$d = (4, -1), \qquad d^R = (-1, 4),$$

compute the following wavelets:

$$a * b, \qquad a^R * b^R,$$
$$c * a, \qquad c * a^R,$$
$$c * a * b, \qquad a * c^R * b^R,$$
$$d * d^R * a, \qquad a * d * d.$$

6. Find the autocorrelation of each of the wavelets in Exercise 5. Which autocorrelations are the same, and why?
7. Find the following convolutions by multiplying z-transforms:

$$(-6, 5) * (3, -2, 1),$$
$$(2, -1, 1, 3, -2) * (1, -1),$$
$$(4, 3, 2, 1) * (2, -1),$$
$$(6, 3) * (1, 2, 3, 4).$$

8. Find the autocorrelations and cross correlations of the following wavelets:

$$(4, -3, 2, 1), \qquad (1, 0, -6, 2),$$
$$(8, -4, 2, -1), \qquad (7, 5, 3, 1).$$

Plot the results. Show that it is necessary to plot only the right halves of the correlation functions.
9. By inspection, say what you can about the autocorrelations and cross correlations of the two wavelets

$$(1, 2, -8, 6), \qquad (6, -8, 2, 1).$$

Confirm by computation.
10. Write out all the mathematical steps to establish that cross correlation can be accomplished by reversing one function in time and convolving

$$\phi_{ab}(k) = a(k) * b(-k).$$

11. If $A(z) = 1 + z + z^2 + z^3 + z^4$, what are the coefficients of $A^2(z)$, $A^3(z)$, and $A^4(z)$? Interpret these results in terms of convolution of waveforms.
12. Let $x(k) = \sin \omega k$ be a sampled sine wave. Convolve this wave with the two-term filter (b_0, b_1) and denote the result by $y(k)$. We think of $x(k)$ as

the input to the filter and $y(k)$ as the output. Show that the output is a sinusoidal wave of the same frequency ω; that is,

$$y(k) = A \sin (\omega k + \theta).$$

Find the amplitude A and the phase θ as a function of ω. The curves $A(\omega)$ and $\theta(\omega)$ plotted against ω are the amplitude spectrum and the phase spectrum of the filter.

13. Show that if $x(k) = \sin \omega k$ and $b = (b_0, b_1, \ldots, b_n)$, then the output $y(k)$ always takes the form $y(k) = A(\omega) \sin [\omega t + \theta(\omega)]$; that is, a sinusoidal wave of the same frequency but different amplitude and phase.
14. A guitar string is stretched from point A to point G. The string is divided into six equal intervals by the points B, C, D, E, F. Paper riders are placed on the string at D, E, and F. The string is held fixed at C and twanged at B. What happens to the riders?
15. We ring a bell by letting the clapper strike its rim. Are all parts of the rim equally loud, or are some parts loud and some almost silent?
16. The ringing of a bell involves a range of frequencies rather than one exact frequency. The frequencies in the range are all quite close and eventually work their way out of phase and interfere destructively with each other, as evidenced by the bell gradually falling silent. Will a bell with the purest tone ring for a long time or for a short time?
17. The greater the amplitude of a wave, the greater its
    (a) frequency
    (b) wavelength
    (c) loudness
    (d) all of these
    (e) none of these
18. A note is sounded when you tap a drinking glass that is partially filled with water. If the glass is filled with more water, the pitch of the sound produced by tapping is
    (a) higher
    (b) lower
19. Resonance occurs when an object is forced to vibrate by external vibrations that are
    (a) at a higher frequency
    (b) at a lower frequency
    (c) of large amplitude
    (d) matched to the natural frequency of the object
20. When a sound source approaches you, there is an increase in the sound's
    (a) speed
    (b) frequency
    (c) wavelength
    (d) all of these
    (e) none of these
21. How does the frequency of the sound change when you fill a container of water and when you empty it?
22. A lifeguard at L on the beach must rescue a drowning person in the water at P. Time is of the essence! Which path from L to P will take the least time?

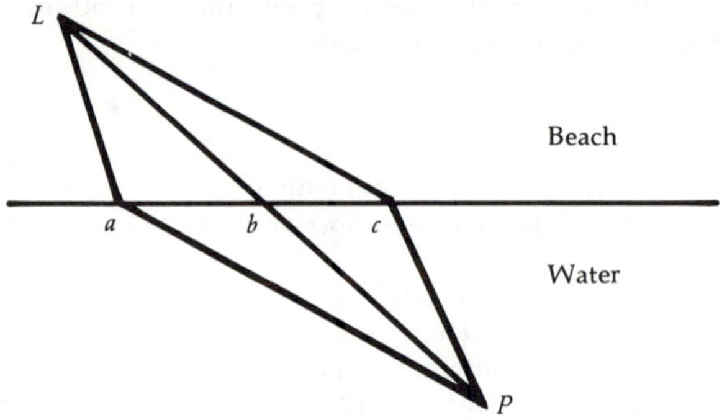

23. Choose the correct answer. If light takes the path of least time between two places, then the way it bends or refracts when it goes from air (or vacuum) into glass suggests that its speed in glass is
    (a) faster than its speed in air
    (b) the same as its speed in air
    (c) slower than its speed in air
24. An upcoming seismic wave usually bends toward the vertical as it approaches the earth surface. What can we say about the surface layer in such a case?
25. Suppose that a pair of toy cart wheels connected by an axle are rolled along a smooth sidewalk and onto a grass lawn. Because of the interaction of the wheels with the grass, they roll slower there than on the smooth sidewalk. If the wheels are rolled onto the lawn at an angle, they will bend from their straight-line path. Which way will the path bend?
26. The cycles of two musical notes A (solid line) and B (dashed line) are shown as follows (two cycles of each):

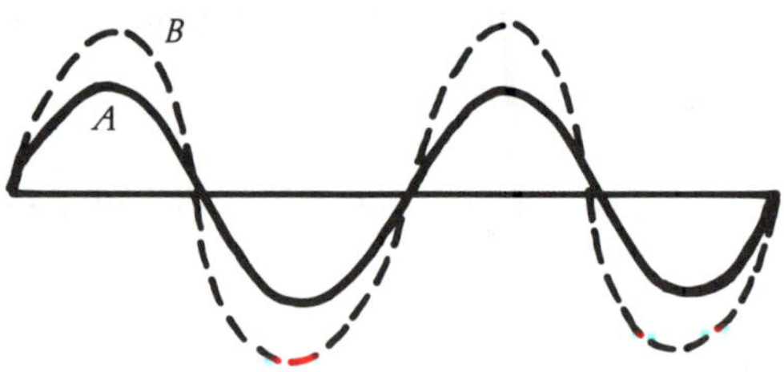

Which note has the higher frequency, and which note is louder?

27. The following waves are called square waves:

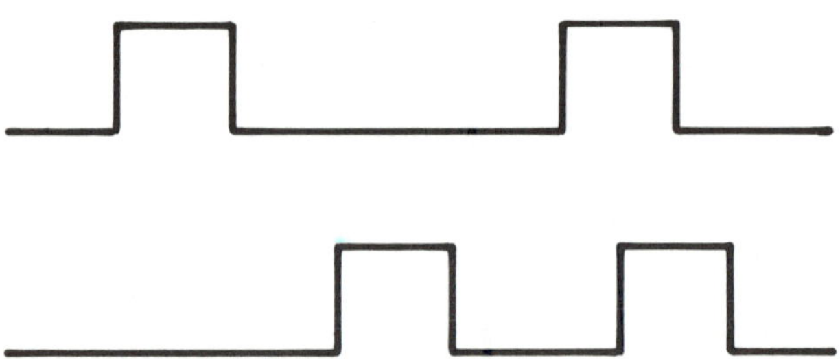

When they are superimposed, they add to make what wave?

28. Find the superposition of the following two sinusoidal (or harmonic) waves:

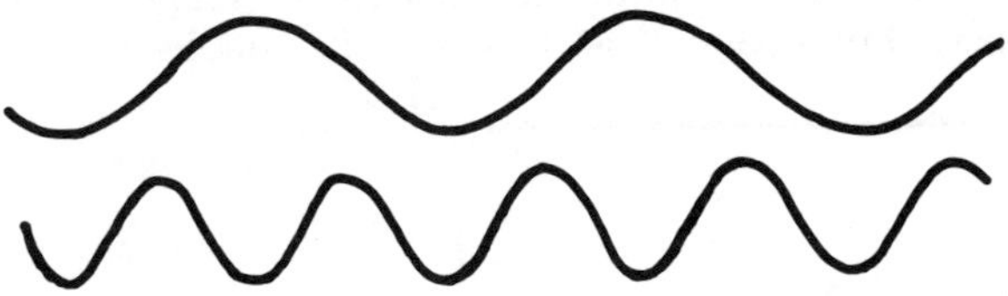

29. Can any waveform be made (built up) by adding (integrating) various pure harmonic (sinusoidal) waves, such as

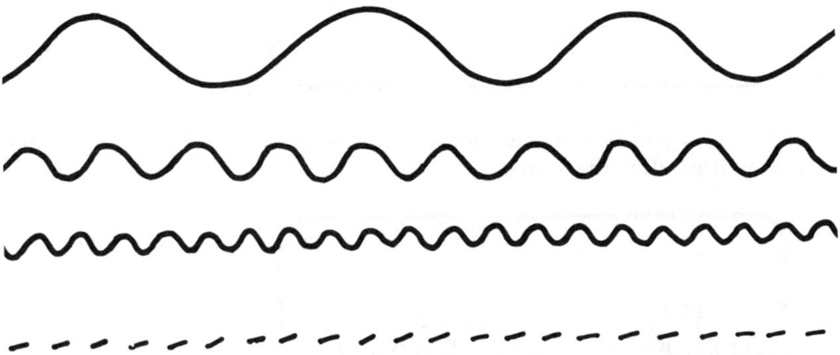

30. The following square wave and sine wave both have the same wavelength.

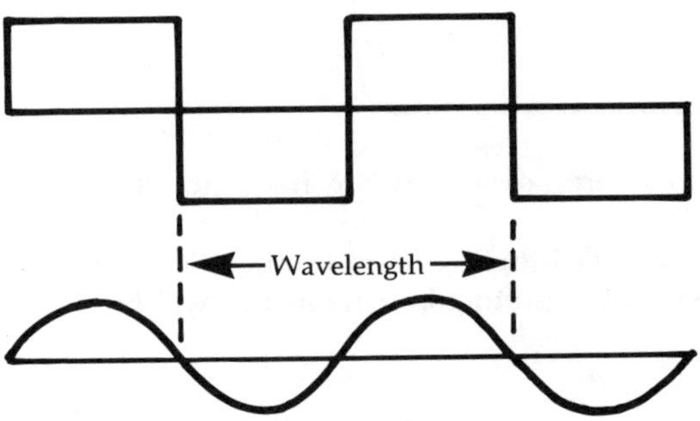

Which of these waves contains the most high-frequency or short-wavelength component waves?

31. Can two light beams ever cancel each other out and result in darkness? Can we add light to more light and come out with no light, or would this violate the law of conservation of energy?

32. The following two waves are generated at the ends of a long rope and travel toward each other.

Will there be any instant at which the amplitude of the rope is everywhere zero?

33. Two different notes are sounded at the same time. Their sounds are added and the sum displayed on oscilloscope A as the following beats:

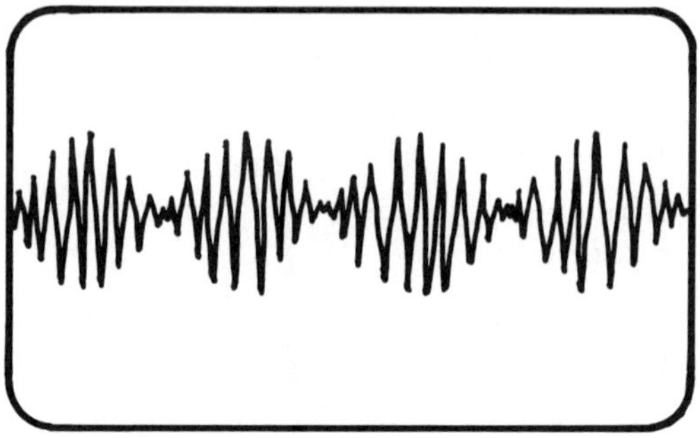

A second pair of notes is sounded and their sum displayed on oscilloscope B as the following beats:

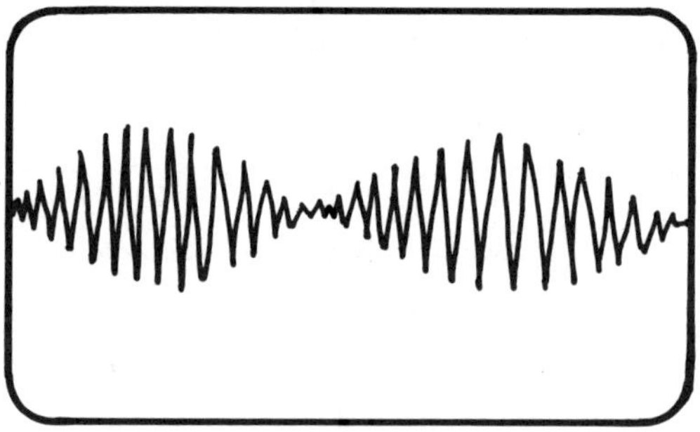

Show that the notes in A are farther apart in frequency than the notes in B.

34. Choose the correct answer. Two tuning forks are sounded, one at 254 Hz and the other at 256 Hz. The resulting beat frequency will be
(a) 2 Hz
(b) 4 Hz
(c) 255 Hz

35. In Fourier synthesis, many sinusoidal waves are added (integrated) to produce a waveform. Does the phase relationships of the component waves make a difference in the result?

36. The autocorrelation function is a measure of the dependence between values of a signal at two different times. What are some properties of the autocorrelation?

37. Show that cross correlation can be performed by reversing the second waveform and convolving.

38. What are some properties of the cross correlation?

39. Show that multiplication of a z-transform by $(1 - z)$ is analogous in discrete time to differentiation in continuous time. Show that dividing a z-transform

by $(1 - z)$ is analogous in discrete time to integration in continuous time.

40. Compute the second-order moment (spread parameter) σ^2 for the following symmetric wavelets (where the center point of the wavelet corresponds to time $k = 0$):

    (0, 0, 1, 0, 0),
    (0, 0.5, 1.25, 0.5, 0),
    (0.25, 0.5, 1, 0.5, 0.25).

41. Show that a three-term symmetric wavelet (a_1, a_0, a_1) is zero-phase if and only if it has the form $(bc, b^2 + c^2, bc)$. Hint: a zero-phase wavelet is an autocorrelation function, and a three-term autocorrelation function must have the form $(b, c) * (c, b)$.
42. Show that convolution is equivalent to replacing each element of one wavelet by an appropriately scaled version of the other wavelet and then summing elements that occur at the same times.

5. *Convolutional models and synthetic seismograms*

Digital models

A model is generally a representation of a physical thing. An analog model, for example, may be a physical structure of electronic components that will react upon an input to produce an output. Such a model has a disadvantage that it may be difficult to modify. It is possible to develop a digital model to accomplish the same purpose. The signals can be represented by a series of values representing uniformly spaced amplitude measurements of a time function over some period of time. The unit-impulse input, for example, would be 0, 0, 0, 1, 0, 0, 0 . . ., with the single value of one representing a single pulse of unit amplitude at time 0. A signal in the form of a series of equally spaced amplitude values is fed into the model as input. The output would be a similarly spaced series of values, which, if plotted and joined with a continuous line, would approximate a continuous response.

An important type of model is one that can be represented by a linear time-invariant digital filter. The actual model requires an impulse response that operates upon the input to produce the output. The mathematical operation that accomplishes the effect of the model is convolution. In the preceding chapter, we studied convolution as a mathematical operation; now we want to apply it to the physical world. Thus, we must find ways to approximate actual physical phenomena as linear time-invariant systems. Although a seismic experiment that produces a seismic trace has essential and significant nonlinear and time-varying characteristics, this approximation is still extremely useful and valuable when carried out with care and understanding.

A model is the embodiment of a concept from which one can deduce effects, which can then be compared to observations. A model is used in understanding the significance of the observations. The model may be conceptual, physical, or mathematical. One might calculate the gravity, magnetic, or seismic effects for an assumed geologic structure, for example, and compare these effects with actual data. Agreement between observations and effects derived from the model can never prove that the model represents the actual situation because of the lack of uniqueness in most geophysical interpretation problems. A block diagram of a model and its effects is given in figure 5.1. For any model, we can define the direct or forward problem as: Given the model, find the effects. The indirect or inverse problem can be described as: Given the effects, find the most plausible model. A typical forward problem in seismology is constructing synthetic traces, given the geologic layering as determined by well logs. Forward problems usually involve *convolution.* As we will see, the synthetic trace is obtained by convolving the reflectivity function with the seismic wavelet. A

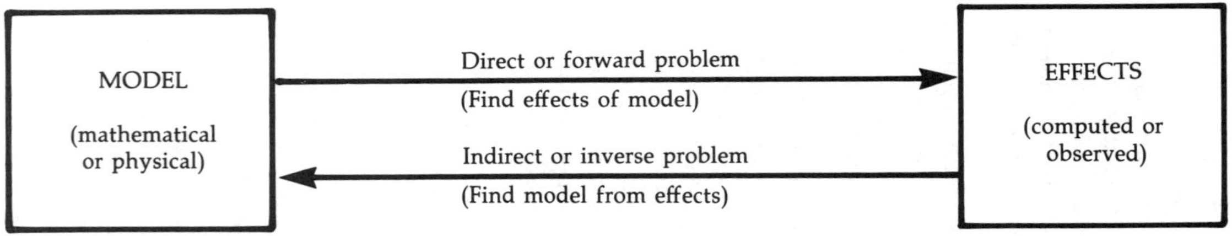

Figure 5.1. A model and the effects it produces.

typical inverse problem in geophysics is finding the reflectivity from knowledge of the seismic traces. As we would expect, inverse problems usually involve the operation of *deconvolution* in one form or another.

We are now in the midst of a universal digital revolution of epic proportions. Digital signal processing is a growing and dynamic field, involving the exploration of new technology and the application of the techniques to diverse fields of science and engineering. The technology has advanced from discrete semiconductor components to very large scale integration (VLSI), with densities above 100,000 components per silicon chip. The availability of fast, low-cost microprocessors and custom, high-density integrated circuits means that increasingly difficult and complex mathematical methods can be reduced to hardware devices. Whereas digital methods originally were used at great expense only because the application demanded high flexibility and accuracy, we have now reached the point in silicon chip technology where the anticipated long-term cost advantages have become a significant factor for the use of digital rather than analog methods.

A digital system is one in which the information is contained and manipulated as a set of discrete numbers, as opposed to an analog system, in which the information is represented as a continuous flow of the quantity constituting the signal.

Digital filters provide a means of filtering data numerically in the time domain by summing weighted samples at successive time increments. Digital filtering is more versatile than conventional electrical (analog) filtering and permits one to filter easily in accordance with arbitrarily chosen characteristics, which might prove difficult to achieve with analog components. Linear digital filtering is carried out by the mathematical operation of convolution. A linear filter may be characterized by its impulse response or, alternatively, by its frequency-domain amplitude and phase response.

Digital filtering may be contrasted with the more conventional analog filtering in that the operation is carried out numerically on the input data as a function of time. Linear digital filtering is done by means of a finite set of numerical filter coefficients. The necessary operations may be implemented by various configurations of equipment, ranging from general-purpose mainframe digital computers to microprocessors. With such equipment, digital filtering becomes a practical operation that can be done on geophysical data economically and accurately.

In the years from the inception of commercial reflection prospecting in 1930 to the 1960s, the electrical analog filter was king. Oil companies spent a great deal of time and money on their design and implementation. A prodigy was growing up in the 1950s, however; it was always described with these words: modern, high-speed, electronic digital computer. Without the modifying words, *computer* meant nothing; a computer was a person who did computations. (In the same way, a typewriter originally was the person who did the typing, not the instrument.) In those early days of computers, IBM would not use the word *computer* and instead called them electronic data-processing (EDP) machines. In fact, everyone then used the word *machine* to mean computer. From the advent of practical solid-state technology in the 1960s to the silicon chip technology of today, the computer came into its own. The number of filtering operations performed on seismic data, as well as the amount of seismic data collected, has increased exponentially during that period. Except for a few strategically placed analog filters, chiefly in the recording equipment, the filters of today are software. They are represented by sequences of numbers and operations in computer programs. The digital filter has replaced the analog filter as king.

We wish to develop here the mathematical properties of finite discrete linear digital filters. The inputs to these linear filters are discrete time series—that is, sequences of observations that are equally spaced with respect to discrete time k. In the context used here, the time series represent digitized readings of geophysical records, such as seismic traces. For convenience, we may take the spacing between each successive observation to be one discrete unit, and thus we may represent the time series as

$$\ldots, x(k-1), x(k), x(k+1), x(k+2), \ldots,$$

where k is an integer.

In the preceding chapter we defined convolution. Let us apply this definition to convolution of a time series with a wavelet. Let the wavelet be

$$a = (a_0, a_1, a_2, \ldots),$$

and let the time series be

$$\ldots, x(-2), x(-1), x(0), x(1), x(2), x(3), \ldots,$$

which is infinitely long in both the positive and the negative directions. Then their convolution is the time series

$$\ldots, y(-2), y(-1), y(0), y(1), y(2), y(3), \ldots,$$

where $y(k)$ is given by the formula

$$y(k) = \sum_{n=0}^{\infty} a_n x(k-n).$$

Like the $x(k)$ time series, the $y(k)$ time series is also infinitely long in both directions.

Now let us illustrate this convolution process schematically. We suppose that the series $x(k)$ is going into a filter. We say that the time series $x(k)$ is the *input* to the filter. Coming out of the filter is the series $y(k)$, so we say that the time series $y(k)$ is the *output* from the filter. The wavelet a is inside the filter and is called the impulse response of the filter. Schematically, we have

Input $x(k)$ → | Impulse response $a = (a_0, a_1, a_2, \ldots)$ | → Output $y(k)$.

Whenever we illustrate an input/output relationship as shown here, we mean the following: The output $y(k)$ is equal to the convolution of the input $x(k)$ and the response u, or

$$y = a * x.$$

A filter that generates its output from its input in this way is called a *time-invariant linear filter.* The coefficients of the impulse response are called the *filter coefficients.* Such time-invariant linear filters form an important class of digital filters. Note that we would have obtained the same result if we had input a into a filter whose impulse response is x. In other words, convolution is commutative:

$$a * x = x * a$$

or

$$\sum_n a_n x(k - n) = \sum_n x(n) a_{k-n}.$$

Consider the situation shown in the diagram where two filters are cascaded:

Input x → [a] → [b] → Output y.

The output of the first filter is $a * x$, which is the input to the second filter. Hence, the output of the second filter is

$$y = b * (a * x).$$

Thus, the two filters may be replaced with one filter whose impulse response is the convolution of the impulse responses of the two filters; that is,

$$c = b * a.$$

The foregoing diagram is thus equivalent to the following diagram:

Input x → [$c = a * b$] → Output y.

We will now discuss the meaning of the concept of the impulse response of the filter. The first concept we need is that of the unit impulse (or spike). This is a simple, positive unit displacement of the type shown in figure 5.2 (top). An automobile has a system of springs and shock absorbers. The purpose of this system is to transmit the ups and downs of a bumpy road into a smoother ride for the passenger. Suppose that we take a very sharp bump of a given size and let that bump represent the unit impulse. We drive over a level road with just this one bump. When we hit the bump, we inject a unit spike into the system. The resulting up-and-down motion of the passenger as a function of time is typically of the form shown in figure 5.2 (bottom). This motion due to a unit spike is the impulse response of the spring/shock-absorber system. In general, the output of any linear system when a spike is used as the input is called the impulse response of the system. It specifies the behavior of the system uniquely.

We shall describe the principle of superposition. Let us suppose that we are driving down a level road at constant speed and that we hit, one after the other,

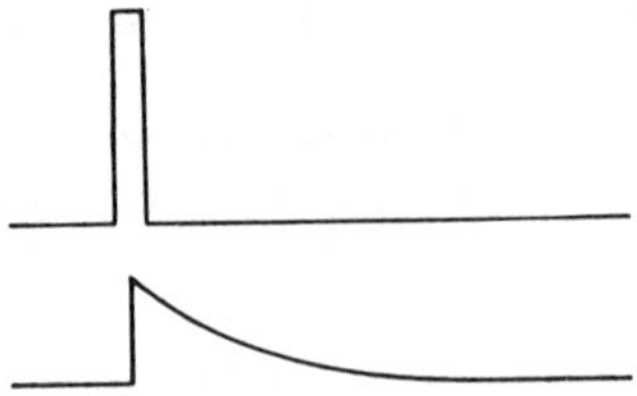

*Figure 5.2. Unit spike input (*top*), and resulting impulse-response function (*bottom*). Reprinted by permission of Gebrüder Borntraeger from Anstey 1970.*

two spike-like bumps, of size 0.50 and 0.75, respectively. Now we invoke the property of linearity. We make use of this powerful theorem: Provided that the system is a linear one, the output obtained from two input spikes is the sum of the two impulse responses produced by the two spikes individually (see fig. 5.3). The usefulness of this result is seen when we note that any road (system input) may be regarded as a contiguous succession of bumps (spikes). Each of these spikes produces an impulse response of an amplitude proportional to the strength of the spike itself. The passenger motion (system output) is the sum or superposition of these impulse responses. In general, any input waveform produces an output waveform given by a superposition of impulse-response functions. The *principle of superposition* can thus be stated as follows: For any linear system and for any input waveform, the output waveform is the superposition of all the impulse responses obtained as if the input waveform were a succession of spikes (see fig. 5.4).

We will now discuss the principle of convolution. In convolution, we want to show that the determination of the output signal at a particular time requires us to add the contributions from all the impulse responses generated by spikes within a time range. This time range is of duration equal to that of the impulse response and extends backward from the given time. If the impulse response is of a decaying type, such as that illustrated, then the contribution from impulses responses generated by spikes in the immediate past is greater than that from more distant spikes. The construction of the output waveform in this manner is an operational way of carrying out the mathematical concept of convolution. The physics behind convolution is exactly the same as the physics of the superposition of impulse responses. Refer to figure 5.2, which depicts the unit spike and the impulse-response function. Figure 5.5 again shows the unit spike but now shows the impulse response reversed in time. Let us keep the spike waveform stationary but proceed to slide the reversed impulse response past it, from left to right. In each new position, we multiply together the strength of the spike and the corresponding ordinate of the impulse response. Thus, in figure 5.5(b), the strength of the spike is unity but the corresponding value of the impulse response is zero, so that the product is zero. We plot this value on the right-hand curve, as shown by the arrow. In (c), the impulse response has moved so that its leading edge coincides with the spike. The product is a positive value, which we plot on the right-hand curve, again as shown by the arrow. Similarly, (d) and (e) show later positions of the impulse response as we slide it by the spike. The arrows show these contributions to the right-hand curve. We see that as the right-hand curve develops, it has the same form as the impulse response itself.

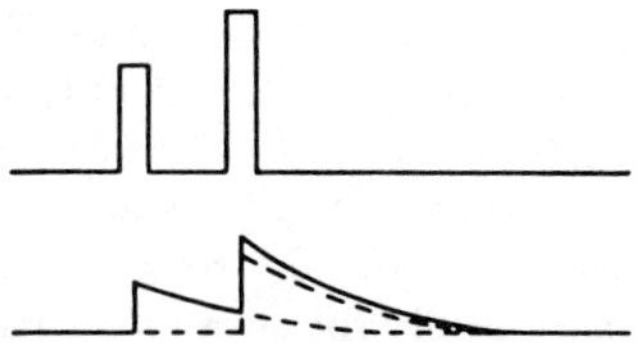

Figure 5.3. The basis of the superposition method of computing the effect of a system on a known input, in the time domain. The top diagram is the input, and the bottom one is the output. Reprinted by permission of Gebrüder Borntraeger from Anstey 1970.

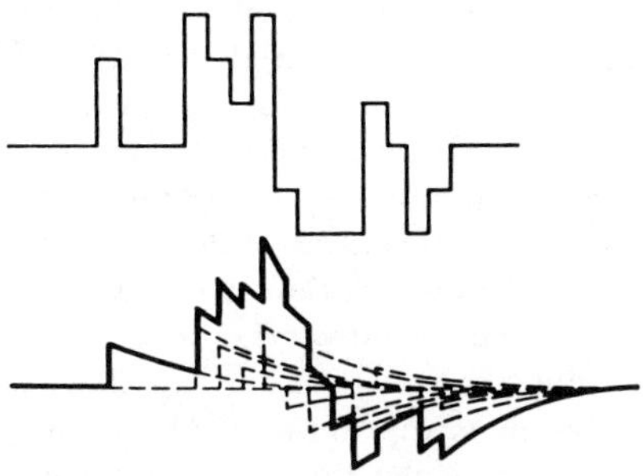

Figure 5.4. The superposition method for a more complicated input. The top diagram is the input, and the bottom one is the output. Reprinted by permission of Gebrüder Borntraeger from Anstey 1970.

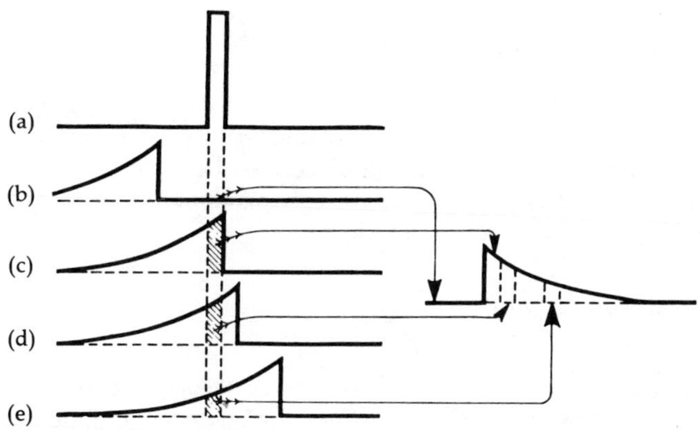

Figure 5.5. Introducing the equivalence of superposition and convolution. The input is the unit spike (a), and the output is the curve to the right. Reprinted by permission of Gebrüder Borntraeger from Anstey 1970.

In figure 5.6, we repeat this sliding process for a double-spike input. The positions of the sliding impulse response in (b) and (c) are such that only the first spike is involved; thus, we are only repeating the operations of the preceding figures. In (d), however, we see that where more than one spike is involved, we must ensure the correct contribution from each. Specifically, for each new position of the sliding impulse response, we must multiply the strength of each spike by the corresponding value of the impulse response and then add the products to obtain the value of the output at the time associated with that position. The output value at the third arrow is seen to be the sum of a large contribution from the second spike and a smaller one from the first spike. Thus, the output waveform is the same as that obtained by superposition in figure 5.3.

In summary, we can think of a filter as an instrument that produces an interaction of an input waveform with an impulse response to produce an output waveform. For any linear system and for any input waveform, the output waveform is the convolution of the system impulse response with the input waveform. However, we can also consider this interaction between impulse response and input waveform in terms of superposition. To repeat, superposition and convolution are merely apparently different ways of describing one physical reality. The concept of convolution, however, brings out that every physical system has a memory whose length is the duration of its impulse response, and that the output obtained at a particular time is a weighted sum of the input at all past times within this memory (see fig. 5.7).

Although we have been expressing convolution as an operation on sampled data, one can also convolve a sample set with a continuous function:

$$y(t) = a * f(t) = \sum_k a_k f(t - k).$$

Each term in the summation represents the function $f(t)$ displaced and scaled; that is, each continuous function $f(t)$ is displaced to the right k units and multiplied by a_k. A special case is that of convolving a continuous function $f(t)$ with a unit impulse located at $t = \tau$. We have

$$\delta_{t-\tau} * f(t) = \sum_k \delta_{k-\tau} f(t - k) = f(t - \tau),$$

since $\delta_{k-\tau}$ is zero except for $k = \tau$, where $\delta_{k-\tau} = 1$. Hence, convolving $f(t)$ with a time-shifted unit impulse displaces the function by the same amount and in the same direction as the unit impulse is displaced. We can also convolve two continuous functions. In this case, the summation becomes the integration:

$$y(t) = x(t) * f(t) = \int_{-\infty}^{\infty} x(\tau) f(t - \tau)\, d\tau.$$

The seismic convolutional model

In his classic work, Norbert Wiener (1949) provided the basis for the statistical theory of communication. Much of the available signal theory of Wiener was developed for radar and telephony purposes. It is based on the assump-

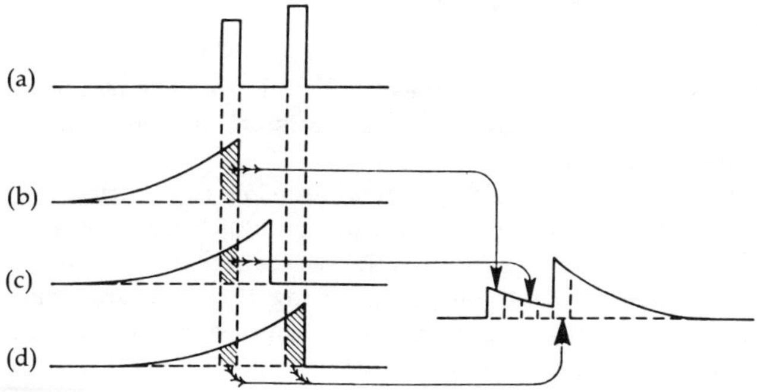

Figure 5.6. The convolution equivalent for the illustration in figure 5.3. Reprinted by permission of Gebrüder Borntraeger from Anstey 1970.

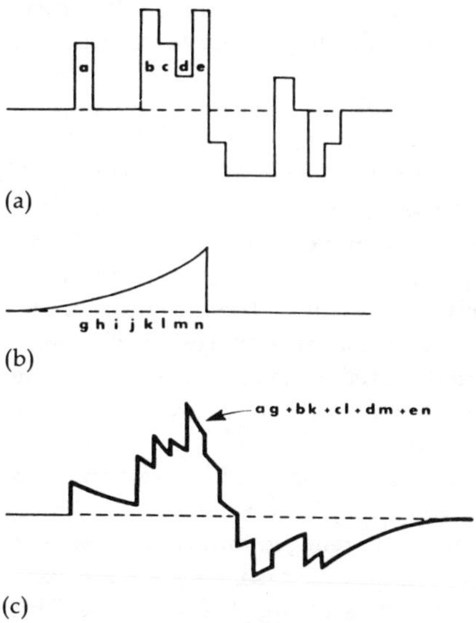

Figure 5.7. The convolution method applied to the example in figure 5.4. We take the complicated waveform (a) and convolve it with the same inpulse response to obtain the output waveform (c). Again, the result is the same as that obtained by superposition in figure 5.4. The reversed impulse response (b) is shown with its initial portion aligned with the input waveform ordinate marked e. *The output at this time is therefore the sum of corresponding ordinate products,* ag + bk + cl + dm + en. *Reprinted by permission of Gebrüder Borntraeger from Anstey 1970.*

tion that a pure signal of some known shape or characteristic is transmitted from a source and is later received, having been contaminated by noise at some distant recording point. The noise is usually assumed to be additive in nature. If $w(k)$ is the known signal, $n(k)$ the noise, and $x(k)$ the received signal, then the model is

$$x(k) = w(k) + n(k).$$

The problem then becomes one of retrieving the known signal $w(k)$ from the received signal $x(k)$. If the signal has been damaged in transmission, we would use instead the model

$$x(k) = c(k) * w(k) + n(k),$$

where $c(k)$ is the transmission filter that modifies the signal. The problem now becomes one of restoring the received signal $x(k)$ to the original shape or character of the transmitted signal $w(k)$.

Unfortunately, most of the operations in petroleum seismic exploration fail to meet the basic assumptions of Wiener's statistical communication theory in its original form. Except for the Vibroseis signal, the source signal is produced with an impulse energy source. On land, weight dropping, thumper, and conventional dynamite all produce a single, relatively sharp impulse, but all sources have in common the lack of a known definitive signal shape or signature that is to be extracted in the processing. Subsequent influences modify the pulse to an uncertain form, so that the processing must proceed under various assumptions or approximations. Despite these serious and far-reaching difficulties, techniques that work extremely well have been developed and form the basis for present-day seismic signal processing.

The reason for this success is that the role played by the components of the model as it is used in telecommunications is completely reversed for seismic signal processing. In telecommunications, as we have seen, the problem is one of extracting a known transmitted signal $w(k)$ from the noise-contaminated received signal $x(k)$. In seismic work, the shot of conventional dynamite develops a very sharp pulse, containing, for all practical purposes, a completely white broad-band signal of all frequencies. Very quickly after the initiation of the pulse, however, it is substantially modified by the elastic properties of transmission through the earth to a severely bandlimited frequency spectrum. The amount and manner of modification is unknown and not readily determined. Hence, the shape or signature of the transmitted waveform is unknown or imperfectly known. To complicate the matter, seismic noise often has properties very similar to those of the signal. All of these factors made Wiener in the 1950s, as well as those who applied his work to radar and telecommunications, believe that his filter theory was not applicable to the seismic case. As we have seen, however, it was a reversal in roles that completely changed this picture. This reversal of thinking opened up a whole new vista of signal processing.

In geophysical exploration, the system includes the earth, the geophones, the amplifiers, and the modulation instruments that place the signal on tape. It is often difficult or impossible to determine the individual contributions of each of the components. The use of the model is complicated by the fact that the response may be somewhat different from trace to trace and from shot to shot. The same general type of broad-band spike signal is transmitted each time, but the conditions of the shot hole and the lithology of the subsurface may change in such a manner as to cause major variations in the received seismic signal. Fortunately, the majority of the components, although they cannot be individually defined, remain constant, and variations in the system are normally due only to some change in the conditions of the transmission medium, the earth, and the coupling of the earth to the instruments. We must exploit the fact that the earth leaves its characteristic imprint on the received seismic records.

The approach used in seismic signal processing, as compared to that used in telecommunications, stems from the recognition that it is not the transmitted signal that we want to recover, but the variations in the lithology of the subsurface. The transmitted seismic wavelet is only an instrument used to obtain the lithology. This approach was developed in the work of the Geophysical Analysis Group in the early 1950s. This group, sponsored by the oil and geophysical industry, had the responsibility of finding a way to apply statistical communication theory to the seismic problem. The oil and geophysical companies not only sent their best research people to participate in this venture but also provided a wide range of seismic data and, if available, well logs. The well logs were few and far between, however, and the seismic data, of course, were analog traces on photographic-paper records.

Seismic theory in the early 1950s regarded a seismic trace as an entity that resulted from the solution of the wave equation with complicated initial and boundary conditions. Earthquake seismologists had obtained solutions in the case of very simple boundary conditions, but no comparable solutions had been obtained by exploration seismologists for any practical situation. As a result, communication theory offered a new approach that was worth trying. The concept that the seismic signal can be represented as the convolution of the seismic wavelet with the earth's reflectivity was developed by Robinson (1954) as a report of the Geophysical Analysis Group. This result is called the *seismic convolutional model,* which takes the form

$$x(k) = c(k) * w(k) + n(k),$$

where k = integer-valued time index
$x(k)$ = received seismic trace
$c(k)$ = reflectivity function (i.e., the sequence of reflection coefficients representing the desired lithology)
$w(k)$ = seismic wavelet (including source effects, absorption, multiple-reflection effects, and receiver effects)
$n(k)$ = additive noise (which generally can be made small by the use of good field techniques)

In the seismic convolutional model, the reflectivity function $c(k)$ is the desired information, not the transmitted signal $w(k)$, as it would be in the case of telecommunications.

The convolution model serves as a bridge that connects and joins the work of field geophysicists with that of research geophysicists. The data experimentally obtained can be used in the convolutional model, and the results can be related with theoretical studies. United Geophysical Company, an active participant in the Geophysical Analysis Group (GAG), used the convolutional model as a forward model to generate synthetic seismograms (Peterson, Fillipone, and Coker, 1955). We will discuss synthetic seismograms in the next section. In the preceding chapter, we presented the development of the Vibroseis method by Continental Oil Company; the convolutional model plays an important part in Vibroseis theory. The first use of the convolutional model, however, was as an inverse model to discover the concept and to develop the theory of predictive deconvolution (Robinson, 1954). Whereas the direct use of the convolutional model (synthetic seismograms) could be applied right away in the 1950s, the inverse use (deconvolution) had to await better quality data and lower cost computing, which finally occurred in the 1960s. The great work of Geophysical

Service, Inc. (GSI), and its close ties, through its chairman Cecil Green, with the Geophysical Analysis Group and the whole MIT academic community, is well known. GSI was the early leader and far-seeing innovator in digital processing (Smith, 1956; Backus, 1959).

In any sort of historical survey of the seismic industry, the work of the geophysical companies always takes a prominent place. Credit for this work is well deserved. Usually, however, equivalent work is done at the same time within oil companies. An oil company, being self-contained, has no reason to disclose the results of its research and especially not to advertise its work. A geophysical company, on the other hand, must advertise its work in order to obtain continuing or new business. The oil companies, through their major and enlightened support of research in geophysics and other areas of petroleum exploration, have made this the era of petroleum, even though the names of many of the company personnel who make significant and far-reaching discoveries and scientific advances are generally unknown to the academic community.

One of the greatest of the oil company geophysicists was Norman Ricker, who upon retirement from Exxon became professor of geophysics at the University of Oklahoma. In the 1940s and 1950s, Ricker developed his wavelet theory (Ricker 1940, 1953) while working at the Exxon Research Center in Tulsa. (This center was given by Exxon to the University of Tulsa in 1965 when Exxon moved to Houston.) Ricker started with the equations of motion of a viscous fluid (the Navier-Stokes equations) and, to the amazement of many physicists, was able to find an analytic solution that represented a traveling seismic wavelet. Such a solution would account for the loss of high frequencies in the source wavelet as it was transmitted through the earth. To test his theory against data, Ricker carried out controlled seismic experiments in the Pierre shale formation in Colorado. Ricker expanded his wavelet theory in many other directions, and the Ricker wavelet became one of the cornerstones of exploration geophysics. I was deeply influenced by the work of Norman Ricker, especially by Ricker's 1953 paper on the control of seismic resolution. However, the Geophysical Analysis Group had only observational data, in the form of field seismograms and well logs, and certainly did not have the funds required to carry out the expensive controlled field experiments as done by Exxon, Mobil, and other oil companies. In order to support and complement the work of Ricker, the GAG had to attack the problem from a completely different point of view. The GAG started with recorded observational data (field traces) and ended with the digital convolutional model of the seismic trace. This was in contrast with Ricker's approach of starting with a theoretical mathematical model (Navier-Stokes equations) and ending with controlled experimental data to confirm the theory. The seismic wavelets found by the two approaches were quite different in shape (fig. 5.8), but no longer could a seismic trace be regarded as an indivisible entity. Instead, it must be looked upon as an elaborate wavelet complex, as originally envisioned by Ricker. This was the beginning of a long friendship. As editor of *Geophysics,* Ricker had the my work published in the 1957 issue. In the autumn of 1964, Professor Ricker visited Sweden, and in his lectures at Uppsala University, in a setting completely removed from the oil industry, the whole beauty of his scientific work came forth to a receptive audience, including several well-known mathematicians and physicists. Within the oil company research

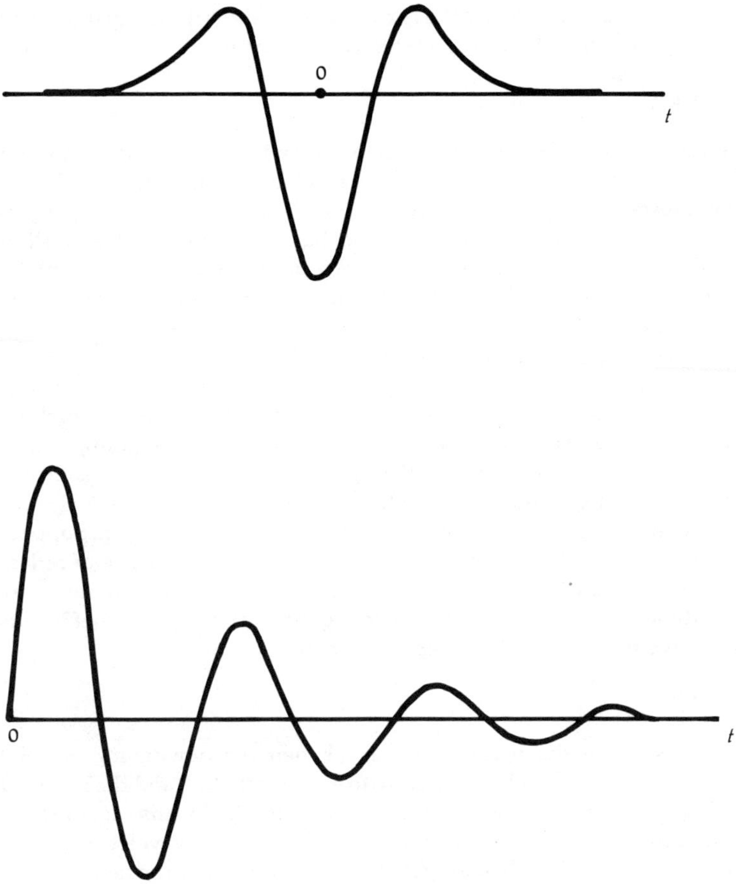

*Figure 5.8. A theoretical Ricker (1940) wavelet (*top*) and a statistically computed wavelet (Robinson, 1954) (*bottom*). The computed wavelet is a causal, minimum-phase wavelet and thus, under proper field conditions (spike-like source), represents the physically transmitted seismic wavelet (field wavelet). The theoretically based Ricker wavelet is a zero-phase wavelet and thus is not a good field wavelet representation, but it is widely used today as an interpreter wavelet and in seismic modeling studies.*

laboratories today, there are other exceptional scientists such as Ricker; unfortunately, however, their work often is not fully brought to the attention of an appreciative academic audience.

The seismic convolutional model forms the starting point of the theory of the filtering of the received seismic data in order to improve the quality of the final display. Such filtering operations form the basis of *wavelet processing.* The objective of wavelet processing is the creation of a visual picture of the subsurface that can be readily and meaningfully interpreted by geologists and geophysicists. At this final stage, the physical properties of transmitted wave motion are no longer important or needed. Instead, a nonphysical wavelet, called the interpreter wavelet $w_i(k)$, is introduced solely for its visual effect. The final seismic display trace—that is, the ultimate processed seismic trace—is called the *inter-*

preter trace $y(k)$. Its model is given by the convolution of the reflectivity (reflection coefficient) function with the interpreter wavelet; that is,

$$y(k) = c(k) * w_i(k).$$

Thus, the interpreter trace is made up of an interpreter wavelet attached to each of the reflection coefficients in the reflectivity function. All the physics of wave transmission present in the received trace $x(t)$, such as source effects, multiple-reflection effects, and receiver effects, should have been removed as well as possible, so that, at this stage, only the lithology in the form of the reflectivity function remains. The artificially introduced interpreter wavelet is used only to make the estimated reflectivity function more pleasing to the eye. The interpreter wavelet contains no physics.

We now will give further discussion and justification of the seismic convolutional model. To make the main ideas clear, consider the simplest example of the reflection seismic process. A source explosion initiates a downgoing seismic wavelet $w(t)$, where t is time. For simplicity, let us suppose that the source is a plane wave and that the interfaces are horizontal. The reflection coefficients of the interfaces make up a sequence of numbers as a function of depth. Suppose that we convert depth to traveltime, so that we now have a sequence of reflection coefficients as a function of two-way traveltime. We denote the two-way vertical traveltime t for interface i by t_i. The sequence of reflection coefficients for the successive interfaces can be represented by

$$c(t_0),\ c(t_1),\ c(t_2),\ \ldots .$$

This sequence is called the *reflectivity function.* When the downgoing wavelet strikes interface n, the wavelet is reflected with amplitude $c(t_n)$, and the reflected wavelet will return to the surface with reflection time t_n. In this ideal model, transmission coefficients are ignored, so that the downgoing wavelet continues to travel down with undiminished amplitude. When it reaches the next boundary, another reflected wavelet is produced. We thus record a sequence of reflected wavelets

$$c(t_n)w(t - t_n)$$

received at the surface, where the interface index n is equal to 0, 1, 2, ... The superposition of all these wavelets forms the seismic trace $x(t)$; that is,

$$x(t) = \sum_{n=0}^{\infty} c(t_n)w(t - t_n).$$

Each of the reflected wavelets represents a primary reflection. The model under discussion does not take any multiple reflections into account. More complex earth models allow for some or all multiples.

Thus, we can describe the convolutional model as follows: If the subsurface consists of plane-parallel homogeneous layers, we can view the reflectivity as the sequence made up of the reflection coefficients associated with the interfaces. In the noise-free case, the reflection seismic trace is the superposition of many wavelets, each having the waveform $w(t)$ but with strengths and arrival times corresponding to the multitude of interfaces encountered in the subsurface. The seismic wavelet $w(t)$ has the wave shape that would be recorded with our actual seismic system for the reflection from a single plane-reflecting boundary in the subsurface. The reflection coefficients $c(t_n)$ correspond to the normal incidence-reflection coefficients. The reflection times t_n correspond to the two-way traveltimes of the reflecting boundaries.

In the convolutional model, the reflection seismic trace $x(t)$ is the convolution of a seismic wavelet $w(t)$ with the subsurface reflectivity $c(t)$. If we use the asterisk as the symbol for convolution, we can write the convolutional model simply as

$$x(t) = c(t) * w(t).$$

In the general case, the recorded seismic trace $x(t)$ is treated as the sum of the seismic signal $w(t) * c(t)$ plus additive noise $n(t)$:

$$x(t) = c(t) * w(t) + n(t).$$

The seismic trace is thus treated as a noisy, filtered version of the subsurface reflectivity. The seismic wavelet is the impulse response of the filter, and the transmission medium (i.e., the reflectivity $c(t)$) is the desired information. This interpretation is the new twist that was given to the seismic convolutional model, as opposed to the conventional convolution models used in telecommunications and other fields. The information theory models devised by Norbert Wiener, Claude Shannon, and others in communication theory in the late 1940s and early 1950s always regarded the transmission medium (which in the seismic case is the reflectivity $c(t)$) as the impulse response of the filter, and the transmitted signal (which in the seismic case is the wavelet $w(t)$) as the desired information. To them, the transmission medium was of no direct informational value, as it only masked the true signal $w(t)$ that was sent; an example would be rain droplets and cloud formations that add clutter to the transmitted radar signal. This reversal of the roles played by the components seems obvious now, but at that time, in the early 1950s, it was the deciding factor that allowed the application of communication theory methods to the seismic exploration problem. More important, this seismic interpretation of the convolutional model opened up the need for deconvolution, which was not even recognized as a problem at that time in radar and telecommunications. This explains why deconvolution originated in the field of exploration geophysics in the early 1950s and actually was in production usage in seismic exploration for about a decade before it was introduced to other disciplines in the late 1960s and early 1970s. Since that time, deconvolution has gained wide acceptance and constant usage in many diverse areas of signal processing.

The convolutional model as written here is for layers with arbitrary reflection times t_n. For digital computation, it is more convenient to have the reflection times occur at equally spaced time points. We can achieve this end by adding hypothetical layers with zero reflection coefficients as necessary, so that the arrival times can be approximated as

$$t_k = k\,\Delta t,$$

where the index k is an integer and Δt is the sampling time interval (such as 4 msec). Thus, the reflectivity function becomes the digital signal $c(k)$ at equally spaced discrete sample times $k = 0, 1, 2, \ldots$. Also, the seismic wavelet $w(k)$, the noise $n(k)$, and the received seismic trace $x(k)$ become digital signals at equally spaced discrete sample times $k = 0, 1, 2, \ldots$. The *convolutional model* simplifies to the discrete convolutional equation given by

$$x(k) = \sum_{n=0}^{\infty} c(n)w(k - n) + n(k),$$

which is the way we wrote it earlier in this section.

The convolutional model is important for an understanding of the seismic method. There has recently been a major resurgence in interest in the model, however, because of its applicability in direct hydrocarbon detection, porosity mapping, and stratigraphic mapping. There is a direct relationship between reflectivity and impedance for the one-dimensional problem. The impedance Z is defined as ρv, where ρ is the density and v is the wave velocity. We recall that the (pressure) reflection coefficient for the interface between layers 1 and 2 is

$$c = \frac{Z_2 - Z_1}{Z_2 + Z_1}.$$

Let us now consider many interfaces and treat the impedance as a function of the two-way traveltime t. We thus write this impedance function as $Z(t)$. Let us now consider a small time interval $t_1 \leq t \leq t_2$. We can thus approximate $Z(t)$ as the average

$$Z(t) = \frac{Z(t_2) + Z(t_1)}{2}$$

and the increment $dZ(t)$ as

$$dZ(t) = [Z(t_2) - Z(t_1)]dt.$$

Thus, the foregoing reflection coefficient becomes approximately

$$c(t)dt = \frac{dZ(t)}{2Z(t)}.$$

From calculus, we know that

$$d \log Z = \frac{dZ}{Z}.$$

Hence, the reflectivity function can be written as the approximation

$$c(t) = \frac{d}{2dt} \log Z.$$

This formula recognizes the existence of continuous variations in acoustic impedance and expresses the reflectivity in terms of the time derivative of the logarithm of impedance versus two-way traveltime. We can also write this equation in terms of density ρ (which does not vary much) and wave velocity $v(t)$ (which varies greatly). Since $Z = \rho v$, the required expression is

$$c(t) = \frac{d}{2dt} \log[\rho v(t)].$$

If we integrate, we obtain

$$\log[\rho v(t)] = \log[\rho v(t_0)] + 2 \int_{t_0}^{t} c(t)\, dt.$$

This relationship between impedance and reflectivity is a linear operation, which may also be treated as a filter. Thus, the seismic signal may be viewed as a noisy filtered version of either the reflectivity or the impedance versus two-way traveltime. In this formulation of the convolutional model, multiple reflections and transmission losses at the reflecting interfaces are neglected. In the context of a properly processed seismic trace (deconvolved and stacked), this approximation is generally valid.

The one-dimensional nature of the convolutional model represents a significant qualification. The presence of curvature of the bedding planes, or variations in impedance parallel to the bedding planes, requires a three-dimensional description of impedance in order to calculate the reflectivity. We should not use the normal incidence-reflection coefficient for angles of incidence greater than about 15°, nor for large offset distances (Hilterman, 1970, 1975). Wavefront divergence and frequency-dependent attenuation may significantly modify the reflectivity from the one-dimensional approximation. Despite all of these approximations, the one-dimensional model can be usefully applied to properly processed seismic data. Furthermore, slant stacking and migration can reduce three-dimensional data to a state at which the one-dimensional convolutional model is useful.

To summarize, in the convolutional model, we think of the earth as a filter of seismic energy. We consider the wave resulting from an explosion as an impulse. To a good approximation, the wave motion at the source of the explosion is zero both before and after the explosion and differs from zero only in an extremely short interval starting at time zero. During this infinitesimal interval, the motion is very large. If nothing were to broaden the impulse, the signal that we record would be simply the sharp impulse convolved with the impulse response of the earth. The result would be zero, except for sharp pulses corresponding to the arrivals of different reflections. However, these reflections would be made up of both primary reflections (which we desire, as they indicate the reflectivity function, or reflection coefficient sequence) and multiple reflections (which we do not desire, as they tend to mask the primary reflections). If our record were of the type consisting of all primary and multiple reflections with an impulse-type wavelet, theoretically we could use dynamic deconvolution (Robinson, 1975) and recover the sequence of reflection coefficients, thus determining the geologic structure with great accuracy. Thus, if this were so, we could easily determine from the recorded data the complete solution to the seismic problem. In practice, however, we get back not only primary and multiple reflections, but also diffractions, surface waves, scattered waves from near-surface irregularities, reflected refractions, three-dimensional effects such as side-swipes, and so on. Moreover, the source impulse quite soon becomes spread out into a broad seismic wavelet, so that the wavelets resulting from all the reflections and other events overlap and interfere with each other on the recorded seismic trace. In addition, the seismic trace is modified by filtering, because of absorption and other causes, and random noise is always superimposed. Despite all this, it is remarkable that we can successfully approximate the recorded seismic trace in the form of successive convolutions. In the convolutional model, the trace is written as the result of the successive convolutions of the shot impulse with the impulse response of the various portions of the earth through which the wave travels. We can arrive at a useful approximate picture by considering the earth to be divided into various zones. Such divisions of the seismic process into zones depend on the purposes to which the models are directed. A typical division would be as follows:

(1) The zone near the shot, where stress levels and the absorption of the highest frequencies is very severe. We write $s(k)$ for the impulse response of this zone and call $s(k)$ the source wavelet. The s in $s(k)$ stands for *source.*
(2) The reflecting sequence of beds whose impulse response $c(k)$ is the reflection coefficient sequence. This is the information we are seeking to discover by seismic exploration. This reflection coefficient sequence $c(k)$ is the traditional symbol used for *coefficient of reflection.*
(3) The multiple reflection generating mechanism whose impulse response is denoted by $m(k)$, where m stands for *multiple reflection*.
(4) The absorption mechanism with impulse response $a(k)$, where a stands for *absorption*.
(5) Finally, the recording instruments and geometry that act in the environment of the near-surface zone. This component exercises considerable filtering action in changing the waveshape. We write $r(k)$ for the impulse response of this last zone, where r stands for *receiver.*

Neglecting additional filtering effects, we thus write the received seismic trace as the (noise-free) convolutional expression:

$$x(k) = s(k) * c(k) * m(k) * a(k) * r(k),$$

where x is the trace, s is the source, c represents the coefficients of reflection, m represents the multiples, a is the absorption, and r is the receiver. Usually, we lump all of the items on the right, except the reflectivity $c(k)$, into the seismic wavelet $w(k)$; that is,

$$w(k) = s(k) * m(k) * a(k) * r(k),$$

so that the (noise-free) convolutional model becomes simply

$$x(k) = c(k) * w(k),$$

where x is the trace, c represents the coefficients of reflection, and w is the wavelet. The objective of seismic processing is to convert the received trace $x(k)$ to the interpreter (or display) trace

$$y(k) = c(k) * w_i(k),$$

where w_i is a spike-like zero-phase interpreter wavelet.

Synthetic seismograms

The seismic reflection trace is a complex montage of events, representing the superposition of a very large number of overlapping reflected pulses from a profusion of geologic interfaces. To understand the seismic trace, the geophysical industry makes use of the synthetic seismogram (Peterson, Fillipone, and Coker, 1955). As we know, the velocities of seismic waves are typically several meters per millisecond. The initial seismic pulse that originated at the shot point travels downward and is reflected. When it returns to the surface, the pulse has been broadened into a wavelet that is stretched across a distance of about 100 m. On the other hand, a typical well log can distinguish features as small as only a few meters in extent. We thus realize that the seismic pulse in the earth is many times longer than the thickness of most of our reservoirs. We realize that the seismic method represents an exploration instrument with a blunt or dull edge in comparison to the reservoirs that we want to discover and map. One of the main ways of sharpening our picture is through the use of *synthetic seismograms*.

An immediate and practical example of the application of the convolutional model is the production of synthetic seismograms (see fig. 5.9). Once the basic wavelet shape is determined, it can be used to synthesize a reliable approximation of a seismic record for any location where an acoustic log is available. A synthetic seismogram is nothing more than a model of the actual seismic trace. The synthetic is developed as follows. First, make a multilayered model of the earth. Each layer represents some small, uniform traveltime increment, usually 1 msec one-way time. From the acoustic log, determine the velocity within each layer. From the velocity function, compute the reflection coefficient at the boundary of each layer. These reflection coefficients will always be numbers between zero and one, positive or negative, and will constitute a small spike at each layer boundary. Let the series of spikes be the input to the system model. As the spikes, in turn, pass through the system, they are convolved with the wavelet (the unit-impulse response). Convolution of the wavelet with the reflection coefficients produces an output trace that is a close approximation to a noise-free seismic trace. The accuracy, of course, depends on whether the wavelet shape has been correctly defined and whether the reflection coefficients computed from the log represent a reliable model of that particular sedimentary region.

The use of the convolutional model not only has the effect of modeling the input and output, but the procedure closely parallels the process of the real earth. The synthetic trace is obtained by convolving a series of spikes (the reflection coefficients) with the seismic wavelet. In the real earth, the reflection coefficients may be considered to be at the layer boundaries. The shot pulse is modified to a broader wavelet, and as it passes through the earth it is reflected by each of the boundaries. The recorded trace thus represents the convolution of the wavelet with the series of reflection coefficients that exist at the boundaries of the individual layers within the earth. The detection and measurement

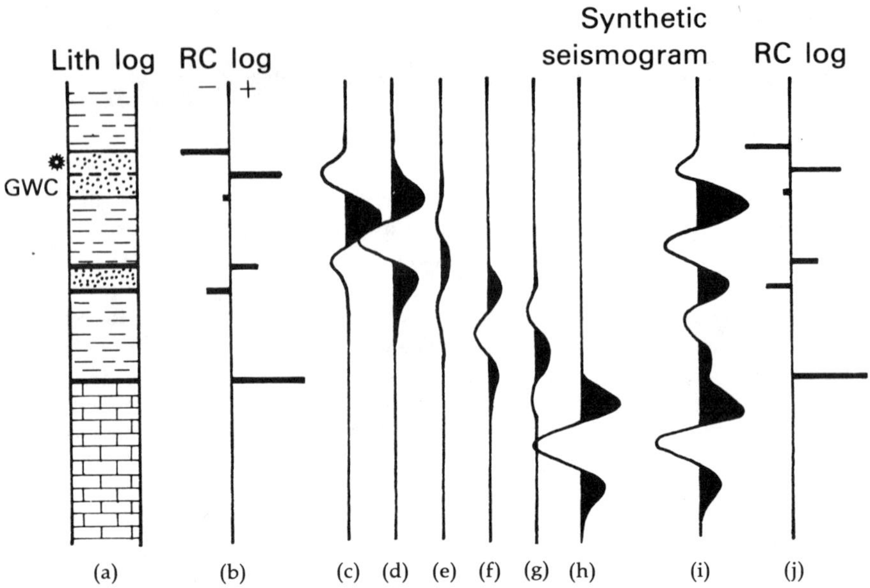

Figure 5.9. The construction of the basic synthetic seismogram. The lithologic log and the RC log (reflectivity function) are shown by (a) and (b). There are six reflectors, represented in time, strength, and sign by the spikes of the RC log. For each of these, in (c) through (h), we draw a reflection pulse starting at the appropriate time, scaled to the appropriate strength, and erect or inverted according to the appropriate sign. They overlap, of course. Then we add them together to obtain the synthetic seismogram shown in (i). This is the seismic response to the given lithologic log for the stated form of the seismic pulse. Finally, (j) is a repeat of the RC log for easy comparison with the synthetic seismogram. The relationship between the two is obscure. First, the obvious part of each pulse is delayed relative to the expected time on the RC log. Furthermore, it is not clear what part of the pulse we should use for our judgments of the time, strength, and sign of the reflections. Reprinted by permission of IHRDC from Anstey 1982.

of these reflection coefficients represent the desired end result of the process of seismic exploration. Geophysicists and interpreters try to find the boundaries and to map them as they carry through the seismic data. We are normally content if we can merely define the location of a boundary within a few milliseconds, since greater precision is usually not attainable with any degree of confidence.

In the foregoing description of the convolutional model, we have used a discrete sequence of reflection coefficients as the reflectivity function. In practice, the reflection coefficients are determined from the acoustic log as a series of spikes at some small interval, such as 1 msec. However, each coefficient may be considered the amplitude of an individual sample taken from a continuous signal sampled at a 1 msec increment. From the standpoint of time-series analysis, the reflection coefficients in fact represent a sampled continuous function. To make this continuous function comparable with the seismic signal, the time scale must be double, since seismic records measure two-way time. The representation of the earth column as a time series of reflection coefficients conforms closely to other operations related to geology. The sedimentary section represents a window extracted from a continuous time series extending through geologic time from minus infinity to infinity. The generation of a seismic record represents the convolution of the seismic wavelet with the time series corresponding to the sedimentary section. As a result, the earth filter obeys all the rules of convolution operations, and the results can be treated by methods of signal analysis. In particular, the process of deconvolution seeks to reverse the effects of convolution of the wavelet with the reflectivity in order to restore the reflectivity function, which is representative of the lithology. A great deal of effort has gone toward the development of improving techniques for deconvolution, and this work has been successful. Virtually every seismic record is deconvolved, and often they are deconvolved twice, both before and after stack.

We will now summarize the main assumption made in the application of the convolutional model to a real seismic trace. The shot pulse, geophone (or hydrophone), and recording system can accurately be characterized as linear time-invariant filters. In the actual data-processing operations, however, we make use of time-variant processes such as spherical divergence correction and normal moveout. The ghosting effect, the near-surface layer effect, and the convolutional effect of source and receiver arrays are all dependent on the emergent angle of the reflected wavefronts. Thus, the single-trace convolutional model is an approximation for these factors, and rapid lateral variations in the near surface produce a noticeable degradation of the approximation. Transmission effects are dependent on the subsurface reflectivity (i.e., the reflection coefficient sequence), and these effects are actually time-varying. The reason is that new reflection coefficients are introduced as the wavelet penetrates deeper, so that the transmission effect depends on seismic traveltime. Thus, the convolutional model for a given seismic trace is most useful when applied to a limited vertical section of the subsurface. Also, the accuracy is better in cases for which the reflections arrive at the surface in waves with a limited emergent angle range.

Let us consider the application of the seismic convolutional model to real situations. There are three basic ways in which the convolutional model is used: (1) data acquisition, (2) data processing, and (3) interpretation.

In data acquisition, the proper control of the seismic source wavelet determines the ultimate quality of the final processed trace. The ratio of signal to noise as a function of frequency is the primary consideration in producing the source wavelet when we activate a seismic source of energy (such as dynamite, air guns, Vibroseis, etc.).

Wavelet processing has the objective of transforming the field data into processed data that best suit the needs of the seismic interpreter (see fig. 5.10). The interpreter wants a seismic section in which the wavelet is of minimum duration compatible with the suppression of additive noise. If the field seismic wavelet is known, then the data can be digitally filtered to produce such an improved wavelet, called the interpreter wavelet. Good processing can produce a record section in which visible lateral variations in amplitude and waveform are due to lateral variations in the subsurface layers, not to lateral variations in the wavelet or additive noise. Also, through data processing, it is possible to convert the trace to an estimate of impedance instead of reflectivity. Such impedance estimates facilitate the tie of seismic data to well log data (see fig. 5.11).

In seismic interpretation, we would like the wavelet to be as short and as sharp as possible so as to facilitate the delineation of thin beds. Wavelet process-

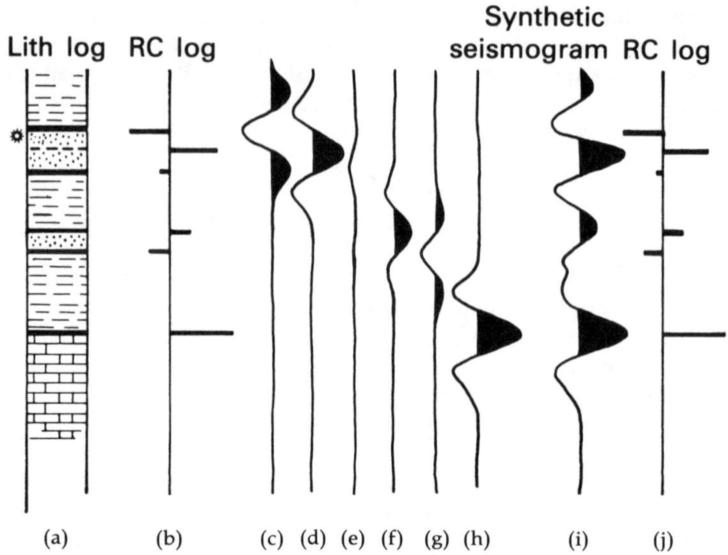

Figure 5.10. The seismogram of figure 5.9 is more directly interpretable, in terms of rock contacts, after wavelet processing. The operation of wavelet processing changes the shape of the pulse in the computer, so that we obtain the record we would have had if the shot had generated this new pulse shape. This new pulse shape is called the interpreter wavelet. In wavelet processing, the computer fixes the time delay and makes the pulse symmetric about the correct time (so that we know where to pick). Also, it makes the center of symmetry black for a positive reflector and white for a negative reflector. After wavelet processing, each pulse is a symmetric interpreter wavelet centered on the correct time. The synthetic seismogram, the result of adding the six interpreter wavelets, now looks much more like the RC log; we have a black peak more or less corresponding to each positive contact and a white trough more or less corresponding to each negative contact. Unfortunately, one black peak and a couple of white troughs do not correspond to a contact and so do not have any useful geologic meaning. By inspection, we can see that these hoaxes are consequences of the wiggly nature of the seismic pulse. Reprinted by permission of IHRDC from Anstey 1982.

ing can shorten the seismic wavelet but can never make it as short as we would like. As a result, the subsurface region of interest must be interpreted on the basis of data in which the diagnostic information has been smeared or spread out to various extents by an imperfect wavelet (figs. 5.12 and 5.13). At this point, models in the form of a synthetic trace come in. In the building of a synthetic seismogram, the subsurface properties are postulated (by use of well logs or other seismic data), and the corresponding reflectivity is calculated. Next, a wavelet, called the interpretive wavelet, is chosen. Convolution of the reflectivity with the interpretive wavelet then provides the interpreter with a synthetic seismic trace, which may be directly compared with the seismic section after digital processing (i.e., after deconvolution, frequency filtering, stacking, and wavelet processing).

The three main inputs to a synthetic seismogram are the v log, the ρ log, and the shape of the seismic pulse. If we have well control, we have the first in the form of the sonic log and the second in the form of the density log. From the velocity and density, we can compute the acoustic impedance ρv. From the acoustic impedance function, we can compute the sequence of reflection coefficients (i.e., the RC log or reflectivity function). As for the seismic wavelet, we can either regard it as a

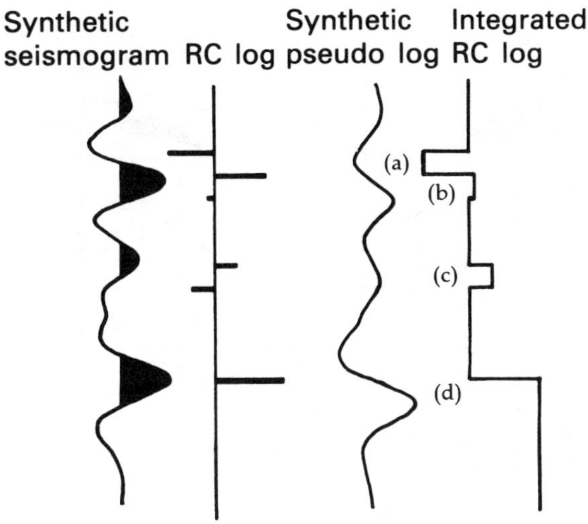

Figure 5.11. An approximate view of the construction of the pseudolog (or seismic log, or acoustic impedance log). The first two traces merely repeat the last two of figure 5.10. The fourth trace is the integrated RC log; this is the log that most clearly represents layers rather than contacts. Then, if we do the same integration (let us think of it, for present purposes, as a simple running sum) on the synthetic trace at the left, we obtain the third trace—the first step toward the synthetic seismic log or synthetic pseudolog. This is our attempt to make the seismic trace represent layers. On comparison, we see that the soft layer (a) clearly deflects to the left; the hard layers (b), (c), and (d) deflect to the right. However, we still have the problem of those outriders on each side of the main indication; on each side of the correct deflection, we have smaller opposite deflections, which are quite spurious. Reprinted by permission of IHRDC from Anstey 1982.

variable or, if we have a seismic line in the area, we can extract it from the seismic data by means of a computer program. Thus, in many oil reservoir situations we can synthesize the expected seismic trace in the neighborhood of the reservoir zone. Such synthetic traces are extremely useful. They allow us to understand why a certain volume of earth gives a certain seismic trace. They also allow us to identify the reflections positively. We must always remember that, because seismic velocities vary from area to area, geophysicists working in a new area cannot always look at the reflections and tell what geologic interfaces they represent. Velocity is so variable that even if the geophysicists know the expected depths, they may find it difficult to deduce the expected times. A synthetic trace tells geophysicists whether or not they will see the reservoir. If not, it tells us why not. If the reason is that the contrast of hardness between the reservoir and its bounding rocks is inadequate, then even excellent processing will not reveal the details. If the reason is that the reservoir is too thin or too close to some very strong reflector that overwhelms it, then we must try to change the shape of the seismic pulse. We must try to change it from its natural spread-out shape into a short zero-phase (i.e., symmetric and quickly damped) pulse centered at the correct reflection time. This wave-shaping process is done by the step in the computational sequence known as wavelet processing. We must always consider the relative cost-effectiveness of this approach and the drilling of holes. The synthetic trace is correct only if the pulse is chosen correctly.

The most successful use of synthetic seismograms is in petroleum reservoir

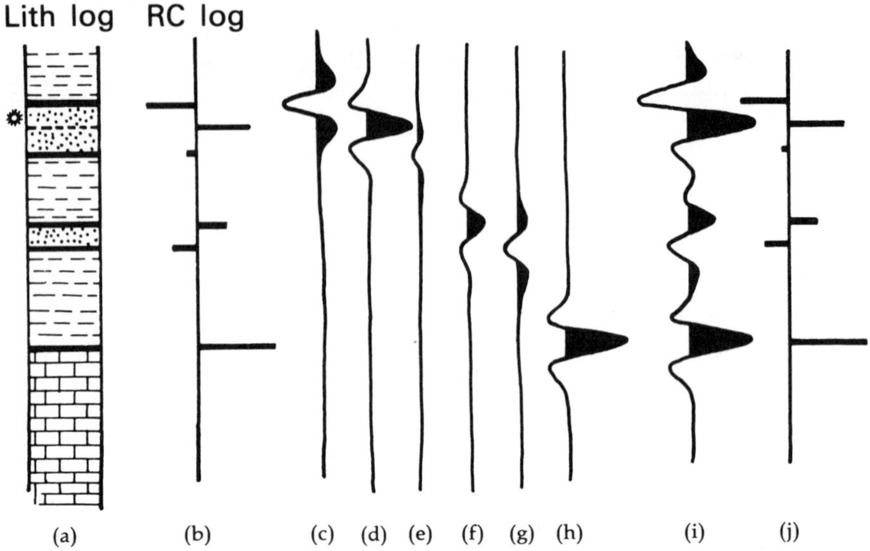

Figure 5.12. The construction of the basic synthetic seismogram with a sharper pulse. We repeat the operations of figure 5.10, but with a sharper pulse. Now we see a clearer and better correlation between the contacts on the RC log and the peaks and troughs of the synthetic trace. We note that the time thickness of the gas column in the upper sand just coincides with the half-period of the pulse, and that this produces constructive interference (or tuning) between the top-sand and gas-contact reflections. This tuning effect exaggerates the gas-water contact. Reprinted by permission of IHRDC from Anstey 1982.

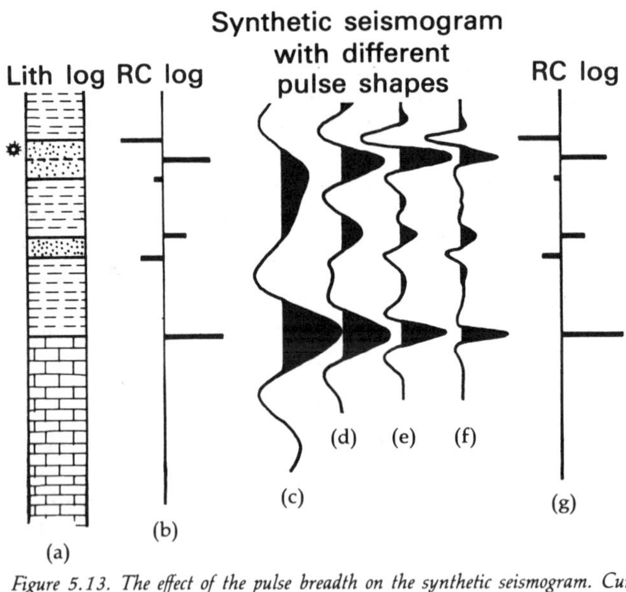

Figure 5.13. The effect of the pulse breadth on the synthetic seismogram. Curves (c), (d), (e), and (f) display the synthetic traces for a range of pulse shapes. The synthetics of (d) and (e) are those we have seen previously, and we now add one with a very poor pulse, trace (c), and one with a very good pulse, trace (f). Reprinted by permission of IHRDC from Anstey 1982.

studies. When we see something on our seismic sections at reservoir level, the synthetic seismogram tells us what part of the reservoir it represents. In particular, the synthetic trace indicates where to look for the all-important gas-liquid contact. A set of synthetics tells us how the seismic response varies as the reservoir changes. We must always be aware of all the fundamental problems whenever we look at a seismic trace. Despite possible pitfalls and ambiguities, the synthetic seismic trace represents one of the most powerful tools available to geophysicists. This is especially true today with the availability in oil companies of large data bases with which geophysicists can construct synthetic seismic sections. The usefulness of synthetics has been demonstrated many times. The synthetic seismogram gives us the seismic trace we can expect to obtain from a defined series of layers in the earth for a given form of seismic pulse. The synthetic is the essence of the modeling step, in which the geophysicist checks the validity of any detailed interpretation of seismic data. From the final interpretation, we synthesize the seismic response and see whether it checks with the original data.

Stationary random processes

We now turn to the statistical approach to time-series analysis. First, we will illustrate the concept of a random process. An observed time series is one specific realization from a stochastic (or random) process. In other words, we consider the received time series to be one sample from a collection of possible signals that might have occurred. The random process is defined as the collection or ensemble of all possible signals, together with their probability of occurrence. We can illustrate the random process graphically by drawing the ensemble of possible discrete time series. As an example, the experiment of repeatedly tossing a coin makes up a random process. Let 1 denote heads, −1 tails, and 0 the coin standing on its edge. (We might suppose that we have a thick coin, not well rounded.) Suppose that the probabilities are

$$\begin{aligned} P(\text{tails}) &= P(-1) = 0.4, \\ P(\text{edge}) &= P(0) \quad\; = 0.2, \\ P(\text{heads}) &= P(1) \quad\; = 0.4. \end{aligned}$$

Suppose that five people each carry out an experiment of five tosses. The results for any one person is a time series, which we write as a horizontal row. We thus have five time series (i.e., five rows), which make up the ensemble, as shown in the following table:

	First toss	Second toss	Third toss	Fourth toss	Fifth toss
First time series	0	−1	0	1	1
Second time series	1	−1	−1	1	0
Third time series	−1	1	0	−1	0
Fourth time series	1	−1	−1	1	1
Fifth time series	−1	0	1	0	−1

Using this table, we shall now discuss the terms *stationary* and *ergodic.* The table should be larger in both directions (i.e., more experiments and more tosses in each experiment), but we want to keep the arithmetic simple.

A random process is said to be stationary if the ensemble properties or averages do not change with time. The ensemble mean value for the third toss is

$$\bar{x}_3 = \frac{1}{5}(0 - 1 + 0 - 1 + 1) = -\frac{1}{5},$$

and the mean square value for the third toss ($k = 3$) is

$$s_3^2 = \frac{1}{5}[0^2 + (-1)^2 + 0^2 + (-1)^2 + 1^2] = \frac{3}{5}.$$

For the fifth toss ($k = 5$), the corresponding averages are

$$\bar{x}_3 = \frac{1}{5}[1 + 0 + 0 + 1 - 1] = \frac{1}{5},$$

$$s_3^2 = \frac{1}{5}[(1)^2 + 0^2 + 0^2 + 1^2 + (-1)^2] = \frac{3}{5}.$$

We say the process is stationary if all the corresponding averages are equal (to within statistical variations). In particular, we must obtain the same mean value at $k = 3$ as at $k = 5$. Actually, $-1/5$ and $1/5$ are not the same, but they are within statistical limits of the ensemble true mean value:

$$\begin{aligned}\bar{x}_{\text{true}} &= (-1)P(-1) + (0)P(0) + (1)P(1) \\ &= (-1)(0.4) + (0)(0.2) + (1)(0.4) = 0.\end{aligned}$$

This coin-tossing process is stationary; its statistical properties do not change with time. A stationary random process is ergodic if the averages computed from one specific time series are the same as the ensemble averages. For example, the mean value of the third time series is

$$\bar{x} = \frac{1}{5}(-1 + 1 + 0 - 1 + 0) = -\frac{1}{5},$$

which, within statistical limits, is the same as the ensemble mean value of zero. The coin-tossing process is ergodic, as the time-averaged parameters obtained from any one series equal the ensemble averages. Ergodicity greatly facilitates analysis, as we need only one times series recorded over a long enough time interval to describe the whole ensemble.

The coin-tossing experiment, as we have seen, represents a stationary, ergodic random process. Because all the tosses are independent of each other, there are no correlations, either between successive tosses or between different experiments. Such a process is an example of *white noise.* Let us compute the autocorrelation function as a time average. The second time series gives

$$\phi(0) = (1)^2 + (-1)^2 + (-1)^2 + (1)^2 + (0)^2 = 4,$$

$$\phi(1) = (1)(-1) + (-1)(-1) + (-1)(1) + (1)(0) = -1,$$

$$\phi(2) = (1)(-1) + (-1)(1) + (-1)(0) = -2,$$

$$\phi(3) = (1)(1) + (-1)(0) = 1,$$

$$\phi(4) = (1)(0) = 0.$$

These values are within statistical limits of the true (i.e., theoretical) values

$$\phi(0) = 4,$$

$$\phi(1) = 0,$$

$$\phi(2) = 0,$$

$$\phi(3) = 0,$$

$$\phi(4) = 0.$$

A much larger experiment would result in empirical autocorrelation values $\phi(1)$, $\phi(2)$, $\phi(3)$, . . ., which would be very small in magnitude relative to $\phi(0)$. A white noise process is defined as one that has zero (theoretical) autocorrelation for all time shifts, except, of course, time shift zero, that is, a spike autocorrelation. For any finite-length time series, however, we cannot expect the empirically computed quantities to be exactly zero because of statistical fluctuations.

The white noise process is a useful concept, as it represents an extreme—namely, the case of no predictability from one observation to the next. The other extreme is a periodic function, such as a sine function, in which past values completely determine the future. Most time series fall in categories that lie between these two extremes, of course, and it is they that we are interested in.

Let us consider a physical system that vibrates. A typical system has a natural rhythm, called the *resonant frequency of vibration.* Many things have resonant frequencies: swings, electric buzzers, horns, and bells. Water in a bathtub or an ocean swishes back and forth at some resonant frequency. There is usually more than one way a thing can vibrate, or resonate; for example, a ball in a bowl can resonate back and forth or it can resonate around a circle. These different ways are called the *resonant modes.* Sometimes the different modes have different resonant frequencies; for example, there are several ways an antenna on a car can whip back and forth. Likewise, there are many ways a string on a musical instrument can resonate, each way with its own resonant frequency.

Consider now a bowl with steep sides, with a small metal ball rolling back and forth in it. Each time the ball goes back and forth, we can give a little push and so add a little energy. Eventually, there will be enough energy added together to get the ball up to the rim and over the top. The trick depends on pushing at the right times and in the right directions. We must match the rhythm of the pushes to the natural rhythm with which the ball rolls back and forth. In other words, our input must be a periodic (sinusoidal) input of the same frequency as the natural frequency (resonance) of the ball in the bowl. The result is a large output; that is, the ball goes over the rim.

A bridge has a resonant frequency. In England, over a hundred and fifty years ago, a company of soldiers once marched over an iron bridge. The soldiers marching in step produced a sinusoidal input. Unfortunately, the input frequency was the same as the natural resonant frequency of the bridge. The vibrations of the bridge built up to such a point that the bridge collapsed. After that, all military commanders were told to have soldiers break step when they cross a bridge. When they break step, the input is no longer sinusoidal but has the properties of white noise. Such input is a sequence of uncorrelated impulses; the bridge still vibrates but not in resonance. The amplitudes of the vibrations do not build up, and the people crossing are safe.

When there is a white noise input, the effects due to interference prevent the output amplitudes from building up to a high value. Suppose that two random impulses are interjected into a system and that they produce two effects. The output is the sum of the two effects. The sum can be zero if the effects are opposite. If the two effects oscillate together, the output oscillation is large. If they oscillate in opposite ways, they cancel each other and the output does not oscillate. This notion extends to all kinds of waves—water waves or sound waves or seismic waves, all of which are produced by something that vibrates. When the effect of several vibrating things or waves comes together, we cannot predict the sum effect until we first know if the things are vibrating together or in opposition. When things vibrate in unison, they are said to be *in phase* or *in sync.* When they vibrate in opposition, they are *in opposite phase* or *180° out of phase.* When vibrations that are out of phase superimpose, they interfere destructively and cancel each other. When they are in phase, they interfere constructively and reinforce each other.

In more general terms, let us consider a physical system that is linear and time-invariant. The system can be described by its impulse-response function. The impulse-response function is defined as the output (response) to a spike input (impulse). As input, we feed into the system a white noise process—that is, a process whose autocorrelation is a spike. What is the output? This result is basic. The output is a stationary random process. This process lies in that large interior region between the two extremes—that is, between the extreme of a white noise process and the other extreme of a periodic process. This basic result can be depicted as

White noise input → [Physical system] → Stationary random output

Let us represent the white noise input as the time series $\epsilon(k)$ and the stationary random output as the time series $x(k)$. Let the impulse response be $w(k)$. Our basic result can now be written as

$$\epsilon(k) * w(k) = x(k);$$

that is, the output is the convolution of the input with the impulse-response function.

What is the autocorrelation of the stationary time series $x(k)$? We know that the autocorrelation of the white noise is a spike, so the autocorrelation must be related in some simple way to the autocorrelation of the impulse response function. What is this relationship? We make use of the fact that the autocorrelation of a signal is equal to the convolution of the signal with its time reverse. Thus, the autocorrelation of $x(k)$ is

$$x(k) * x(-k) = [\epsilon(k) * w(k)] * [\epsilon(-k) * w(-k)].$$

Since convolution is commutative, we can rearrange the right-hand side as

$$x(k) * x(-k) = [\epsilon(k) * \epsilon(-k)] * [w(k) * w(-k)].$$

The first expression in brackets on the right-hand side is the autocorrelation of the white noise. This autocorrelation is a spike, and a unit spike acts as unity in convolution. Hence, we have the important result that

$$x(k) * x(-k) = \sigma^2 \, w(k) * w(-k),$$

where σ^2 is a scale factor that represents the power (energy) of the white noise. This result says that the autocorrelation of the stationary time series is equal to the autocorrelation of the impulse-response function, except for the scale factor σ^2. Thus, the white noise does not (and cannot) influence the resonances of the system.

The random reflectivity model

The seismic trace can be visualized as the response of the earth to a seismic source, such as an explosion. As we have seen, the convolutional model represents the received trace $x(k)$ as the convolution of a seismic wavelet $w(k)$ with the reflectivity function $c(k)$. We recall that the reflectivity function is defined as the sequence of reflection coefficients of the subsurface interfaces. For computational purposes, we must use a discrete model, so we pick some sampling time interval, such as 4 msec, as the spacing (in two-way time) of the (mathematically defined) interfaces represented by the $c(k)$.

We now come to the random reflectivity hypothesis. A hypothesis is an assertion subject to verification; it is a conjecture that, within a theory or ideational framework, accounts for a set of facts and that can be used as a basis for further investigation. It is a proposition stated as a basis for argument and reasoning. The random reflection hypothesis represents a working proposition. It does not represent absolute truth. The random reflection hypothesis does, however, represent a picture that is supported by the data in enough cases so that it is useful. The random reflection hypothesis is a benchmark as to the statistical distribution of the reflection coefficients used as a reference point in a seismic survey of a completely unknown region.

The random reflectivity hypothesis states that the reflectivity function is a sample from a white noise process. In other words, the reflection coefficients $c(k)$ form a sequence of uncorrelated random variables. This hypothesis would be expected to hold in areas where the various sediments were laid down over geologic time in a nonpredictable, random manner. As a result, the reflection coefficients would be unrelated to each other and thus would form a random white noise sequence. Our notation for a white noise sequence is $\epsilon(k)$, so the random reflectivity hypothesis can be written as

$$c(k) = \epsilon(k).$$

Let us now look at some of the properties of a white noise process $\epsilon(k)$. Suppose that we have N observations $\epsilon(1), \epsilon(2), \ldots, \epsilon(N)$. These would correspond to the reflection coefficients of N interfaces at equally spaced two-way traveltimes. The autocorrelation would be

$$r(k) = \frac{1}{N} \sum_{n=1}^{N-k} \epsilon(n + k)\epsilon(n) = r(-k)$$

for $k = 0, 1, 2, \ldots, N - 1$. We assume tht the random numbers $\epsilon(k)$ have zero mean, variance σ^2, and are independent. The theoretical autocorrelation is thus

$$\rho(k) = \begin{cases} \sigma^2 & \text{for } k = 0 \\ 0 & \text{otherwise.} \end{cases}$$

Because of statistical fluctuations, the empirical autocorrelation $r(k)$ will differ from the theoretical autocorrelation $\rho(k)$. However, there is a probable range around the theoretical values in which we can expect the empirical values to lie most of the time. From statistical reasoning, we know that the extent of this probable range would decrease as the sample size increases. Thus, as N becomes large, $r(0)$ tends to the theoretical value σ^2. The other empirical autocorrelations should be zero, but they also have a statistical fluctuation due to the finite sample size N. Their probable range would be centered around the theoretical value of zero. As N becomes large, the value of each $r(k)$ will tend to the theoretical value of zero (see fig. 5.14).

When we say that a reflectivity function is random, we do not mean that its autocorrelation for nonzero shifts is identically zero, but that they more or less lie in a band given by the probable range. Some will lie outside this range, of course, but most should lie within it. An actual reflectivity function can be computed from the well logs from an oil well, and then the autocorrelation of the empirical reflectivity function can be computed. We would plot this autocorrelation to see how much it deviates from the theoretical spike autocorrelation. If the deviation is within statistical limits, then we say that the random reflectivity hypothesis is upheld for the geologic sedimentary column represented by that particular oil well. This computation has been done for oil wells in many different oil fields, and the results show that random reflectivity represents a good working hypothesis for the explorationist who is working in an unexplored area. The situation corresponds to card games. A good working hypothesis for card games is that the card shuffle is random. The shuffling mixes the playing cards together to change their order of arrangement. The shuffled deck of cards lies on the table before us, even as the layers of an unknown sedimentary basin appear as random and unknown as cards in a shuffled deck. If we drew the cards one by one and found them black, red, black, red, black,

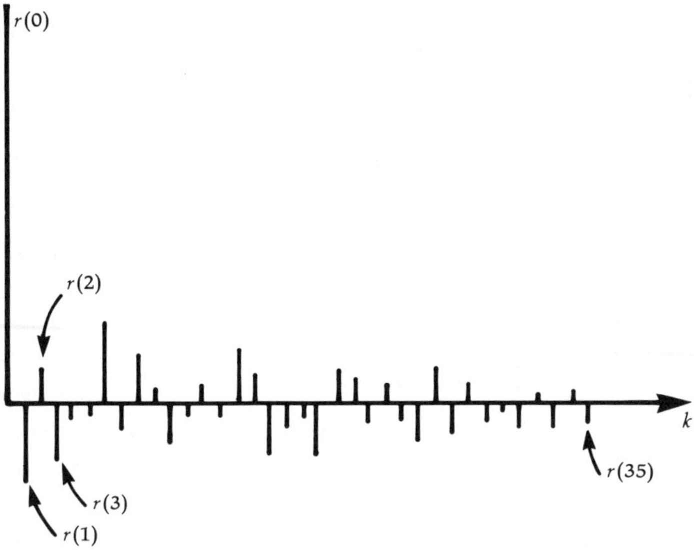

Figure 5.14. The right-hand side of the empirical autocorrelation function of a white noise sequence of random variables. The true autocorrelation function is a spike at k = 0.

red—all the way to the last—we would be surprised, even as we would be surprised if the reflection coefficients were plus, minus, plus, minus, plus, minus —all the way to the deepest interface.

In playing with a particular dealer, or in exploring a particular basin, as more experience is gained, certain idiosyncrasies might come out as significant, definite biases. For such a dealer, we would then want to modify our hypothesis. In the same way, as the explorationist learns more about an area, he would want to modify the hypothesis in accord with what he has learned about the actual reflection coefficient distribution present in the sedimentary column.

In exploration, we are always working against time; time is money. We must start with a noncommittal hypothesis and then modify it only to the extent that time and money will allow. The random reflectivity hypothesis is the best such initial hypothesis.

The random reflectivity convolutional model (Robinson, 1954) is

$$x(k) = w(k) * \epsilon(k),$$

where the reflectivity function $\epsilon(k)$ is white noise (see fig. 5.15). Thus, the trace is a realization of a stationary random process; in particular, the autocorrelation of the trace is approximately equal to the autocorrelation of the seismic wavelet. This result is important, because from a received trace we cannot directly measure the wavelet because of the mutual overlapping of wavelets and the resulting interference with each other. Although we cannot measure the wavelet directly, we can obtain its autocorrelation, for, as we have shown in the preceding section, the autocorrelation of the received seismic trace is equal to the autocorrelation of the wavelet, except for a scale factor. This determination of

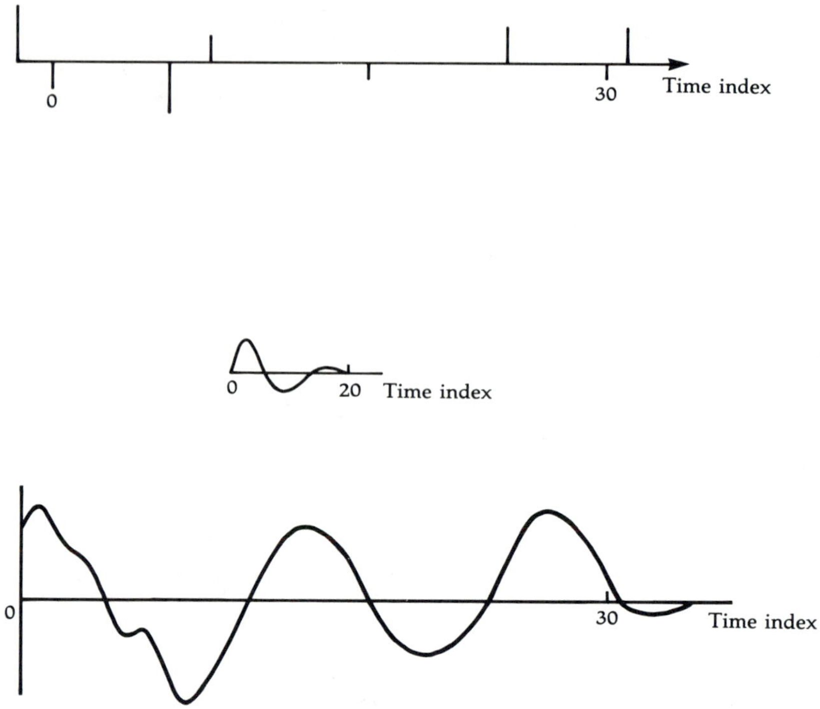

*Figure 5.15. (*top*) Random sequence of reflection coefficients. (*middle*) The seismic wavelet. (*bottom*) The seismic trace given by their convolution. Reprinted by permission of Geophysics from Robinson 1954 (1967).*

its autocorrelation does not determine the wavelet, of course, since many wavelets have the same autocorrelation. However, we have made an important first step in the determination of the seismic wavelet.

To summarize, the random reflectivity convolutional model states that a given discrete, finite-length seismic trace is additively composed of wavelets with random amplitudes. Because of this randomness, the autocorrelation of the trace is equal to the autocorrelation of the wavelet, within a scale factor. This equality is only approximate, of course, because of statistical fluctuations, especially for large time shifts where the number of cross products is few. Hence, we have the basic result that we can find the autocorrelation of a wavelet, at least approximately, without any direct measurement of wavelet. This result is one of the basic building blocks of the statistical use of deconvolution (Robinson, 1954).

The fact that the autocorrelation of the trace gives the autocorrelation of the basic wavelet should seem reasonable, as the trace is assumed to be generated by the convolution of a white reflectivity function with the wavelet. The autocorrelation of white noise is zero at all shifts except shift zero, and hence the contributions to the autocorrelation of the trace must be due to the wavelet. The important result concerning the autocorrelation of a seismic trace being the autocorrelation of the wavelet is illustrated in figures 5.16 and 5.17.

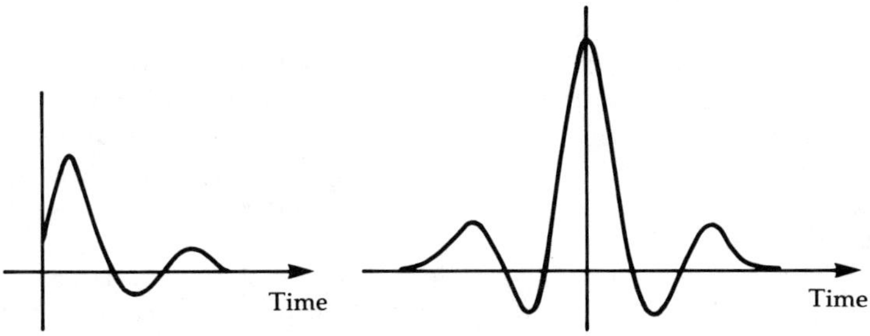

*Figure 5.16. Seismic wavelet (*left*) and its autocorrelation (*right*). Reprinted by permission of Geophysics from Robinson 1954 (1967).*

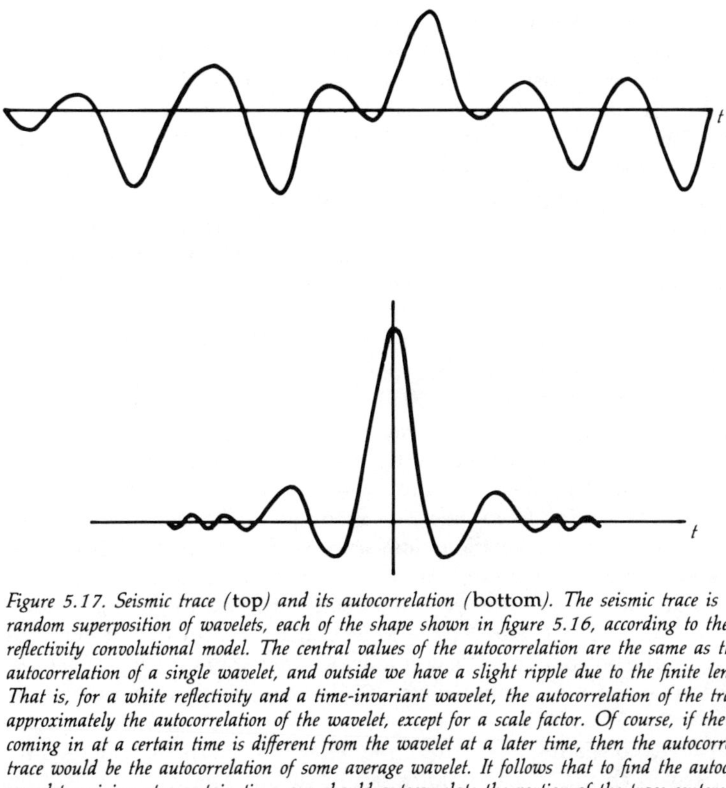

*Figure 5.17. Seismic trace (*top*) and its autocorrelation (*bottom*). The seismic trace is made up of a random superposition of wavelets, each of the shape shown in figure 5.16, according to the random reflectivity convolutional model. The central values of the autocorrelation are the same as the autocorrelation of a single wavelet, and outside we have a slight ripple due to the finite length of data. That is, for a white reflectivity and a time-invariant wavelet, the autocorrelation of the trace is approximately the autocorrelation of the wavelet, except for a scale factor. Of course, if the wavelet coming in at a certain time is different from the wavelet at a later time, then the autocorrelation of the trace would be the autocorrelation of some average wavelet. It follows that to find the autocorrelation of a wavelet arriving at a certain time, one should autocorrelate the portion of the trace centered about that time instant. The portion of trace used in computing the autocorrelation should be up to ten times the length of the wavelet in order to cancel out statistical variations. Reprinted by permission of Geophysics from Robinson 1954 (1967).*

Multiple reflections

Under the assumption of a random reflectivity (i.e., a white noise sequence of reflection coefficients), the autocorrelation of that trace, within a scale factor, is simply the autocorrelation of the basic wavelet. This assumption of randomness is generally valid if the occurring wavelets are the result of a succession of primary reflections. However, a certain lack of randomness is present when multiple reflections occur.

In this section, we will consider the multiple reflections known as ghosts (Lindsey, 1960) and those known as water or near-surface reverberations (Backus, 1959). Surface-to-bed and interbed multiples can be handled in a similar fashion. We want to determine their characteristics from their effects on the autocorrelation. Our purpose here is to use the autocorrelogram only to identify the problem.

The ghost reflection is caused by some of the explosive energy traveling up and being reflected, thereby producing a secondary spike. Suppose that the ghost-producing interface has reflection coefficient -0.5 and that the time delay is 40 msec. The succession of wavelets recorded by the geophone due to primary reflections will be assumed to be randomly spaced. We assume that, following each basic wavelet by a time delay of 40 msec, there will be a similar wavelet reduced 50 percent in amplitude and inverted because of upward-traveling energy being reflected by a negative reflection coefficient. Thus, the seismic wavelet is a wavelet complex composed of the direct wavelet and the ghost. This wavelet complex is randomly spaced down the trace (Robinson, 1954). When we autocorrelate the trace, we average out the random reflectivity, and we are left with the autocorrelation of the wavelet complex. The autocorrelation of the wavelet complex is made up of a center portion due to the direct wavelet and prominent side lobes at time shifts -50 msec and $+50$ msec. Such side lobes can be interpreted as ghosts on the seismic trace (see fig. 5.18).

Water reverberations are a type of multiple reflections that occur on a marine seismic trace. They are produced by the following mechanism. Water reverberations are multiples resulting from reflection at the top and bottom of the water layer. Let $n\Delta t$ be the two-way traveltime from surface to water bottom and back, where Δt is the time interval between successive values of the sampled data and n is an integer. Let the reflection coefficient of an upgoing incident wave at the surface be -1, and let the reflection coefficient of a downgoing incident wave at the bottom be c. We will consider here only the zero-offset case (coincident source and receiver positions), so that the wave motion is straight up and down (vertical raypaths). The source in the water sets off a reverberation in the water layer consisting of the multiple reflections bouncing between the top and bottom. Each round-trip bounce results in the product of the bottom reflection coefficient and the surface reflection coefficient—that is, the product $c(-1)$, which is $-c$. Thus, the sequence of multiples (0, 1, 2, 3, . . . round-trip bounces) will have amplitudes $(-c)^0, (-c)^1, (-c)^2, (-c)^3, \ldots$, each spaced $n\Delta t$ apart. Thus, this multiple-generating (or reverberation) wavelet is

$$m_1 = (1, 0, \ldots, 0, -c, 0, \ldots, 0, c^2, 0, \ldots, 0, -c^3, 0, \ldots),$$

where there are $n - 1$ zeros between each nonzero entry. A reflection from a deep interface must pass through the water layer twice, once on the way down

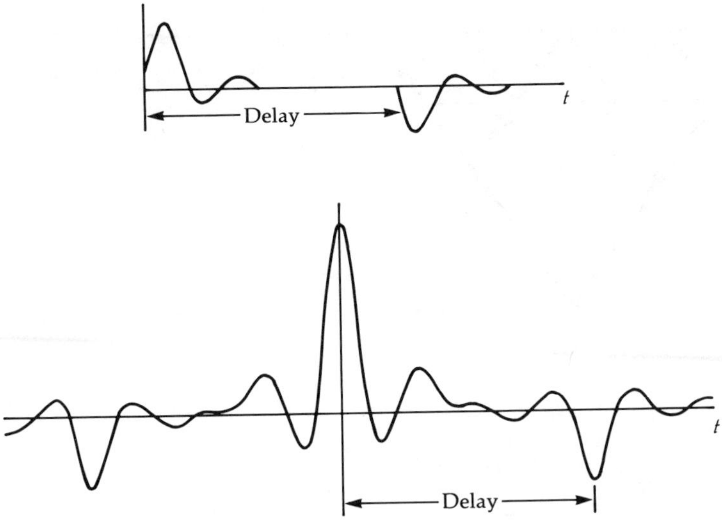

Figure 5.18. (top) *The wavelet complex made up of the direct wavelet and the ghost.* (bottom) *Autocorrelation of the seismic trace.*

and once again on the way up. Thus, the total effect of the water layer is given by convolution of the foregoing reverberation wavelet with itself; that is, the complete multiple-generating wavelet is

$$m = m_1 * m_1 = (1, 0, \ldots, 0, -2c, 0, \ldots, 0, 3c^2, 0, \ldots, 0, -4c^3, 0, \ldots).$$

Thus, the impulse response of the water layer for various water depths $d = vn\Delta t/2$, where v is the wave velocity (about 5000 ft/sec) in the water (see fig. 5.19), are

$$\begin{aligned} &(1, -2c, 3c^2, -4c^3, 5c^4, \ldots) && \text{when } n = 1, \\ &(1, 0, -2c, 0, 3c^2, 0, -4c^3, \ldots) && \text{when } n = 2, \\ &(1, 0, 0, -2c, 0, 0, 3c^2, \ldots) && \text{when } n = 3. \end{aligned}$$

Let us now consider this model (Robinson, 1954). The seismic trace x is the convolution of the random reflectivity ϵ, the complete multiple m, and the seismic wavelet w; that is,

$$x(k) = m(k) * w(k) * \epsilon(k).$$

The autocorrelation of the seismic trace is therefore

$$x(k) * x(-k) = [m(k) * m(-k) * w(k) * w(-k)]\sigma^2,$$

where σ^2 is a constant representing the power of the white noise reflectivity function. Except for the scale factor σ^2, we see that the autocorrelation of the

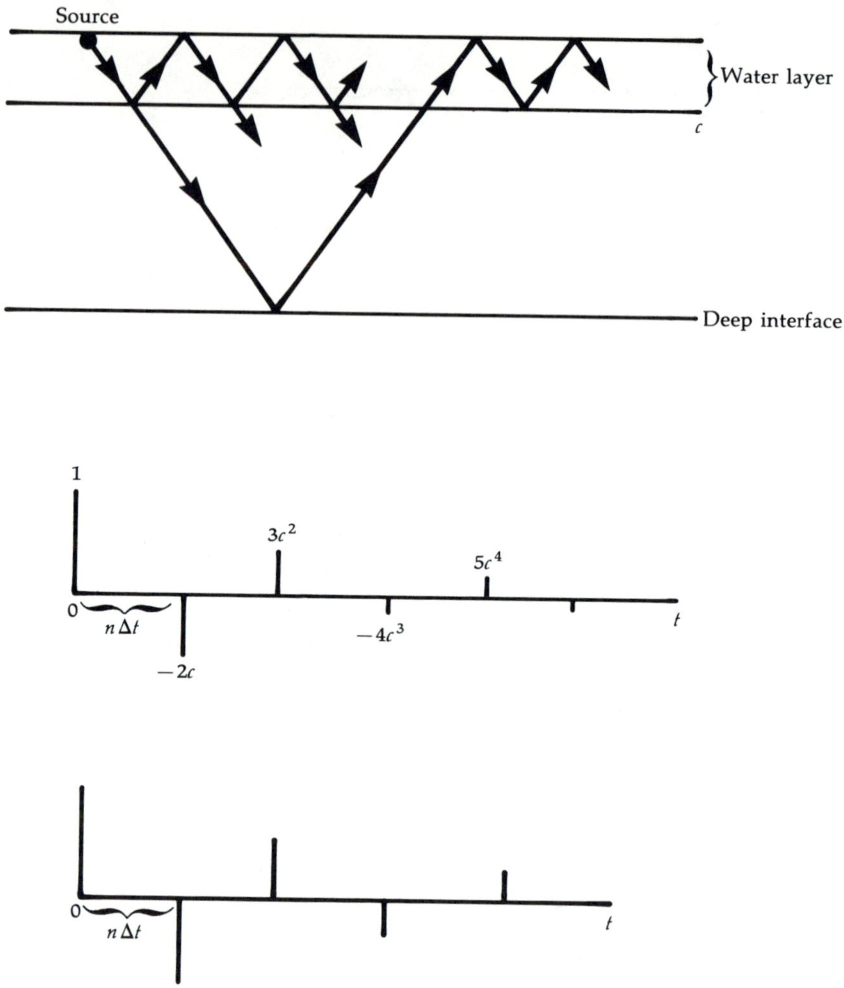

Figure 5.19. (top) *The water reverberation mechanism with bottom reflection coefficient* c. (middle) *The two-way reverberation wavelet* m *(for* $c > 0$*).* (bottom) *The right half of the autocorrelation of* m.

trace is the convolution of the autocorrelation of the multiple m with the autocorrelation of the wavelet w.

To explain this situation further, one component of the seismic trace is a succession of primary wavelet reflections, assumed to have random amplitudes and to be randomly spaced. Following each basic wavelet by a time of $n\Delta t$, however, is a similar wavelet inverted with respect to the basic wavelet. Here we assume a hard bottom ($c > 0$). Following each second wavelet by the same time $n\Delta t$ is a similar wavelet that is not inverted with respect to the basic wavelet, and after this another, and so on, because of the bounces in the water layer. Thus, the seismic trace can be considered to be additively composed of the composite wavelet

$w(k) * m(k),$

with random amplitudes and arrival times as given by the reflectivity $\epsilon(k)$. In other words, the seismic trace $x(k)$ is the superposition of such composite wavelets with random amplitudes and randomly spaced down the trace. The autocorrelation of the trace would average out the random part and would simply be the autocorrelation of the composite wavelet, except for a slight ripple due to statistical fluctuations (see figs 5.20 and 5.21).

We can therefore conclude the following about the autocorrelation of a marine seismic trace. On the autocorrelation function, a succession of prominent side lobes, each separated by $n\,\Delta t$ seconds where $n\,\Delta t = 2d/v$, could be interpreted as reverberations on the seismic trace. From the autocorrelation, we can obtain the reverberation time $n\,\Delta t$. This parameter is needed when we wish to apply predictive deconvolution. Also from the autocorrelation, we can obtain the sign of the reflection coefficient c of the bottom. If the central peak of autocorrelation is followed by a trough, the reflection coefficient is positive, but if the central peak is followed by another peak, it is negative.

Exercises

1. Let the source wavelet be (5, −4, 2, 1).
   (a) If the direct wave is followed by a ghost of half-strength two time units later, what is the composite wavelet?
   (b) Instead, suppose that the direct wave is followed by a half-strength negative multiple one time unit later and then followed by a quarter-strength positive multiple two time units later. What is the composite wavelet? (Solve the problem by use of z-transforms.)
2. Consider a filter with impulse response (2, −1, −1). Find its z-transform. By factoring the z-transform, show that the filter is equivalent to two cascaded filters with impulse responses (2, 1) and (1, −1).
3. Given the seismic wavelet (10, 9, −7, −8, 0, 5, 3), plot the composite wavelet made up of the wavelet plus 0.5 times its delayed replica for delays of 1, 2, 3, 4, 5, 6, 7, and 8 time units. Interpret these composite wavelets as being caused by reflections along a thin bed that pinches out. Plot the composite wavelets in the form of a Widess diagram. Repeat in the case when the delayed replica is multiplied by −0.5 instead of 0.5. At what delay is resolution possible for the 0.5 case? For the −0.5 case?
4. Most of the energy in a seismic wavelet is contained in a band of frequencies centered about the dominant frequency. The dominant period can be defined as the time between two major crests. The dominant frequency is the reciprocal of the dominant period. If T is the dominant period, then the dominant frequency is $f = 1/T$. Let the wavelet (5, 4, −3, −4, 0, 2, 1) represent a seismic wavelet digitized at 4 msec intervals. What is the dominant frequency of this wavelet? Assume a sandstone layer in shale, with a reflection coefficient of the top interface given by 0.1 and the bottom interface −0.1. If the two-way traveltime through the sandstone is 12 msec, compute and graph the composite reflection waveshape. Do the same in the case when the two-way time is 16 msec. (In doing this exercise, take into

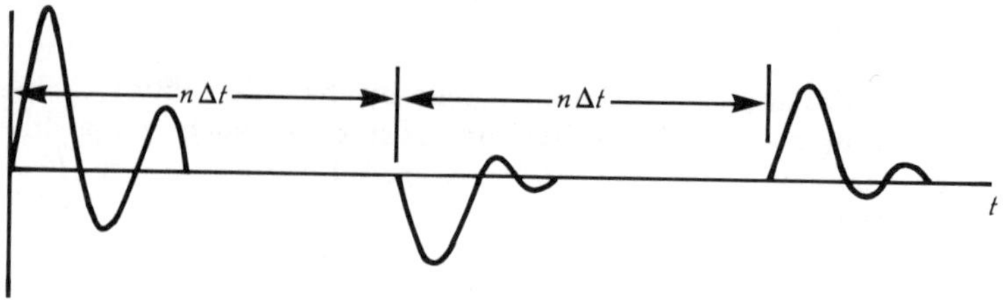

Figure 5.20. The composite wavelet m(k) * w(k).

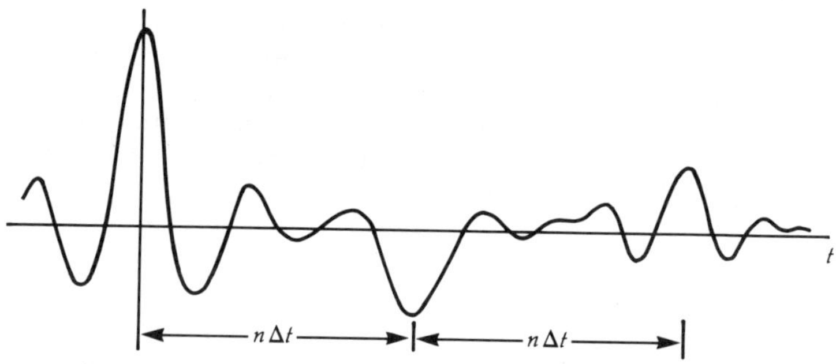

Figure 5.21. The right-hand side and some of the left-hand side of the autocorrelation function of the seismic trace. Except for statistical fluctuations and a scale factor, this curve is the same as the autocorrelation of the composite wavelet in figure 5.20. This result holds under the random reflectivity hypothesis.

account only primary reflections; that is, neglect transmission effects by setting all transmission coefficients equal to unity, and neglect multiple reflections).

5. Let a sequence of reflecting layers each have two-way traveltime of 4 msec and let the reflection coefficients of the successive interfaces be (0.05, −0.1, 0.05, 0.03, −0.07, 0.10). Let the seismic wavelet digitized at a sampling rate of 250 samples per second be the wavelet (5, 4, −4, −5, 0, 2). Plot the composite reflected waveform (neglecting transmission effects and multiples) and graph the results. Also convolve the reflectivity sequence with the wavelet and compare with your answer.
6. Let a sandstone layer containing gas in its upper portion and water in its lower portion be sandwiched in shale. Two-way time through the gas sand is 8 msec and through the water sand is 16 msec. Assume a shale-to-gas-sand reflection coefficient of −0.2, a gas-sand-to-water-sand reflection coefficient of 0.3, and a water-sand-to-shale reflection coefficient of −0.1. For the wavelet of Exercise 5, find the composite reflection wavelet (neglecting transmission effects and multiples). How can we obtain the same result by convolution? Plot all results.
7. Let a sandstone layer be 24 m thick, with a velocity of 3000 m/sec. A downgoing wave strikes the top of the sandstone and is reflected with reflection coefficient 0.1. The part that is transmitted into the sandstone is reflected at the bottom of the sandstone layer with reflection coefficient

−0.1. Neglect amplitude loss in transmission through the top of the sandstone (i.e., set transmission coefficients equal to one), and neglect intrabed multiples.

(a) If the basic wavelet has amplitudes at 4 msec sampling given by 8, 7, −7, −5, 0, 4, 2, find the composite wavelet shape that is reflected by the sandstone layer. Explain why in this case we can say that the top and base of the sandstone are resolved.

(b) If the basic wavelet has amplitudes at 4 msec sampling rates given by 4, 8, 10, 7, 0, −7, −6, −5, −1, 0, 2, 4, 1, 2, find the composite wavelet shape. Is the sand layer resolved in this case?

(c) What is the dominant frequency in each of the wavelets in (a) and (b)? How does resolution depend on this frequency?

(d) Let a shale bed be in the middle of the sandstone layer so that there is 12 m of sand above the shale and 12 m below. Let the thickness of the shale be such that the traveltime through it is the same as through 12 m of sand. Let the reflection coefficient at the top of the shale be −0.1 and at the bottom of the shale be 0.1. For the wavelet in (a), find the composite reflected wavelet. This example shows that component reflections from several reflectors can interfere so as to give a tuned response. The ringing character results from the natural resonance.

8. The basic equation for wavelength is

$$\lambda = \frac{\text{velocity}}{\text{frequency}} = \frac{v}{f},$$

where v is wave velocity (in meters per second or feet per second) and f is frequency (in hertz). The wavelength λ will then be in units of meters or feet. Calculate wavelengths for the following cases:

(a) Shallow rocks: 2000 m/sec, 50 Hz.

(b) Deep rocks: 6000 m/sec, 25 Hz.

9. What is the impedance contrast, or reflectivity, of the following geologic model?

shale: v = 3.0 km/sec, ρ = 2.3 gr/cc

---

limestone: v = 6.0 km/sec, ρ = 2.6 gr/cc

Suppose that the interface is buried at a depth of 3 km beneath a homogeneous section of shale. What is the normal incidence seismogram that would be recorded at the surface if we used an ideal impulse signal with a pressure of one unit. (Neglect any attenuation due to propagation effects.) Use the following plotting scales: traveltime — 1 in. = 1 sec; amplitude — 1 in. = 1 amplitude unit. Plot the cases of both particle velocity and pressure. Repeat, but with a composite signal waveform consisting of a unit pressure pulse followed 0.25 sec later by another pressure pulse with a magnitude of −0.5 units.

10. Given the source wavelet $w(k)$ = 0, 1, 2, 1, 0, −1, −2, −1, 0 with sampling interval 10 msec, compute and plot the synthetic seismogram $x(k)$ obtained

by convolving this wavelet with the reflection coefficient sequence $c(k)$ corresponding to the following velocity structure:

	Surface
$\Delta z_1 = 10$ m, $v = 1000$ m/sec, $\rho = 2.5$ gr/cc	
	Base of layer 1
$\Delta z_2 = 400$ m, $v = 5000$ m/sec, $\rho = 2.5$ gr/cc	
	Base of layer 2
$v = 3000$ m/sec, $\rho = 2.5$ gr/cc	

Plot $c(k)$ and the synthetic seismogram $x(k)$ at 10 msec intervals. Use amplitude scale of 1 in/2 amplitude units, time scale of 1 in./50 msec. Repeat, but with $\Delta z_2 = 100$ m.

11. The relationship between velocity v, frequency f, and wavelength λ is $f = v/\lambda$. We would like to clearly resolve the top and bottom reflections from certain geologic beds. To do so, we decide to use a seismic signal with a dominant frequency corresponding to a wavelength no greater than the bed thickness. Assume constant density. Determine this frequency for the following beds (where b is the thickness of the bed):

(a) $v = 1$ km/sec

$v = 3$ km/sec $\quad b = 500$ m

$v = 1$ km/sec

(b) $v = 4$ km/sec

$v = 1$ km/sec $\quad b = 100$ m

$v = 4$ km/sec

(c) $v = 2$ km/sec

$v = 4$ km/sec $\quad b = 250$ m

$v = 5$ km/sec

At what frequencies would you expect *constructive* interference for these beds? At what frequencies would you expect *destructive* interference?

12. The dominant period of a wavelet is defined as the time T between two major crests, and the dominant frequency is $f = 1/T$. If T is in seconds, then f is in hertz. Compute the primary reflection amplitudes that would be measured at the surface of the following geologic model if one used 10 Hz, 50 Hz, and 100 Hz signals, respectively. Take into account transmission losses but not spherical divergence and anelastic attenuation. Normalize amplitudes with respect to the amplitude of the first reflection. Plot the $\log_{10}$ of amplitude versus traveltime (two-way, of course).

$v = 0.5$ km/sec, $\Delta z = 0.1$ km

$v = 2$ km/sec, $\Delta z = 5$ km

$v = 6$ km/sec, $\Delta z = 10$ km

$v = 8$ km/sec

Here v is compression wave velocity and Δz is layer thickness. Repeat, but include spherical divergence.

13. Plot and compare the reflection amplitudes for a 10 Hz signal versus time for the following geologic models. Normalize to the amplitude of the first primary. Include transmission losses but not spherical divergence and anelastic attenuation effects:

(a) $v = 1$ km/sec, $\Delta z = 2$ km

$v = 3$ km/sec

Plot primary and first 4 order multiple amplitudes. Repeat, but include spherical divergence.

(b) $v = 1$ km/sec, $\Delta z = 2$ km

$v = 3$ km/sec, $\Delta z = 6$ km

$v = 5$ km/sec, $\Delta z = 10$ km

$v = 6$ km/sec, $\Delta z = 12$ km

$v = 7$ km/sec

Plot amplitudes of primaries only. Repeat, but include spherical divergence.

To answer the following exercises, refer to books in the geological library.

14. Briefly define the following:
    (a) Seismic sequence
    (b) Geophone array
    (c) Peglegs
    (d) Sideswipe
    (e) Stack
    (f) Chronostratigraphic chart
    (g) Flat spot
    (h) Seismic facies
    (i) Pinnacle reef
    (j) Deconvolution
    (k) Migration
15. How and why does the seismic velocity of rocks in a typical sedimentary basin depend on each of the following?

(a) Depth
(b) Geologic age
(c) Porosity

16. Why are anticlines more likely to be discovered by a reconnaissance reflection survey than are synclines? (Hint: It is not because they are more common. Think 3-D!)
17. Present three general but reasonable geologic situations that might give rise to a bright spot.
18. What is the maximum dip that can be observed on an unmigrated seismic section, and why?
19. What is meant by the terms *direct* and *indirect hydrocarbon detection?*
20. How can one differentiate between high, low, and variable energy depositional environments on a seismic section?
21. What is a petroleum independent?

References

Al-Chalabi, M., 1973, Series approximations in velocity and traveltime computation: Geophys. Prosp., v. 21, p. 783–795.

——— 1974, An analysis of stacking, rms, average, and interval velocities over a horizontally layered ground: Geophys. Prosp., v. 22, p. 458–475.

Al-Sadi, H. N., 1980, Seismic exploration: Boston, Birkhauser Boston, Inc.

Anstey, N. A., 1970, Signal characteristics and instrument specifications, Vol. 1 of Seismic prospecting instruments: Berlin, Gebrüder Borntraeger.

——— 1977, Seismic interpretation: the physical aspects: Boston, IHRDC.

——— 1980, Seismic exploration for sandstone reservoirs: Boston, IHRDC.

——— 1982, Simple seismics: Boston, IHRDC.

Backus, M.M., 1959, Water reverberations, their nature and elimination: Geophysics, v. 24, p. 233–261.

Berkhout, A.J., 1977, Least squares inverse filtering and wavelet deconvolution, Geophysics, v. 42, p. 1369–1383.

Cosgrove, J.J., Edwards, C.A.M., and Grigsby, J.K., 1969, Velocity determinations and the digital computer: Houston, Olympic Geophysical Co.

Crawford, J.M., Doty, W.E., and Lee, M.R., 1960, Continuous signal seismograph: Geophysics, v. 25, p. 95–105.

Dix, C.H., 1955, Seismic velocities from surface measurements: Geophysics, v. 20, p. 68–86.

——— 1981, Seismic prospecting for oil: Boston, IHRDC.

Dobrin, M.B., 1976, Geophysical prospecting: New York, McGraw-Hill.

Duncan, J.W., and Levin, F.K., 1973, Effect of normal moveout on a seismic pulse: Geophysics, v. 38, p. 635–642.

Fourier, J.B.J., 1807, Theorie analytique de la chaleur: Paris, Académie des Sciences.

Geyer, R.L., 1969, The Vibroseis system of seismic mapping: Tulsa, Seismograph Service Corp.

Hilterman, F.J., 1970, Three-dimensional seismic modeling: Geophysics, v. 35, p. 1020–1037.

——— 1975, Amplitudes of seismic waves. A quick look: Geophysics, v. 40, p. 745–762.

Levorsen, A.I., 1967, Geology of petroleum: San Francisco, W.H. Freeman.

Lindsey, J.P., 1960, Elimination of seismic ghost reflections by means of a linear filter: Geophysics, v. 25, p. 130–140.

Mayne, W. H., 1962, Common reflection point horizontal stacking techniques: Geophysics, v. 27, p. 927–938.

Morley, L., and Claerbout, J., 1983, Predictive deconvolution in shot-receiver space, Geophysics, v. 48, p. 515–531.

Neidell, N.S., and Taner, M.T., 1971, Semblance and other coherency measures for multichannel data: Geophysics, v. 36, p. 482–497.

Peterson, R.A., Fillipone, W.R., and Coker, F.B., 1955, The synthesis of seismograms from well log data: Geophysics, v. 20, p. 516–538.

Ricker, N., 1940, The form and nature of seismic waves and the structure of seismograms: Geophysics, v. 5, p. 348–366.

——— 1953, Wavelet contraction, wavelet expansion, and the control of seismic resolution, Geophysics, vol. 18, pp. 769–792.

Robinson, E. A., 1954, Predictive decomposition of time series with applications to seismic exploration: Cambridge, M.I.T. Geophysical Analysis Report No. 7 (Reprinted in Geophysics, v. 32, p. 418–484, 1967).

——— 1957, Predictive decomposition of seismic traces: Geophysics, v. 22, p. 767–778.

——— 1975, Dynamic predictive deconvolution: Geophys. Prosp., v. 23, p. 779–797.

——— 1980, Physical applications of stationary time series: High Wycombe, Bucks, England, Charles Griffin and Co.

——— 1983, Migration of geophysical data: Boston, IHRDC.

Robinson, E.A., and Treitel, S., 1980, Geophysical signal analysis: Englewood Cliffs, N.J., Prentice-Hall.

Schramm, Jr., M.W., Dedman, E.V., and Lindsey, J.P., 1977, Practical stratigraphic modeling and intepretation: Tulsa, AAPG Mem. 26, p. 477–502.

Sengbush, R.L., 1983, Seismic exploration methods: Boston, IHRDC.

Sheriff, R.E., 1973, Encyclopedic dictionary of exploration geophysics: Tulsa, Society of Exploration Geophysicists.

——— 1978, A first course in geophysical exploration and interpretation: Boston, IHRDC.

——— 1980, Seismic stratigraphy: Boston, IHRDC.

Smith, M.K., 1956, Noise analysis and multiple seismometer theory: Geophysics, v. 21, p. 337–360.

Taner, M.T., and Koehler, F., 1969, Velocity spectra: Geophysics, v. 34, p. 859–881.

Taner, M.T., Koehler, F., and Alhilali, K.A., 1974, Estimation and correction of near-surface time anomalies: Geophysics, v. 39, p. 441–463.

Tatham, R.H., and Stoffa, P., 1976, A potential hydrocarbon indicator: Geophysics, v. 41, p. 837–849.

Telford, W.M., Geldart, L.P., Sheriff, R.E., and Keys, D.A., 1976, Applied geophysics: Cambridge, England, Cambridge University Press.

Waters, K. H., 1978, Reflection Seismology: New York, John Wiley.

Wiener, N., 1949, Extrapolation, interpolation, and smoothing of stationary time series with engineering applications: Cambridge, M.I.T. Press.

Bibliography

Velocity

Cook, E.E., and Taner, M.T., 1969, Velocity spectra and their use in stratigraphic and lithologic differentiation: Geophys. Prosp., v. 18, p. 433–448.

Cunningham, A.B., and Heffring, H.H., 1980, Interpretation of velocity spectra: Geophysics, v. 45, p. 1741–1752.

Douze, E.J., and Laster, S.J., 1979, Statistics of semblance: Geophysics, v. 44, p. 1999–2003.

Garotta, R., 1978, Velocities, models applied to seismic interpretation: Oil and Gas Journal, Oct. 30, p. 126–134.

Garotta, R., and Michon, D., 1967, Continuous analysis of the velocity function and of the move out corrections: Geophys. Prosp., v. 15, p. 584–597.

May, B.T., and Straley, D.K., 1979, Higher-order moveout spectra: Geophysics, v. 44, p. 1193–1207.

Naess, O.E., and Bruland, L., 1981, Velocity analysis using iterative stacking: Geophys. Prosp., v. 29, p. 1–20.

Neidell, N.S., and Taner, M.T., 1971, Semblance and other coherency measures for multichannel data: Geophysics, v. 36, p. 482–497.

Pollet, A. 1974, Simple velocity modelling and the continuous velocity section: Presented at 44th Annual International Meeting of SEG, Nov. 10–14, Dallas.

Robinson, J.C., 1969, HRVA—A velocity analysis technique applied to seismic data: Geophysics, v. 34, p. 330–356.

Schneider, W.A., and Backus, M.M., 1968, Dynamic correlation analysis: Geophysics, v. 33, p. 105–126.

Schultz, P.S., 1982, A method for direct estimation of internal velocities: Geophysics, v. 47, p. 1657–1671.

Sherwood, J.W.C., and Poe, P.H., 1972, Continuous velocity estimation and seismic wavelet processing: Geophysics, v. 37, p. 769–787.

Starichenko, N.D., and Meshby, V.I., 1975, Utilization of the horizontal velocity spectra for a determination of the nature of interfering waves: Neftegazovaya Geologika Geofizika, no. 7, p. 36–40.

Stoffa, P.L., Diebold, J.B., and Buhl, P., 1982, Velocity analysis for wide aperture seismic data: Geophys. Prosp., v. 30, p. 25–57.

Taner, M.T., and Koehler, F., 1969, Velocity spectra-digital computer derivation and applications of velocity functions: Geophysics, v. 34, p. 859–881.

Ursin, B., 1977, Seismic velocity estimation: Geophys. Prosp., v. 25, p. 658–666.

Moveout

Al-Chalabi, M., 1973, Series approximation in velocity and traveltime computations: Geophys. Prosp., v. 21, p. 783–795.

——— 1974, An analysis of stacking, rms, average, and interval velocities over a horizontally layered ground: Geophys. Prosp., v. 22, p. 458–475.

Diebold, J.B., and Stoffa, P.L., 1981, The traveltime equation, tau-p mapping, and inversion of common midpoint data: Geophysics, v. 46, p. 238–254.

Gangi, A.F., and Yang, S.J., 1976, Traveltime curves for reflections in dipping layers: Geophysics, v. 41, p. 425–440.

Hajnal, Z., and Sereda, I.T., 1981, Maximum uncertainty of interval velocity estimates: Geophysics, v. 46, p. 1543–1547.

Kaila, K.L., and Krishna, V.G., 1979, A new computerized method for finding effective velocity from reversed reflection traveltime data: Geophysics, v. 44, p. 1064–1076.

Shah, P.M., 1973, Use of wavefront curvature to relate seismic data with subsurface parameters: Geophysics, v. 38, p. 812–825.

Shaw, P.M., and Levin, F.K., 1973, Gross properties of time-distance curves: Geophysics, v. 38, p. 643–656.

Ray tracing

Blackburn, G., 1980, Errors, stacking velocity-true velocity conversion over complex geologic situations: Geophysics, v. 45, p. 1465–1488.

Cerveny, V., and Ravindra, R., 1971, The theory of seismic head waves: University of Toronto Press.

Daley, P.F., and Hron, F., 1977, Reflection and transmission coefficients for transversely isotropic media: Bull. SSA, v. 67, p. 661–676.

Deregowski, S.M., and Rocca, F., 1981, Geometrical optics and wave theory of constant offset sections in layered media: Geophys. Prosp., v. 29, p. 374–406.

Gerritsma, P.H.A., 1977, Time-to-depth conversion in the presence of structure: Geophysics, v. 42, p. 760–772.

Grant, F.S., and West, G.F., 1965, Interpretation theory in applied geophysics: New York, McGraw-Hill.

Hron, F., 1973, A numerical ray generation and its application to the computation of synthetic seismograms for complex layered media: Geophys. J. Roy. Astr. Soc., v. 35, p. 345–349.

Hron, F., Daley, P.F., and Marks, L.W., 1977, Numerical modelling of seismic body waves in oil exploration and crustal seismology: Proc. A.S.M.E. Sympos. on Computing Methods in Geophysical Mechanics, Atlanta, p. 21–42.

Hron, F., and Kanasewich, E.R., 1971, Synthetic seismograms for deep seismic sounding studies using asymptotic ray theory: Bull. SSA, v. 61, p. 1169–1200.

Hron, F., Kanasewich, E.R., and Alpaslan, T., 1974, Partial ray expansion required to suitably approximate the exact wave solution: Geophys. J. Roy. Astr. Soc., v. 36, p. 607–625.

Hubral, P., 1975, Locating a diffractor below plane layers of constant interval velocity and varying dip: Geophys. Prosp., v. 23, p. 313–322.

——— 1976, Interval velocities from surface measurements in the three-dimensional plane layer case: Geophysics, v. 41, p. 233–242.

——— 1976, CDP ray modeling in the presence of 3-D plane isovelocity layers of varying dip and strike: Geophys. Prosp., v. 24, p. 478–491.

——— 1976, Some remarks on computing travel times in plane layered media: Geophys. Prosp., v. 24, p. 719–724.

——— 1977, Time migration-some ray theoretical aspects: Geophys. Prosp., v. 25, p. 738–745.

——— 1980, Computation of the normal moveout velocity in 3D laterally inhomogeneous media with curved interfaces: Geophys. Prosp., v. 28, p. 221–239.

——— 1980, Wavefront curvatures in three-dimensional laterally inhomogeneous media with curved interfaces: Geophysics, v. 45, p. 905–913.

——— and Krey, T., 1980, Interval velocities from seismic reflection time measurements: Tulsa, Soc. Expl. Geophysicists.

Kaufman, H., 1953, Velocity functions in seismic prospecting: Geophysics, v. 18, p. 289–297.

Kleyn, A.H., 1977, On the migration of reflection time contour maps: Geophys. Prosp., v. 25, p. 125–140.

Krey, T., 1976, Computation of interval velocities from common reflection point moveout times for *n* layers with arbitrary dips and curvatures in three dimensions when assuming small shot-geophone distances: Geophys. Prosp., v. 24, p. 91–111.

Limond, W.Q., and Patriat, P., 1975, The accuracy of determination of seismic interval velocities from variable angle reflection (disposable sonobuoy) records: Geophys. J. Roy. Astr. Soc., v. 43, p. 905–938.

May, B.T., and Covey, J.D., 1981, An inverse ray method for computing geologic structures from seismic reflections—zero-offset case: Geophysics, v. 46, p. 268–287.

May, B.T., and Hron, F., 1978, Synthetic seismic sections of typical petroleum traps: Geophysics, v. 43, p. 1119–1147.

Merland, J., 1974, Application of seismic modelling to velocity studies: Presented at CSEG National Convention, Calgary, April.

Miller, M.K., 1974, Stacking of reflections from complex structures: Geophysics, v. 39, p. 427–440.

Psencik, I., 1974, Ray amplitudes of waves propagating in inhomogeneous media with curved interfaces: Proc. XIV ESC Assembly, Trieste, p. 171–175.

Robinson, E.A., and Mendel, J.M., 1983, Geo-signal processing: input-output and state-variable approaches: In preparation.

Sattlegger, J., 1964, Series for three-dimensional migration in reflection seismic interpretation: Geophys. Prosp., v. 12, p. 115–134.

——— 1965, A method of computing true interval velocities from expanding spread data in the case of arbitrary long spreads and arbitrarily dipping plane interfaces: Geophys. Prosp., v. 13, p. 306–318.

——— 1969, Three-dimensional seismic depth computation using space-sampled velocity logs: Geophysics, v. 34, p. 7–20.

——— 1982, Migration of seismic interfaces: Geophys. Prosp., v. 30, p. 71–85.

Sattlegger, J.W., Rohde, J., Egbers, H., Dohr, G.P., Stiller, P.K., and Echterhoff, J.A., 1981, INMOD-Two dimensional inverse modeling algorithm based on ray theory: Geophys. Prosp., v. 29, p. 229–240.

Sattlegger, J.W., and Stiller, K.P., 1974, Section migration, before stack, after stack, or in between: Geophys. Prosp., v. 22, p. 297–314.

Sattlegger, J., Stiller, P.K., and Echterhoff, J.A., 1976, Dip selective migration velocity determination: Geophys. Prosp., v. 24, p. 650–659.

Synthetic seismograms and the convolutional model

Anstey, N. A., 1960, Attacking the problems of the synthetic seismogram: Geophys. Prosp., v. 8, p. 242–260.

——— 1970, Seismic prospecting instruments, v. 1: Berlin, Geopublication Associates.

Arya, V. K., and Holden, H. D., 1979, A geophysical application: deconvolution of seismic data. Digital Signal Processing: N. Hollywood, Calif., Western Periodicals, p. 324–338.

Baranov, V., 1959, Synthetic seismograms with and without multiple reflections: Bull. AFTP, July.

——— 1960, The computation and use of synthetic seismograms: Erdol Zeitschrift, no. 3, March.

——— 1960, Film synthetique avec reflexions multiples-theorie et calcul pratique: Geophys. Prosp., v. 8, p. 315–325.

Baranov, V., and Kunetz, G., 1958, Calcul des sismogrammes synthetiques avec des reflexions multiple: C.R. Ac. Sc., t. 247, p. 1886–1889.

Bois, P., Chauveau, J., Grau, G., and Lavergne, M., 1960, Sismogrammes synthetiques: possibilities, techniques de realisation et limitations: Geophys. Prosp., v. 8, p. 260–314.

Bois, P., and Hemon, C., 1963, Etude statistique de la contribution des multiples aux seismogrammes synthetiques et reels: Geophys. Prosp., v. 11, p. 313–349.

Born, W. T., 1941, The attenuation constant of earth materials: Geophysics, v. 6, p. 132–148.

Collins, F., and Lee, C. C., 1950, Seismic wave attenuation characteristics from pulse experiments: Geophysics, v. 21, p. 16–40.

Delaplanche, J., 1961, Quelques exemples d'utilisation des films synthetiques: Geophys. Prosp., v. 9, p. 427–443.

Delaplanche, J., Hagemann, R. F., and Bollard, P. G. C., 1963, An example of the use of synthetic seismograms: Geophysics, v. 28, p. 842–854.

Delaplanche, J., and Ledoux, Y., 1960, L'analyse des films reels et leur comparison avec les films synthetiques: Geophys. Prosp., v. 8, p. 326–346.

Dennison, A. T., 1960, An introduction to synthetic seismogram techniques: Geophys. Prosp., v. 8, p. 231–241.

Durschner, H., 1958, Synthetic seismograms from continuous velocity logs: Geophys. Prosp., v. 6, p. 272–284.

Faust, L. Y., 1953, A velocity function including lithologic variation: Geophysics, v. 18, p. 271–288.

Futterman, W. I., 1962, Dispersive body waves: J. Geophys. Res., v. 67, p. 5279–5291.

Gardner, G. H. F., Gardner, L. W., and Gregory, A. R., 1974, Formation velocity and density—the diagnostic basics for stratigraphic traps: Geophysics, v. 39, p. 770–780.

Gerritsma, P. H. A., 1977, Time to depth conversion in the presence of structure: Geophysics, v. 42, p. 760–772.

Godfrey, L. M., Stewart, J. D., and Schweiger, F., 1968, Application of dinoseis in Canada: Geophysics, v. 33, p. 65–77.

Goupillaud, P. L., 1961, An approach to inverse filtering of near-surface layer effects from seismic records: Geophysics, v. 26, p. 754–760.

Hilterman, F. J., 1970, Three-dimensional seismic modelling: Geophysics, v. 35, p. 1020–1037.

——— 1975, Amplitudes of seismic waves—a quick look: Geophysics, v. 40, p. 745–762.

Kelly, K. R., Ward, R. W., Treitel, S., and Alford, R. M., 1976, Synthetic seismograms: a finite difference approach: Geophysics, v. 41, p. 2–27.

Lavergne, M., 1975, Pseudo-diagraphies de vitesses en offshore profond: Geophys. Prosp., v. 23, p. 695–711.

Lavergne, M., and Willm, C., 1977, Inversion of seismograms and pseudo velocity logs: Geophys. Prosp., v. 25, p. 231–250.

Levin, F. K., and Lynn, R. D., 1959, Deep-hole geophone studies: Geophysics, v. 23, p. 662–671.

Lindseth, R. O., 1971, Recent advances in digital processing of geophysical data —A review: Tulsa, SEG, private publication, 4th edition.

——— 1976, Seislogs: Presented to Geophysical Society of Houston, November 10–11.

McDonal, F. J., Angona, P. A., Mills, R. L., Sengbush, R. L., Van Nostrand, R. G., and White, J. E., 1958, Attenuation of shear and compressional waves in Pierre Shale: Geophysics, v. 23, p. 421–440.

Neidell, N. S., Stratigraphic modelling and interpretation: geophysical principles and techniques: Tulsa, AAPG Special Publication of Seismic Stratigraphy.

Newman, P., 1973, Divergence effects in a layered earth: Geophysics, v. 38, p. 481–488.

Nikolic, Z. J., 1975, A recursive time-varying band-pass filter: Geophysics, v. 40, p. 520–526.

O'Brien, P. H. S., 1961, A discussion on the nature and magnitude of elastic absorption in seismic prospecting: Geophys. Prosp., v. 9, p. 261–275.

——— 1967, The use of amplitudes in seismic refraction survey, *in* Seismic refraction prospecting: Tulsa, Soc. Expl. Geophysicists.

O'Doherty, R. F., and Anstey, N. A., 1971, Reflections on amplitudes: Geophys. Prosp., v. 19, p. 430–458.

Peterson, R. A., Fillipone, W. R., and Coker, F. B., 1955, The synthesis of seismograms from well log data: Geophysics, v. 20, p. 516–538.

Ricker, N., 1940, The form and nature of seismic waves and the structure of seismograms: Geophysics, v. 5, p. 348–366.

——— 1953, The form and laws of propagation of seismic wavelets: Geophysics, v. 18, p. 10–40.

——— 1953, Wavelet contraction, wavelet expansion and the control of seismic resolution: Geophysics, v. 18, p. 769–792.

Robinson, E. A., 1954, Predictive decomposition of time series with application to seismic exploration: Ph.D. dissertation, Dept. of Geology and Geophysics, M.I.T. (Reprinted in Geophysics, v. 32, p. 418–484, 1967).

——— 1957, Predictive decomposition of seismic traces: Geophysics, v. 22, p. 767–778.

——— 1966, Multichannel z-transforms and minimum-delay: Geophysics, v. 31, p. 482–500.

——— 1967, Statistical communication and detection with special reference to digital data processing of radar and seismic signals: High Wycombe, Bucks, England, Charles Griffin and Co.

——— 1967, Multichannel time series analysis with digital computer programs, second edition, 1983, Boston, IHRDC.

——— 1975, Dynamic predictive deconvolution: Geophys. Prosp., v. 23, p. 779–797.

Robinson, E. A., and Treitel, S., 1967, Principles of digital Wiener filtering: Geophys. Prosp., v. 15, p. 311–333.

Rockwell, D. W., 1967, The digital computer's role in the enhancement and interpretation of North Sea seismic data: Geophysics, v. 32, p. 259–281.

Schneider, W. A., and Backus, M. M., 1968, Dynamic correlation analysis: Geophysics, v. 33, p. 105–127.

Schoenberger, M., and Levin, F. K., 1974, Apparent attenuation due to intrabed multiples: Geophysics, v. 39, p. 278–291.

Shannon, C. E., 1948, A mathematical theory of communication. Bell Sys. Tech. J., v. 27, p. 379.

Sengbush, R. L., 1962, Stratigraphic trap study in Cottonwood Creek field, Big Horn Basin, Wyoming: Geophysics, v. 27, p. 427–444.

Sengbush, R. L., Laurence, P. L., and McDonal, F. J., 1961, Interpretation of synthetic seismograms: Geophysics, v. 26, p. 138–157.

Sheriff, R. E., 1973, Encyclopedic dictionary of exploration geophysics: Tulsa, Soc. Expl. Geophysicists.

——— 1973, Factors affecting amplitudes—a review of physical principles, in Lithology and direct detection of hydrocarbons using geophysical methods: Symposium of the Geophysical Society of Houston (see also Geophys. Prosp., v. 25, p. 125–138).

——— 1976, Inferring stratigraphy from seismic data: AAPG Bull., v. 60, p. 528–542.

——— 1977, Limitations on the resolution of seismic reflections and the geological detail derivable from them: AAPG Memoir 26, Tulsa, American Association of Petroleum Geologists.

Sherwood, J. W. C., 1962, The Seismoline, an analog computer of theoretical seismograms: Geophysics, v. 27, p. 19–34.

Shugart, T. R., and Barry, K. M., 1975, Zero phase seismic sections: Presented at 45th meeting of SEG, Denver.

Shugart, T. R., and Mathews, J. D., 1977, A marine seismic wavelet and its component parts: Presented at 47th annual meeting of SEG, September 1977, Calgary.

Silverman, D., Lash, C. C., and Hadley, C. F., 1963, Murac, a multiple reflection analog computer: Geophysics, v. 28, p. 975–989.

Taner, M. T., Cook, E. E., and Neidell, N. S., 1970, Limitations of the reflection seismic method: Lessons from computed simulations: Geophysics, v. 35, p. 551–573.

Taner, M. T., and Koehler, F., 1969, Velocity spectra—digital computer derivation and applications of velocity functions: Geophysics, v. 34, p. 859–881.

Tranter, C. E., and Kerns, C. W., 1972, Raven Creek Field, Campbell County, Wyoming: stratigraphic oil and gas fields: SEG Special Publications No. 10, Tulsa, Soc. Expl. Geophysicists.

Treitel, S., and Robinson, E. A., 1966, The design of high resolution digital filters: IEEE Trans. on Geoscience Electronics, v. GE-4, no. 1, p. 25–38.

——— 1966, Seismic wave propagation in layered media in terms of communication theory: Geophysics, v. 31, p. 17–32.

Trorey, A. W., 1962, Theoretical seismograms with frequency and depth dependent absorption: Geophysics, v. 27, p. 766–785.

——— 1970, A simple theory for seismic diffractions: Geophysics, v. 35, p. 762–784.

White, J. E., 1965, Seismic waves: radiation, transmission and attenuation: New York, McGraw-Hill Book Co.

Wiener, N., 1949, Extrapolation, interpolation and smoothing of stationary time series: Cambridge, Mass., M.I.T. Press.

Williamson, E. D., and Adams, L. H., 1923, Density distribution in the earth: J. Wash. Acad. Sci., v. 13, p. 413.

Wu, C. S., 1971, Method of geophysical prospecting by measuring the attenuation of seismic waves in the earth: U.S. Patent #3,662,965, issued November 23.

Wuenschel, P. C., 1960, Seismogram synthesis including multiples and transmission coefficients: Geophysics, v. 25, p. 106–219.

Wyllie, M. R. J., Gardner, G. H. F., and Gregory, A. R., 1962, Studies of elastic wave attenuation in porous media: Geophysics, v. 27, p. 569–589.

Deconvolution

Berkhout, A. J., 1973, On the minimum length of one-sided signals: Geophysics, v. 38, p. 657–672.

——— 1974, Related properties of minimum-phase and zero-phase time functions: Geophys. Prosp., v. 22, p. 683–703.

——— 1977, Least squares inverse filtering and wavelet deconvolution: Geophysics, v. 42, p. 1369–1383.

Deregowski, S. M., 1971, Optimum digital filtering and inverse filtering in the frequency domain: Geophys. Prosp., v. 19, p. 729–768.

De Voogd, N., 1974, Wavelet shaping and noise reduction: Geophys. Prosp., v. 22, p. 354–369.

Douze, E. J., 1971, Prediction error filters, white noise and orthogonal coordinates: Geophys. Prosp., v. 19, p. 253–264.

Ekstrom, M. P., 1973, A spectral characterization of the ill conditioning in numerical deconvolution: IEEE Trans. Audio Electroacoust, v. AU-21, p. 344–348.

Ford, W. T., 1978, Optimum mixed delay spiking filters: Geophysics, v. 43, p. 125–132.

Ford, W. T., and Hearne, J. H., 1966, Least-squares inverse filtering: Geophysics, v. 31, p. 917–926.

Fourmann, J. M., 1980, Signature processing—review and comparison of some processing principles: Presented at the 42nd Annual EAEG Meeting, Istanbul.

Jury, E. I., 1964, Theory and application of the *z*-transform method: New York, John Wiley and Sons.

Kailath, T., 1974, A view of three decades of linear filtering theory: IEEE Trans. Inf. Theory, v. IT-20, p. 146–181.

Kolmogorov, A. N., 1941, Stationary sequences in Hilbert space (Russian): Bull. Moscow State Univ. Math., v. 2, p. 40.

Kuo, B. C., 1964, Automatic control systems: Englewood Cliffs, N.J., Prentice-Hall, Inc.

Lackoff, M. R., and LeBlanc, L. R., 1975, Frequency-domain seismic deconvolution filtering: Acoust. Soc. Am. J., v. 57, p. 151–159.

Lacoss, R. T., 1971, Data adaptive spectral analysis methods: Geophysics, v. 36, p. 661–675.

Lee, Y. W., 1960, Statistical theory of communication: New York, John Wiley and Sons.

Levinson, N., 1947, The Wiener rms (root mean square) error criterion in filter design and prediction: J. of Math. and Phys., v. 25, p. 261–278.

MIT Geophysical Analysis Group, 1952–1957, GAG Reports No. 1 through No. 11: Cambridge, Mass., Massachusetts Institute of Technology.

Paley, R. E. A. C., and Wiener, N., 1934, Fourier transforms in the complex domain: Am. Math. Soc. Colloq. Publ., v. 19, ch. 1.

Rice, R. B., 1962, Inverse convolution filters: Geophysics, v. 27, p. 4–18.

Robinson, E. A., 1954, Predictive decomposition of time series with application to seismic exploration: Ph.D. thesis, MIT, Cambridge, Mass. (Reprinted in Geophysics, v. 32, p. 418–484, 1967).

——— 1957, Predictive decomposition of seismic traces: Geophysics, v. 22, p. 767–778.

——— 1962, Random wavelets and cybernetic systems: High Wycombe, Bucks, England, Charles Griffin and Co.

——— 1963, Structural properties of stationary stochastic processes with applications, Ch. 12, Time series analysis, ed. M. Rosenblat: New York, John Wiley and Sons.

——— 1964, Wavelet composition of time series, Ch. 2; Recursive decomposition of stochastic processes, Ch. 3; Econometric model building; ed. H. O. Wold: Amsterdam, North Holland Publishing Co.

——— 1966, Multichannel *z*-transforms and minimum delay: Geophysics, v. 31, p. 482–500.

——— 1967, Statistical communication and detection with special reference to digital data processing of radar and seismic signals: High Wycombe, Bucks, England, Charles Griffin and Co.

——— 1967, Multichannel time series analysis with digital computer programs, second edition: Boston, IHRDC, 1983.

——— 1978, Waves propagating in random media as statistical time series, *in* Applied time series analysis, ed. D. F. Findley: New York, Academic Press.

——— 1980, Physical applications of stationary time series: High Wycombe, Bucks, England, Charles Griffin and Co., and New York, Macmillan.

Robinson, E. A., and Treitel, S., 1964, Principles of digital filtering: Geophysics, v. 29, p. 395–404.

——— 1965, Dispersive digital filters. Rev. Geophys., v. 3, p. 433–461.

——— 1967, Principles of digital Wiener filtering: Geophys. Prosp., v. 15, p. 311–333.

——— 1969, Robinson-Treitel reader: Tulsa, Seismograph Service Corp.

Schneider, W. A., Larner, K. L., Burg, J. P., and Backus, M. M., 1964, A new data-processing technique for the elimination of ghost arrivals on reflection seismograms: Geophysics, v. 29, p. 783–805.

Schoenberger, M., 1974, Resolution comparison of minimum-phase and zero-phase signals: Geophysics, v. 39, p. 826–833.

Simpson, S. M., Jr., and Robinson, E. A., 1960–1964, Advanced Research Projects Agency, Project VELA UNIFORM, Reports No. 1 through No. 8, Contract No. AF(604) 7378.

Solodovnikov, V. V., 1952, Introduction to the statistical dynamics of automatic control systems, Ch. 1, trans. ed. J. B. Thomas and L. A. Zadeh: New York, Dover Publications, Inc., 1960.

Stone, D. G., 1981, Seismic filter design from borehole data: Technical Papers, 51st Annual Meeting, Society of Exploration Geophysicists, Tulsa.

Titchkosky, K. A., 1969, Seismic deconvolution: Calgary, published privately.

Treitel, S., 1970, Principles of digital multichannel filtering: Geophysics, v. 35, p. 785–811.

Treitel, S., and Robinson, E. A., 1964, The stability of digital filters: IEEE Trans. Geoscience Electronics, v. 2, p. 6–18.

——— 1966, The design of high resolution digital filters: IEEE Trans. Geoscience Electronics, v. 4, p. 25–38.

——— 1966, Seismic wave propagation in layered media in terms of communication theory: Geophysics, v. 31, p. 17–32.

Vaage, S., Haugland, K., and Utheim, T., 1983, Signatures from single airguns, Geophys. Prosp., v. 31, p. 87–97.

Wiener, N., 1949, Extrapolation, interpolation, and smoothing of stationary time series: Cambridge, Mass., The Technology Press, and New York, John Wiley and Sons.

Wiggins, R. A., and Robinson, E. A., 1965, Recursive solution to the multichannel filtering problem: J. Geophys. Res., v. 70, p. 1885–1891.

Wood, L. C., Heiser, R. C., Treitel, S., and Riley, P. L., 1978, The debubbling of marine source signatures: Geophysics, v. 43, p. 715–729.

Ziolkowski, A. M., 1982, An air gun model which includes heat transfer and bubble interactions: Technical Program Abstracts, 52nd Annual Meeting, Society of Exploration Geophysicists, Tulsa.

Ziolkowski, A. M., Lerwill, W. E., March, D. W., and Peardon, L. G., 1980, Wavelet deconvolution using a source scaling law: Geophys. Prosp., v. 28, p. 872–901.

Migration

Alford, R.M., Kelly, K.R., and Boore, D.M., 1974, Accuracy of finite-difference modeling of the acoustic wave equation: Geophysics, v. 39, p. 834–842.

Baker, B.B., and Copson, E.T., 1950, The mathematical theory of Huygens principle: Oxford, Oxford University Press.

Berkhout, A.J., 1981, Wave field extrapolation techniques in seismic migration, a tutorial: Geophysics, v. 46, p. 1638–1656.

Berryhill, J.R., 1977, Diffraction response for nonzero separation of source and receiver: Geophysics, v. 42, p. 1158–1176.

Bleistein, N., and Cohen, J.K., 1979, Direct inversion procedure for Claerbout's equations: Geophysics, v. 44, p. 1034–1040.

——— 1982, The velocity inversion problem—present status, new directions: Geophysics, v. 47, p. 1497–1511.

Castle, R.J., 1982, Wave-equation migration in the presence of lateral velocity variations: Geophysics, v. 47, p. 1001–1011.

Chun, J.H., and Jacewitz, C.A., 1981, Fundamentals of frequency domain migration: Geophysics, v. 46, p. 717–733.

Claerbout, J.F., 1970, Coarse grid calculations of waves in inhomogeneous media with application to delineation of complicated seismic structure: Geophysics, v. 35, p. 407–418.

Claerbout, J.F., and Doherty, S.M., 1972, Downward continuation of moveout-corrected seismograms: Geophysics, v. 37, p. 741–768.

Clayton, R.W., and Engquist, B., 1980, Absorbing boundary conditions for wave-equation migration: Geophysics, v. 45, p. 895–904.

Clayton, R.W., and Stolt, R.H., 1981, A Born-WKBJ inversion method for acoustic reflection data: Geophysics, v. 46, p. 1559–1567.

Cohen, J.K., and Bleistein, N., 1979, Velocity inversion procedure for acoustic waves: Geophysics, v. 44, p. 1077–1087.

Doherty, S.M., and Claerbout, J.F., 1976, Structure independent velocity estimation: Geophysics, v. 41, p. 850–881.

Emerman, S.H., Schmidt, W., and Stephen, R.A., 1982, An implicit finite-difference formulation of the elastic wave equation: Geophysics, v. 47, p. 1521–1526.

French, W.S., 1974, Two-dimensional and three-dimensional migration of model-experiment reflection profiles: Geophysics, v. 39, p. 265–277.

———, 1975, Computer migration of oblique seismic reflection profiles: Geophysics, v. 40, p. 961–980.

Gardner, G.H.F., French, W.S., and Matzuk, T., 1974, Elements of migration and velocity analysis: Geophysics, v. 39, p. 811–825.

Gazdag, J., 1978, Wave equation migration with the phase shift method: Geophysics, v. 43, p. 1342–1351.

——— 1980, Wave equation migration with the accurate space derivative method: Geophys. Prosp., v. 28, p. 60–70.

——— 1981, Modeling of the acoustic wave equation with transform methods: Geophysics, v. 46, p. 854–859.

Greenberg, M.D., 1971, Applications of Green's functions in science and engineering: Englewood Cliffs, N.J., Prentice-Hall.

Hatton, L., Larner, K., and Gibson, B.S., 1981, Migration of seismic data from inhomogeneous media: Geophysics, v. 46, p. 751–767.

Herman, A.J., Anania, R.M., Chun, J.H., Jacewitz, C.A., and Pepper, R.E.F., 1982, A fast three-dimensional modeling technique and fundamentals of three-dimensional frequency-domain migration: Geophysics, v. 47, p. 1627–1644.

Hilterman, F.J., 1975, Amplitudes of seismic waves—a quick look: Geophysics, v. 40, p. 745–762.

Judson, D.R., Lin, J., Schultz, P., and Sherwood, J., 1980, Depth migration after stack: Geophysics, v. 45, p. 361–375.

Kuhn, M.J., and Alhilali, K.A., 1977, Weighting factors in construction and reconstruction of acoustical wave fields: Geophysics, v. 42, p. 1183–1198.

Larner, K.L., Hatton, L., Gibson, B., and Hsu, I-C., 1981, Depth migration of imaged time sections: Geophysics, v. 46, p. 734–750.

Raz, S., 1981, Direct reconstruction of velocity and density profiles from scattered field data: Geophysics, v. 46, p. 832–836.

——— 1981, Three dimensional velocity profile inversion from finite-offset scattering data: Geophysics, v. 46, p. 837–842.

——— 1982, A procedure for multidimensional inversion of seismic data: Geophysics, v. 47, p. 1422–1430.

Reynolds, A.C., 1978, Boundary conditions for the numerical solution of wave propagation problems: Geophysics, v. 43, p. 1099–1110.

Robinson, E.A., 1983, Migration of geophysical data: Boston, IHRDC.

Sattlegger, J.W., 1975, Migration velocity determination: Part I. Philosophy: Geophysics, v. 40, p. 1–5.

——— 1975, Migration velocity determination: Part II. Applications: Geophysics, v. 40, p. 6–16.

Schneider, W.A., 1978, Integral formulation for migration in two and three dimensions: Geophysics, v. 43, p. 49–76.

Schultz, P.S., and Claerbout, J.F., 1978, Velocity estimation and downward continuation by wavefront synthesis: Geophysics, v. 43, p. 691–714.

Stoffa, P.L., Buhl, P., Diebold, J.B., and Friedemann, W., 1981, Direct mapping of seismic data to the domain of intercept time and ray parameters—a plane wave decomposition: Geophysics, v. 46, p. 255–267.

Treitel, S., Gutowski, P.R., and Wagner, D.E., 1982, Plane-wave decomposition of seismograms: Geophysics, v. 47, p. 1375–1401.

Trorey, W.A., 1977, Diffractions for arbitrary source-receiver locations: Geophysics, v. 42, p. 1177–1182.

Ursin, B., 1982, Quadratic wavefront and traveltime approximations in inhomogeneous layered media with curved interfaces: Geophysics, v. 47, p. 1012–1021.

Weglein, A.B., Boyse, W.E., and Anderson, J.E., 1981, Obtaining three-dimensional velocity information directly from reflection seismic data: an inverse scattering formalism: Geophysics, v. 46, p. 1116–1120.

Weglein, A.B., and Gray, S.H., 1983, The sensitivity of Born inversion to the choice of reference velocity: a simple example: Geophysics, v. 48, p. 36–38.

Yilmaz, O., and Claerbout, J.F., 1980, Prestack partial migration: Geophysics, v. 45, p. 1753–1779.

Answers to selected exercises

Chapter 1

1. Geophysics is the study of the earth by quantitative physical methods, especially by seismic reflection and refraction and by gravity, magnetic, electrical, and radiation methods. It includes the branches of (a) seismology, (b) geothermometry (heating of the earth, heat flow, and volcanology and hot springs), (c) hydrology (ground and surface water and sometimes including glaciology), (d) oceanography, (e) meteorology, (f) gravity and geodesy (the earth's gravitational field and the size and form of the earth), (g) atmospheric electricity and terrestrial magnetism (including ionosphere, Van Allen belts, telluric currents, and so on), (h) tectonophysics (geologic processes in the earth), and (i) exploration and engineering geophysics. Geochronology (the dating of earth history) and geocosmogony (the origin of the earth) are sometimes included in the foregoing list. Exploration geophysics is the use of seismic, gravity, magnetic, electrical, or electromagnetic methods in the search for oil, gas, minerals, water, and so on, for economic exploitation.
2. An electrical survey makes measurements at or near the earth's surface of natural or induced electrical fields to map mineral concentrations or for geological or basement mapping. Geothermal prospecting takes measurements of variations in earth temperature that are not attributable to daily variations in solar heating. The objectives may be to locate sources of geothermal energy, geological features that affect heat flow (salt domes, dikes, faults, and so on), or groundwater variations. A gravity survey takes measurements of the gravitational field at a series of different locations over an area of interest. The objective in exploration work is to associate variations with differences in the distribution of densities and, hence, of rock types. In a magnetic survey, measurements of the magnetic field or its component (such as the vertical component) are taken at a series of different locations over an area of interest, usually with the objective of locating concentrations of magnetic materials or of determining depth to basement. Differences from the normal field are attributed to variations in the distribution of materials having different susceptibility and perhaps also remanent magnetization. A reflection survey is a program to map geologic structure employing the seismic reflection method. Measurements are made of the arrival time of events attributed to seismic waves that have been reflected from interfaces where the acoustic impedance changes. The objective usually is to map variations in the depth and attitude of the interfaces, which usually are parallel to the bedding. A second objective is to define stratigraphic

variations from normal moveout (NMO) measurements or from the amplitude and character of reflection events. A refraction survey is a program to map geologic structure by using head waves. Head waves involve energy that enters a high-velocity medium (refractor) near the critical angle and travels in the high-velocity medium nearly parallel to the refractor surface. Head-wave arrivals are identified in terms of time after the shot and distance from the shot. The objective is to determine the arrival times of the head waves so as to map the depth to the refractors in which they travel.

3. *Electric log* is a generic term including all electrical borehole logs (SP, normal, lateral, laterologs, induction, microresistivity logs). The term is also used for records obtained in surface resistivity surveying.
4. *Anisotropy* is the variation of a physical property that depends on the direction in which it is measured. Geophysical usage of *anisotropy* is sometimes qualified as apparent anisotropy or effective anisotropy to distinguish it from the point property possessed by crystals. The general elasticity tensor relating stress and strain in anisotropic media contains 21 independent constants. If properties are the same in two directions (transversely isotropic medium), these reduce to five independent constants. Isotropic media possess only two independent elastic constants. Anisotropic media have two modes of S and one mode of P body-wave propagation; wavefronts are not necessarily orthogonal to the directions of wave propagation; and Snell's law requires modification. The term *aeolotropy* is synonymous with *anisotropy*.
5. A homogeneous, isotropic medium is a medium with properties independent of the location of a seismic wave and its direction of travel. Homogeneous means the same at all points. Isotropic means having the same physical properties regardless of the direction in which they are measured. Strictly, isotropic applies only to an arbitrarily small neighborhood surrounding a point and to single properties.
6. Ground roll is surface-wave energy that travels along or near the surface of the ground. It is usually characterized by relatively low velocity and low frequency but high amplitude. Ground roll tends to mask desired signals; shot and geophone patterns, filtering, and stacking are used to discriminate against it. Rayleigh waves are usually the main component of ground roll.
7. Hole fatigue is the effect involved where there is a delay between the detonation of a shot and the initiation of the seismic impulse from it because of changes in the shot environment (usually formation of a cavity) produced by an earlier shot in the same hole.
8. Hole noise from a shot is caused by hole blow, or discharge of the gases resulting from the explosion. Hole noise may last for several seconds. It is often excessively strong on geophone groups near the shothole, and it attenuates rapidly with distance.
9. An unplugged shot hole would be dangerous to people, animals, and equipment. A hole plug is used to plug a shot hole after shooting. Usually, the plug is pushed far enough into the hole to prevent its being dislodged, and earth is shoveled over it level with the surrounding ground.
10. An implosion is a collapse into a region of very low pressure. The creation of such a region under water (as with the Hydrosein or Flexichoc) causes the water to rush in with great force, and the collision of the inrushing water

with itself generates a seismic shock wave. The outrush of water propelled by a bubble of high-pressure gas (as from an underwater explosion or from an air gun) leaves behind a region of the vapor pressure of the water, into which the water subsequently collapses, causing the bubble effect.

11. A horizon, or an interface, is an underground surface separating two different rock layers. In this sense, horizon refers to the limit of one's experience. A map based on a reflection event may be called a horizon map, sometimes contrasted to a phantom map. A phantom on a seismic section is a line drawn parallel to the dip of nearby reflection events. Phantoms are mapped where one cannot follow one reflection event far enough to develop a horizon map based on that event alone.
12. As a wave travels outward from the source, there is, at any given time, a surface that joins all the points where motion is just about to start. Such a surface is called a wavefront. In an isotropic, homogeneous material, the wavefronts are a set of concentric spheres with the center at the source. In isotropic, nonhomogeneous media, the form of wavefronts is determined by the wave velocity distribution. The energy can be conceived as traveling down a very large number of pyramids of infinitesimal cross section. The centerline of any one pyramid is regarded as a ray, and its direction is normal to the wavefront at any time. If the wavefront is distorted by a nonuniform velocity distribution, the ray is bent so that it always stays perpendicular to the instantaneous wavefront. In many situations, the ray convention is a convenient one to follow, since it is usually simpler than dealing with the complexities of wavefront construction. It is merely a convenience, however, and it leads to difficulties whenever the normal to the wavefront cannot be uniquely defined.
13. This refraction wave corresponds to the case of critical incidence—the ray that hits an interface at the critical angle, propagates along the lower side of the interface, and then travels from the interface into the upper medium, also at the critical angle. This wave is normally called the head wave. The head wave is thus a wave that enters and leaves the high-velocity layer at the critical angle.
14. Refraction in a strong velocity-gradient zone may reverse the component of seismic rays in the downward direction and return them to the surface, where they may be observed (at offset) as refraction arrivals, even though they have no appreciable path through a distinctive refractor. Thus, diving waves are waves whose raypaths are curved because of vertical velocity gradient. Thus, diving waves return to the surface, even as a person who dives into the water.
15. An isotime map gives contours of reflection time, time intervals, and the like. The time interval between two reflections is mapped in order to locate stratigraphic changes in the interval, reef buildup, variation in salt thickness, and so forth. The isotime may vary because of variation in the velocity or the thickness, or both.
16. Fermat's principle states that seismic energy follows a raypath such that the first-order variation with all neighboring raypaths is zero. It is sometimes described as that path for which the traveltime is stationary (which could be either a minimum, a maximum, or a stationary value) compared to all neighboring paths. If the medium is not homogeneous, the raypath will not

be straight but will bend or curve so that, in the usual case, the overall traveltime is a minimum. The resulting raypath is called the least-time path, or brachistochrone. Snell's law is a consequence of Fermat's principle.

17. A marker bed is a bed or sequence of beds that yields characteristic reflections over a more or less extensive area.
18. Mode conversion is conversion of P-wave energy into S-wave energy, or vice-versa, by oblique incidence on an interface. The resulting wave is called a converted wave.
19. Multiplicity refers to the number of independent raypaths that combine to provide a single output trace. It is the number of common midpoint traces that sample essentially the same portion of reflector but with different offsets. A 12-fold CMP recording, for example, involves recording each subsurface point 12 times, once from each of 12 different offset distances.
20. First-break data and the refraction wavetrains that follow the first breaks usually are so strong that they have to be excluded from the stack to avoid degrading the quality of shallow reflections. This is done by muting, which involves arbitrarily assigning zero values to traces during the period when the first breaks and following wavetrains are arriving.
21. A diversity stack is used with CMP stacking to discriminate against ground roll and similar high-amplitude wavetrains. In some applications, a diversity stack is a stack in which amplitudes that exceed some threshold are excluded completely instead of merely being given small weight. Amplitudes less than this threshold may not be affected.
22. Wind noise is random noise attributed principally to ground unrest caused when the wind moves plants, trees, and so forth, and shakes their roots. Often, any seismic background noise (in the absence of a shot), regardless of the natural source, is referred to as wind noise.
23. *Acoustic log* is a generic term for well logs that display any of several aspects of acoustic-wave propagation. In some acoustic logs (sonic log, continuous-velocity log), the traveltime of the compressional wave between two points is measured. In others (amplitude log), the amplitude of part of the wavetrain is measured. Other acoustic logs (character log, 3-D log, VDL-log, microseismogram log, signature log) display part of the wavetrain in wiggle or variable-density form. Still others (cement-bond log, fracture log) are characterized by the objective of the measurements, rather than by their form. A borehole televiewer is also an acoustic log.
24. Slowness is the reciprocal of velocity. Transit time over an interval, as measured by the sonic log, is slowness.
25. Vertical exaggeration is the use of a vertical scale that is larger than the horizontal scale. Exaggeration makes subtle effects more evident but distorts structural relationships. Seismic time sections involve variable vertical exaggeration because the velocity varies with depth. The selection and interpretation of significant features on record sections is greatly affected by vertical exaggeration. The migration of data is another factor involved in the direct interpretation of record sections.
26. Divergence is the decrease in amplitude of a wavefront because of geometrical spreading. With body waves, the energy spreads out as the spherical wavefront expands, causing the energy density to vary inversely as the square of the distance (spherical divergence), hence causing wave amplitude

to vary inversely as the distance. With surface waves,the energy density varies inversely as the distance (cylindrical divergence), hence causing wave amplitude to vary inversely as the square root of the distance. Tube waves do not suffer energy loss because of divergence. Note that energy density decreases for other reasons also.

27. Frequency (denoted by the symbol f) is the repetition rate of a periodic waveform. The frequency f is the reciprocal of the period T; that is, $f = 1/T$. Frequency is usually measured in cycles per second, or hertz. The angular frequency ω, measured in radians per second, is given by

$$\omega = 2\pi f.$$

The *dominant frequency* of a wavelet is an approximate repetition (the reciprocal of the peak-to-peak time interval), even though the entire wavelet does not repeat. The *frequency content* of a waveform is the amplitudes of the sinusoidal components into which the waveform can be decomposed by Fourier analysis, even where there is nothing repetitive about the waveform itself.

28. Gain control is the variation of the amplification or attenuation of an amplifier to compensate for variations in input signal strength. Types of gain control used in seismic amplifiers include *automatic gain control* (AGC), in which the gain of each channel is controlled automatically and independently of other channels; *ganged automatic gain control,* in which the gain of all channels is the same although automatically determined, being based either on an average of the energy levels of a number of the channels or on a single channel; *preset, or programmed, gain control,* in which the gain as a function of record time is determined arbitrarily prior to the shot; and *binary gain control,* in which gain is allowed to vary only by factors of two but the time at which the gain changes is determined automatically. The times at which the changes occur are recorded so that the gain effects can be removed in processing. The *gain trace* is the trace on a seismic record that indicates the gain (amplification) used on one or more channels.

29. Because surface information is essential in the processing of the observed reflection data, an integral part of the field activity in a seismic reflection survey is the determination of the thickness and the velocity of the surface layer. One method is the up-hole survey, in which drill holes about 50 to 100 m deep are drilled along the seismic line at points 5 to 10 km apart. Small charges are fired at a series of points down the drill hole. Traveltimes of the generated waves from source to surface are recorded. From a plot of traveltime (reduced to vertical path) versus depth, both the average velocity and the depth of the surface layer are determined. Another method is the special short-spread refraction survey. It is used when it is difficult to drill reasonably deep holes for up-hole surveying. This is a working method, since the surface layer is almost always of a lower velocity than the underlying medium. A third method makes use of first arrivals on the reflection records. The usual reflection spreads generally are not suitable for giving useful information on a relatively thin surface layer. First arrivals on these records can be used, however, to get an idea of the velocity of the layer below the surface layer. A normal reflection record includes the upward transmission

arrival for a wave recorded by a special geophone placed on the surface close to the shot hole. This vertical time, which is usually available for each shot, provides a direct means of computing the velocity in the medium that lies above the shot.

30. The weathering layer is the low-velocity layer on the earth's surface. Below the weathering layer, the velocity abruptly increases. The base of weathering is the boundary between the weathering layer of low seismic velocity and the underlying layer of appreciably higher velocity. The seismic weathering layer may or may not correspond to the geologic weathering or to the water table. The boundary is involved in deriving time corrections for seismic records. To avoid confusion with the geologic weathering, the term *low-velocity layer* (LVL) is often used in place of *seismic weathering layer.*

Chapter 2

1. The most important parameter is interval velocity. During the first years of seismic exploration, the velocities determined for each reflection from normal moveout (NMO) times represented averages over the entire section between the surface and the reflecting boundary. It was not until fairly recently that recording and processing techniques were introduced that made it practical to determine velocities over intervals of depth comparable to the thickness of a single formation. The long spreads now in common use make it possible to compute such interval velocities with an accuracy that can provide much more useful geologic information than could have been derived from the shorter spreads previously employed.
2. One method is the one discussed in the text. It involves the determination of NMO velocities based on the relationship between the traveltimes for reflections and the distances at which they are recorded. A knowledge of these for adjacent reflections allows computation of the velocities within the layers bounded by the respective reflecting surfaces. The other method, inverse modeling, involves construction of a layered model and assignment of velocities to each layer. Ray tracing is used to determine times for each model, and the velocities are iteratively adjusted by comparisons of the times thus computed with observed times.
3. An isovelocity surface is one on which the seismic velocity is constant. In the absence of structural uplift and lateral variations of lithology, isovelocity surfaces are apt to be horizontal planes. Where structural uplift occurs, however, isovelocity surfaces tend to follow structure, but with less relief than the structure. Changes in velocity bend seismic rays and hence alter apparent structure. Curvature of velocity surfaces can act toward seismic rays as lenses act toward light rays, distorting both structure and measurements, such as velocity calculated from normal moveout. Knowledge of how the velocity varies is essential for correct interpretation.
4. A multiple represents seismic energy that has been reflected more than once. If this definition is strictly adhered to, virtually all seismic energy involves multiples, and primary energy that has been reflected only once is scarcely observable. The important distinction is between long-path and short-path multiples. A long-path multiple arrives as a distinct event, whereas a short-path multiple arrives so soon after the primary that it merely adds tailing

to the primary. Usually, it does not matter whether the reflections being mapped are primary but rather whether their attitude is indicative of the attitude of the appropriate portion of the section, so that the tailing produced by the many short-path pegleg multiples does not detract seriously. The attitude of long-path multiples is not likely to be representative of the portion of the section associated with their arrival time. Usually, the long-path multiples have traveled more in the slower (shallower) part of the section than primaries with the same arrival time, so that they ordinarily show more normal moveout and can be attenuated by stacking.

5. The required expression is

$$t^2(x) = \left(\frac{2d}{v}\right)^2 + \left(\frac{x}{v}\cos\phi\right)^2.$$

Note that this equation differs from that of the flat-layer model in that $2d/v$ replaces $t(0)$ and $v/\cos\phi$ replaces v. This equation may be described as a hyperbolic equation in which the velocity estimated from the reflection time differences (the NMO) Δt is an apparent velocity v_{app} related to the true velocity v by

$$v_{\mathrm{app}} = v/\cos\phi.$$

Thus, we must also know the dip angle of the reflecting interface to estimate the velocity. The effect of dip angle on velocity can easily be computed; for dip angles less than approximately 12°, the apparent velocity exceeds the true velocity by no more than 2 percent.

6. The stacking time t_s is no longer the vertical two-way time from the common surface point to the reflecting horizon, but rather for the raypath normal to the dipping interface. The stacking velocity is the apparent velocity $v_{\mathrm{app}} = v/\cos\phi$. Thus, to obtain the true velocity from the measured apparent velocity v_{app} associated with the best-fit hyperbola, it is necessary to know the dip angle ϕ. The dip angle cannot be determined from a single CMP trace gather, but may be estimated from two or more adjacent gathers. It must be emphasized, however, that the dip is assumed to be constant in this procedure; that is, the reflecting horizon is planar. The stacking velocity, which in this case is the estimated apparent velocity v_{app}, is used to correct the reflection events by the NMO correction to the stacking time $t_s = 2d/v$, which is the traveltime for $x = 0$. This NMO correction is done prior to summation of these CMP traces. It should be noted that this stacking time aligns the events, although the events do not have coincident reflection points.

7. If the seismic traverse is in the strike direction ($\theta = 90°$), the apparent velocity equals the true velocity. The greatest difference occurs in the dip direction ($\theta = 0°$). For dip angles $\theta < 10°$, the velocity difference usually can be ignored.

8. For complex subsurface configurations, inverse modeling provides a viable means of determining interval velocities. Basically, inverse modeling involves the construction of a model consisting of a cross section of the subsurface geometry (depth vs. horizontal distance) based on the observed

reflection times on a record section. The configuration of each reflecting surface and the interval velocities between the surfaces are determined by working progressively downward into deeper layers. Iterative ray-tracing procedures are used to find the position of each successively deeper reflecting surface and the interval velocity in the layer immediately above it that gives the best agreement with observed reflection time. This permits a refinement of the interval velocities and results in a depth section that correctly represents subsurface structures. The oil companies and geophysical companies make use of other such approaches, and the good ones are generally kept company-confidential.

10. Multiples have the same type of (x^2, t^2) relationship as primaries. Thus, someone must interpret velocity analysis plots and determine which values will emphasize primaries as opposed to multiples. Multiples have not traveled as deep in the earth as primaries with the same arrival time and so usually are associated with stacking velocity values that are smaller than those for primaries. Thus, a velocity analysis interpreter usually assumes that the larger velocity values should be honored in order to emphasize primaries. The fact that velocity usually increases with depth is the reason for multiples having smaller stacking velocities; under some circumstances, however, velocity does not increase with depth, and this criterion may therefore be wrong in such instances.
16. (a) The average velocities are 5000, 5676, 6846, 6695, 7534. The rms velocities are 5000, 5695, 6951, 6758, 7712.
    (b) The two-way traveltimes are as follows:

		Offset				
Interface		0	1800	3000	4200	6000
1	Exact	0.1600	0.3940	0.6210	0.8551	1.2105
	Approx.	0.1600	0.3940	0.6210	0.8551	1.2105
2	Exact	0.493	0.586	0.721	0.884	1.154
	Approx.	0.493	0.586	0.722	0.887	1.163
3	Exact	0.993	1.027	1.083	1.161	1.311
	Approx.	0.993	1.027	1.083	1.163	1.316
4	Exact	1.763	1.783	1.818	1.869	1.973
	Approx.	1.763	1.783	1.818	1.869	1.974
5	Exact	2.363	2.374	2.394	2.424	2.487
	Approx.	2.363	2.374	2.394	2.424	2.487

Chapter 3

1. (a) 1
   (b) $1/(1 - z)$
   (c) $1/(1 - cz)$
   (d) $z/(1 - z)^2$
   (e) $(1 - az \cos bT)/(1 - 2az \cos bT + a^2z^2)$
   (f) $az \sin bT/(1 - 2az \cos bT + a^2z^2)$
6. Yes.

7. We are led to the necessity for reflection. Not all the incident energy in the shale can be transmitted into the limestone, and the amount that is reflected is just what provides equality between the particle velocities on the two sides of the interface. The process of reflection does not involve a loss of energy; the sum of the transmitted and reflected energy is equal to the incident energy (assuming no loss to heat).
8. In seismic work, we do not ordinarily measure energy; we measure amplitude—pressure amplitude if we are at sea, particle velocity amplitude if we are on land.
9. $(A_r/A_i) = c_{\text{pressure}} = (Z_2 - Z_1)/(Z_2 + Z_1)$. Note that both the incident and reflected amplitudes are in the same material (the Z_1 material). Also note that measurements of pressure amplitude are independent of the direction of travel of the wave. A compression is a compression, and a rarefaction is a rarefaction, whichever way the wave is going. The reflection amplitude is decided primarily by the numerator of the reflection coefficient expression (i.e., the contrast of acoustic impedance across the interface). If Z_2 is greater than Z_1 (as with the example of shale followed by limestone), the reflection coefficient is positive, and an incident compression is reflected as a compression. If Z_2 is less than Z_1 (which would be the case if the wave is incident from the limestone side), the reflection coefficient is negative, and an incident compression is reflected as a rarefaction. If the reflection coefficient of an interface is positive when viewed from one side, it is negative when viewed from the other side.
10. Two limiting values are implicit in the reflection coefficient formula: Z_2 is zero (a free surface) and Z_2 is infinite (a fixed surface). The first is approximated by the interface between earth (or sea) and the air. The pressure reflection coefficient is virtually -1, so that an upcoming seismic wave is sent back down into the earth by the free surface, substantially undiminished in amplitude but inverted in sign. As a corollary of this, to a wave in the air, the earth appears as a fixed surface. It is not possible, therefore, to generate a seismic signal in the earth by shooting in the air.
11. Suppose that the incident medium is the one with the greater impedance. The incident wave is maintaining a certain balance between the pressure acting and the motion produced in harmony with the acoustic impedance of the hard material. The incident wave hits the soft material, in which motion is much more readily produced. As a result, a rarefaction opens up behind the wave. The rarefaction then propagates in the backward direction as the reflected wave.
12. If the bottom end is left free, slow vibration of the top end of the rope produces an equal and opposite motion at the bottom end. No forces are exerted at the bottom end, but there is a maximum of motion. If the bottom end is anchored, however, the inevitable zero of motion at the anchor point is accompanied by significant forces exerted by the rope on the anchoring mechanism.
13. The pressure vanishes at a free surface.
14. $(A_t/A_i) = t_{\text{pressure}} = (2Z_2)/(Z_2 + Z_1)$.
15. In the case of a positive reflection coefficient, the fact that the transmitted pressure amplitude is larger than the incident pressure amplitude is physically reasonable. If there were no reflection, and the whole of the incident

energy were transmitted from the soft into the hard medium, then we know that the pressure amplitude in the hard medium would be increased to a large notional value (although it would be offset by a corresponding reduction of particle velocity). In the real world, there is reflection, and not all the incident energy is transmitted. The consequence is that the pressure amplitude of the transmitted signal is less than the aforementioned notional value but still greater than the incident amplitude. The key point about the transmission coefficient formula is that the measurements are made in different materials. Note, also, that the process of transmission cannot change the sign of the transmitted signal; it cannot turn a compression into a rarefaction.

16. The signal transmitted into a harder material has a particle velocity smaller than that of the incident signal (the reverse of the situation obtained with a pressure geophone). The sum of the reflected energy and the transmitted energy is equal to the incident energy. The sum of the reflected and transmitted amplitudes is not equal to the incident amplitude. However, the sum of the incident and reflected amplitudes is equal to the transmitted amplitude; this is the continuity equation that prevents rupture at the interface.
17. The answer depends on the type of geophone. If it is a pressure-sensitive geophone, then the output increases as the geophone passes into the harder material. If it is a velocity-sensitive geophone, the output decreases in the harder material.
18. $T = (4Z_1Z_2)/(Z_2 + Z_1)^2$. The two-way transmission coefficient implies measurements made in the same material; it is always less than one and so always represents a diminuation of amplitude introduced by transmission (as we would expect). Furthermore, the two-way transmission coefficient is unaffected by the type of receiver used (particle velocity-sensitive or pressure-sensitive).
19. If Z_2 is greatly different from Z_1, the reflection coefficient will be large in magnitude and the transmission coefficient small. A smooth transition between the two impedances (Z_2 close to Z_1) transmits the signal best. This is impedance matching.

Chapter 4

4. $(4 - 2z + z^2)(1 - 2z + 4z^2) = z^2(4 - 2z + z^2)(z^{-2} - 2z^{-1} + 4)$.
14. The riders at D and F jump off, but not the one at E.
15. When the bell is struck by the clapper, the circular rim is driven into the form of an oval, which elastically rebounds into another oval. The rim vibrates continuously from one oval form to the other as long as the bell rings. However, there are four points along the rim that do not vibrate—the intersection points of the two ovals. No sound comes from these points. The silence extends away from the bell because the sound waves that arrive in the vicinity of these points from other parts of the bell arrive out of phase and so interfere destructively. If the points of rest gradually travel around the rim, the ring of the bell is wavy.
21. When we run water into a container, the air space becomes smaller. The frequency of the sound that comes out of the container increases. By listening, we can almost tell when the container is full without looking. When

we run water out of the container, the size of the air space gets larger. The large air space, or cavity, has a lower resonant frequency. The liquid flow pulses at the resonant frequency of the air cavity, and that frequency decreases as the liquid comes out. The glugging sound frequency emitted on emptying is much lower, however, than the hissing sound emitted on filling. The reason is that when the container is emptying, the water must vibrate with the air, but when the container is filling, the air alone vibrates. On emptying, a larger mass must vibrate, and it is harder to accelerate a larger mass. Thus, when emptying the container, the mass of the water slows down the rate of vibration.

22. The least-time path is *LcP.* The lifeguard does not run directly from *L* to *b* to *P* in one straight line, because the guard can go faster running over the beach than swimming through the water. The guard's speed on the *L* to *c* run and the reduced distance of the slow *c* to *P* swim more than compensate for the increased distance traveled. By running to *c* and then swimming to *P*, the distance the guard must travel is not minimized, but the time required for the rescue is minimized. If a dolphin were employed as a lifeguard, the dolphin would take the path *LaP* (the least-time path for the dolphin). The dolphin makes good time in the water but can hardly move on the beach. This principle of least time (Fermat's principle) is important in seismology. It happens that whenever a wave travels from one point to another, it will do so along the path requiring the least time.
23. The correct answer is (c). Light bends toward the normal when it goes from air into glass. Because it is the path of least time, the speed of light is slower in the glass than in the air.
24. The surface layer is a low-velocity medium.
25. The path of the wheels will bend toward the normal. Suppose that the left wheel is the first to encounter the grass, where the greater interaction results in slowing. This acts as sort of a pivot, because the right wheel remains at its greater speed on the sidewalk until it meets the grass. When both wheels are in the grass, they again move in a straight line. The same occurs with wavefronts of waves incident at an angle upon a low-velocity medium.
26. Both notes have the same period, so the frequency does not differ. The height (amplitude) of the wave *B* is greater, however, which indicates a greater vibratory energy. Therefore, *B* is the louder note.
27. The vertical displacements of waves are added together to form their composite, as shown in the following diagram. This is a simple case, because the displacements for both waves are everywhere either zero or a single positive value.

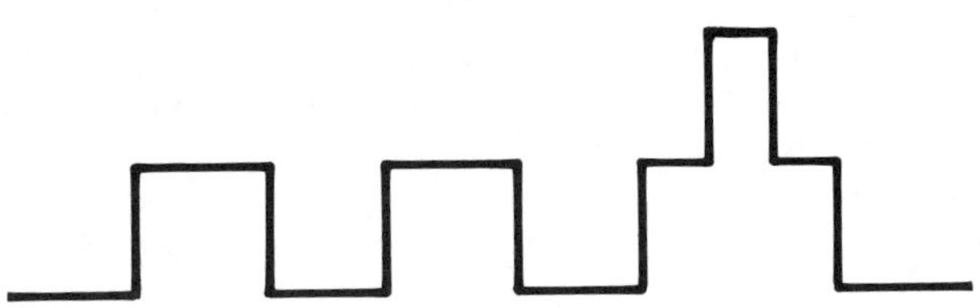

28. The vertical displacement everywhere along the composite wave is simply the algebraic sum of the displacements of the two sinusoidal waves:

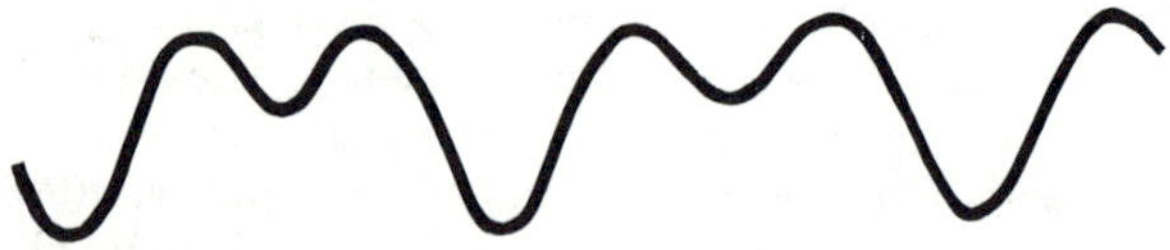

29. Fourier (1807) showed that a waveform of any shape can be represented as a sum (integral) of many pure waves. If the desired waveform has small bumps or sharp corners, then many short-wavelength (high-frequency) waves will be required to make it.
30. Because the square wave has the sharpest corners, it is composed of waves of the highest frequencies. A square wave is built up with many high-frequency sinusoidal waves. If we send a square wave through an amplifier or transmission line that is not capable of handling high frequencies, the square wave will have its high-frequency components robbed and will come out with round shoulders.
31. Light is a wave, and waves can combine so as to reinforce each other or cancel each other. The same can be done with water waves, where the crest of one wave overlaps the trough of the other wave, or with seismic waves or any other kinds of waves. This is called destructive interference. It is not a violation of the law of conservation of energy. If light cancels light, where does the energy go? Every time you make a setup where light cancels light at one location, there is another location (usually very near) where light reinforces light, and all the energy that is missing from the canceled location shows up at the reinforced location.
33. If the two notes had periods 1 and 2, respectively, and if they started out in phase, they would be in phase again after two periods of the first (corresponding to one period of the second). The frequency difference would be $1 - 0.5 = 0.5$. If the two notes had periods 1 and 1.25, however, then five periods of the first would correspond to four periods of the second, so that it would take longer for them to get in phase again. The frequency difference in this case is $1 - 0.8 = 0.2$. When the two notes are in phase, we have a beat. Thus, the greater the difference in the two frequencies, the shorter the interval between places where matching occurs. If the frequencies are only very slightly different, then matching places are fewer and farther apart. If the two frequencies are the same, of course, they never get out of phase, and they either match or mismatch everywhere. We can hear the beats as throbs from a twin-engine boat. If the engines run exactly in sync, there are no beats, but if one runs a little faster than the other, it has a slightly higher frequency than the other and beats occur. As the difference in frequency increases, the beat frequency increases. In *A*, the beats occur twice as often as in B, so the difference in frequency of the notes fed into *A* is twice the difference in frequency of the pair of notes fed into B. Thus, the notes that make up *B* are closer in frequency.
35. The Fourier synthesis of a waveform involves adding together sinusoidal waves of different frequencies and different phases. If the same components are added together with different phase relations, different waveforms will result.
36. At $k = 0$, the autocorrelation is a maximum that gives the energy (power) of the waveform. About this central peak, the autocorrelation is symmetric.

Because the autocorrelation is a zero-phase function, it damps out more rapidly (i.e., it is more spikey) than any other symmetric function with the same energy (power) spectrum. The autocorrelation of a waveform contains no information on the phase relations between frequency components in the waveform.

37. We have

$$\phi_{xy}(k) = \phi_{yx}(-k) = \sum_n y_{n-k}x_n = \sum_n x_n y_{-(k-n)} = x_k * y_{-k},$$

where y_{-k} is the reverse of y_k.

38. Some properties of the cross correlation $\phi_{ab}(k)$ are as follows:
   (a) It is not symmetric.
   (b) It is a measure of correlation between two different signals.
   (c) An equivalent definition is

$$\phi_{ab}(k) = \sum_n a_n b_{n-k}.$$

   (d) $\phi_{ab}(k) \neq \phi_{ba}(k)$, but $\phi_{ab}(k) = \phi_{ba}(-k)$.
   (e) The cross correlation preserves information on the phase difference between any frequency components present in both the signals.

Chapter 5

21. The U.S. Internal Revenue Service definition of a petroleum independent is as follows:

> An "independent producer" with respect to any quarter, is any person other than a person to whom subsection (c) of Section 613A does not apply by reason of Paragraph 2, (relating to certain retailers) or Paragraph 4, (relating to certain refiners) or Section 613A(d) or to whom subsection (c) would not apply if Subsection (D) (2) or (4) were applied, on a quarterly rather than an annual basis.

Index